THE DATACAD 11 PROJECT BOOK

Created by:
Dr. Leonard O. Nasman

&

Michael R. Smith

Distributed by:
Tech Ed Concepts

550 Pembroke Street
Pembroke, NH 03275
800-338-2238

Published by:
Microcomputer Education Systems Inc.

copyright 1997, 1998, 2000, 2001, 2004

The DataCAD 11 Project Book

by Leonard O. Nasman and Michael R. Smith

Copyright 1997, 1998, 2000, 2001, 2004 by Microcomputer Education Systems Inc.

All rights reserved. No part of this book may be reproduced or used in any form or by any means, such as: graphic, electronic, or mechanical including photocopying, recording, videotaping, or information storage and retrieval systems without written permission of the publisher.

Distributed by: Tech Ed Concepts
550 Pembroke St.
Pembroke, NH 03275
800-338-2238
www.tecedu.com

DataCAD is a registered trademark of DATACAD LLC Inc., 20 Tower Lane, Avon Park South, Avon, CT, 06001. (www.datacad.com)

Printed in the United States of America.

10 9 8 7 6 5 4 3 2 1

•• **NOTE:** This printing has been updated for DataCAD 11.05.

ISBN Number 0-9740796-5-0

Nasman, Leonard O.
Smith, Michael R.
The DataCAD 11 Project Book / Leonard O. Nasman, Michael R. Smith

1. DataCAD (Computer system)
2. Computer aided design
3. Architectural design
4. Computer aided drawing
5. Architectural drafting
6. Architecture

To The Student

Preface

This project book is designed to introduce DataCAD to new users, and to introduce architectural CAD through the use of the DataCAD software. The book is aimed at students in technology education, "tech prep", or introductory level architectural CAD courses, vo-tech courses, college architectural design classes, and architects and designers just starting with DataCAD. The book is divided into a series of drawing projects developed to help new users explore and appreciate three dimensional architectural design and drawing.

DataCAD 11 for Windows requires an IBM PC compatible running Windows 98/Me/NT/2000/XP, a hard disk drive with 200MB of available space (for a full install), 128MB of memory, a CD ROM drive, a 3 button mouse, and a good graphics display adapter. Also, if you are buying a computer, get the latest and fastest processor, the largest hard disk, the most RAM, and the best color monitor you can afford. It is quite amazing how much time can be saved (especially when removing hidden lines from complex pictorial drawings) with a faster machine. If you are a professional, some sort of removable media (like a CD R drive) is desirable. DataCAD drawing files can get quite large, although with Version 11 all files are compressed when stored, so that they take up 1/10th as much disk space. And it is also very important to manage the organization, storage, and backup of important files.

To The Student

As you work through the projects in this book, you will see places to check off the steps as they are completed. This will help you keep track of where you left off whenever you are interrupted or take a break. Because the power to do wonderful drawings with DataCAD is so great, you will be tempted to jump ahead and skip steps. ***Don't do it***. The sequence of steps within the exercises presented in this workbook has been carefully designed to give you as much power as possible in as short a time as possible. If you skip steps, you will miss something that will cause you to waste time and become frustrated. Remember, it is the hasty person who takes the longest. So read the information very carefully. Maybe even go through an exercise several times so that you milk it for all it is worth. The reward will come when you can use DataCAD to create more architectural drawing in minutes than your friends can do in days with conventional tools. But be careful, drawing with DataCAD can be so much fun that you might become addicted.

The DataCAD 11 Project Book

To The Teacher

The projects in this book include complete step-by-step instructions. Information is presented between the steps to help the students understand the concepts and processes being used so that they will be able to create their own designs. These materials can be used for either group or individualized instruction.

For introductory programs of very short duration, it is recommended that projects 1, 2, 3, and 10 be completed (in that order). For a more complete introduction to architectural CAD add projects 4 through 8. (Note that projects 4 through 8 must be completed in sequence since each project uses drawing files from earlier projects.) Project 9 provides a brief introduction to creating 3D cityscapes. In Project 9 the students will create a "block city" and then move around the city through the use of 3D views. Project 11 introduces the **FrameIt** macro for creating floor, wall, and roof framing.

It is desirable to integrate architectural CAD into a traditional architectural drawing course. After the students are introduced to basic CAD concepts, they can apply those concepts to traditional architectural drawing projects. For example, after completing projects 1 through 3, the students can be given an assignment to create small emergency shelters, or perhaps a small vacation cottage. After completing the motel room projects the students will have enough experience to create a single family residence, or town house design. An important point is that DataCAD is not a subject, but simply a new tool for doing traditional architectural designs.

If you will be introducing both architectural and mechanical CAD to your students, it is recommended that architectural CAD be covered first. Because the students are already familiar with the terminology of architecture (door, window, wall, etc.) there are very few vocabulary stumbling blocks. Because students will be able to create exciting, understandable 3D drawings with DataCAD within the first hour or so, they are more likely to be "turned on" to CAD. After they have some CAD experience, it will be easier for them to deal with the relatively abstract and much more formal concepts of mechanical drawing.

You can have each student save files on floppy disks. However, since it is possible to create large files easily, a floppy disk may be inadequate for some work. You might want to establish a folder on the hard drive for each student. Drawing file security is always an issue to consider.

To The Teacher

A Personal Note From the Authors:

Dr. Leonard O. Nasman:

Dublin, Ohio, Winter 2001

For a long time I shared the opinion of many of my fellow teachers that students should develop significant manual drawing skills before starting with CAD. With the advent of software such as DataCAD, I have changed my mind. I now believe that it is appropriate to introduce students to 3D architectural CAD the very first day of class. The DataCAD software is so easy to use, and so powerful, the students can get a lot farther into architectural design in a much shorter time than is possible using traditional tools. However, I would have students *extensively* practice some traditional drawing techniques (such as sketching, shading, lettering, or applying media) to prepare preliminary design ideas and to enhance plots of 3D views of architectural designs for the purpose of creating presentation drawings.

Michael R. Smith:

Maynard, Massachusetts, Autumn, 2004

Dr. Nasman was kind enough to allow me to update and add to the text of the previous version of this book for DataCAD 10. As the author of two previous DataCAD books (*The Official DataCAD User's Guide*, and *DataCAD 10 Advanced Construction Drawing & Visualization*), and an avid user of DataCAD, it has been my pleasure to keep the tradition of Dr. Nasman's DataCAD Project Book alive.

Because so much of the book involves the use of symbols I decided to add Chapter 16 so you can learn to make your own symbols. I chose to continue using the Relative Cartesian input method that Dr. Nasman has used in the chapters of his previous books, but since I and many users don't use this method I felt it was important to add Appendix D, which describes the other methods of input in DataCAD. And finally, also in Appendix D, I added information about DataCAD's *Identify/Set All* feature. DataCAD just would not be DataCAD without this feature, and you should get to know it well.

This version created: Thursday, November 11, 2004.

The DataCAD 11 Project Book

The DataCAD 11 Project Book
- To The Student iii
- To The Teacher iii

Chapter 1
Beginning Architectural Drawing 1
- Starting a New Drawing 2
- Selecting Menu Options 8
- Turning On Walls 10
- Drawing and Erasing Lines 14
- Creating Rectangular Polygons 17
- Adjusting the Display 21
- Adding a Door 23
- Adding Windows 26
- Quitting DataCAD 29

Chapter 2
Adding Furniture and Text 33
- Loading a File 34
- Adding Furniture (Using Symbols) 35
- Creating and Naming Layers 41
- Shortcuts 44
- Adding Text 45
- Moving Text 48

Chapter 3
3D Drawing 55
- Creating 3D Views 58
- Using the Viewing Sphere 60
- Hidden Line Removal 62
- Moving the Pictorial Image 64
- Changing Enlargement Factors 67
- Saving the Drawing 71
- Creating Perspective Views 72

Preface, Page vii

The DataCAD 11 Project Book

Chapter 4
Ferd's Motel Room, Phase 1 81
Creating and Naming Drawing Layers 82
Creating Walls 84
Setting Snap Size 85
Adding Plumbing 86
Adding Interior Walls 89

Chapter 5
Ferd's Motel Room, Phase 2 101
Adding a Standard Door 103
Centering a Door 105
Centering and Cleaning a Wall 107
Adding a Cut Wall Opening 108
Adding a Bi-fold Door 109
Adding a Sliding Door 110
Creating a New Drawing File 112

Chapter 6
Ferd's Motel Room, Phase 3 115
Adding Furniture 116
Changing the Symbol Scale 121
Using Dynamic Rotation 122
Changing the Z-Offset 123

Chapter 7
Ferd's Motel 8, Mirror & Group 133
The Mirror Function 135
The Rotation Function 141

Chapter 8
Editing and Dimensioning 147
Erasing Layers 148
Removing a Door 149
The Stretch Function 151
Adding Dimensions 152
Design Changes 160

Preface, Page viii

Chapter 9
3D Modeling, Ferd City **163**
Changing the Z-base 165
Creating 3D Blocks... 165
The Copy Function 166
Creating a Dome 168
Adding Block Buildings 171

Chapter 10
3D Modeling, Using Macros **179**
Using the AEC_MODL Macro....................................... 183
Fixing Wall Cutouts... 187
Adding a Roof... 189

Chapter 11
Using the Frameing Macro................................. **197**
Creating Wall Framing 198
Adding Openings 202
Editing a Frame... 206
Adding Floor Framing to Your Plan.............................. 209
Roof Framing 215

Chapter 12
Using Shader and Object Viewer........................ **221**
Moving the Playhouse to 0,0 222
Creating a Slab.. 223
Using GoToView.. 225
Using the Shader Function... 229
Adding a Street Lamp... 236
Adding Background Images 238
Adjusting Model Colors.. 240
Using the Object Viewer... 243

Preface, Page ix

The DataCAD 11 Project Book

Chapter 13
Detail Drawings 253
    Using the Layer Manager 254
    Creating a 2D Detail Drawing 256
    Changing Line Type 258
    Adding a Circle 259
    Adding Hatching 260
    non-Associative Hatching 263

Chapter 14
Typical Wall Section 269
    Selecting a Line Type 271
    Trimming Lines 276
    Adding Blocking 277
    Adding Insulation 278
    Adding the Floor 281
    Adding Text 285

Chapter 15
My Old House 291
    Defining Layers 292
    Creating the First Floor Slab 293
    Creating First Floor Walls 296
    Creating the Second Floor Slab 300
    Adding Second Floor Walls 302
    Adding Windows 305
    Adding Doors 306
    Adding the Roof 307
    Adding Stairs 310

Chapter 16
Creating and Modifying Symbols 315
    Create a 2D Symbol 316
    Create a 3D Symbol 322
    Explode a Symbol 329
    Redefining Symbols 331
    Replacing Symbols 333

Appendix A
The DataCAD 11 Project Book Quick Keys 345
Appendix B
Using Selection Sets 347
    The Selection Sets Menu 348
    The Mask Funcion 353
Appendix C
Using Object Viewer 355
    Setting the DataCAD Rendering Settings.................. 359
    Using the Object Viewer.. 360
Appendix D
Input Methods and Identify/Set All 365
    Drawing With Relative Cartesian 366
    Drawing With Absolute Cartesian 369
    Drawing With Relative Polar..................................... 370
    Drawing With Absolute Polar.................................... 372
    Drawing With Direction/Distance 375
    Using Identify / Set All.. 376

The DataCAD 11 Project Book
_____

CHAPTER 1

Beginning Architectural Drawing

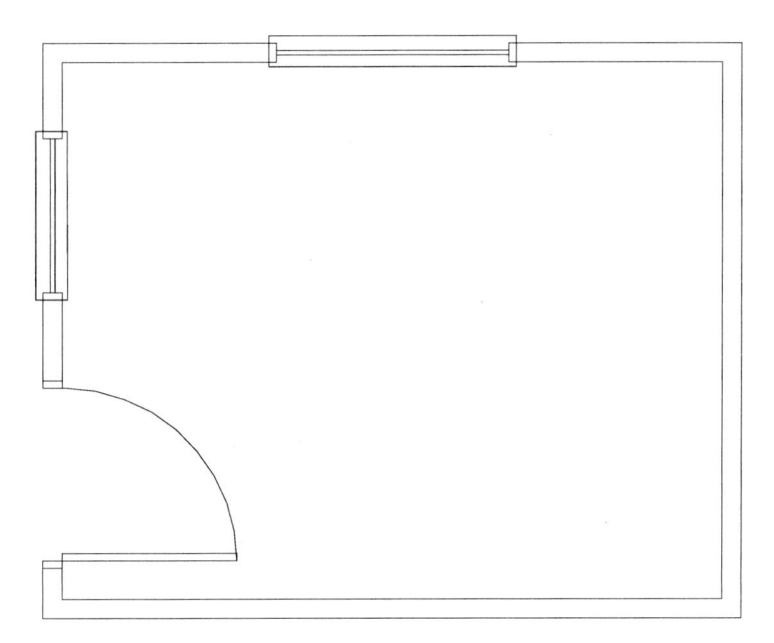

Project Goals

When you have completed this project, you will be able to:

1. Start the DataCAD software.
2. Set the disk drive path for saving drawings.
3. Name and start a new drawing.
4. Create a floor plan with double line walls.
5. Create a door swing.
6. Insert a window in a wall.
7. Save the drawing and exit the DataCAD software.

The DataCAD 11 Project Book

Prerequisite: None

Introduction

This project is designed to have you create your first floor plan with DataCAD in a minimum amount of time. Unlike some software that requires you to adjust a large number of system settings before you can start drawing, DataCAD pre-sets most system settings for standard architectural work so that you can start drawing right away. The DataCAD user interface is very clever in that you don't have to know very much about the available functions to start doing useful architectural drawing and design. However, as you gain experience, you will find that you have complete control over almost every available function. The projects in this book will discuss the various DataCAD functions as they are needed to do common architectural design operations. This "learn by doing" approach has been found to help new users do productive work in a minimum amount of time.

> •• **NOTE:** If DataCAD has not yet been installed on your computer, follow the instructions found in the DataCAD Up & Running manual that is supplied with the software.

Starting a New Drawing

1.__ **Turn on your computer.** (If you don't know how, ask your teacher for help.)

> •• **NOTE:** Check to see if your system has a shortcut icon on the **Windows Desktop**. If it has, *double click* on the **DataCAD 11** icon instead of following the next step.

DataCAD 11

Chapter 1, Page 2

Starting a New Drawing

2.__ From the Windows **TASK BAR**, select **Start, Programs, DataCAD, DataCAD 11** (see Figure 1-1).

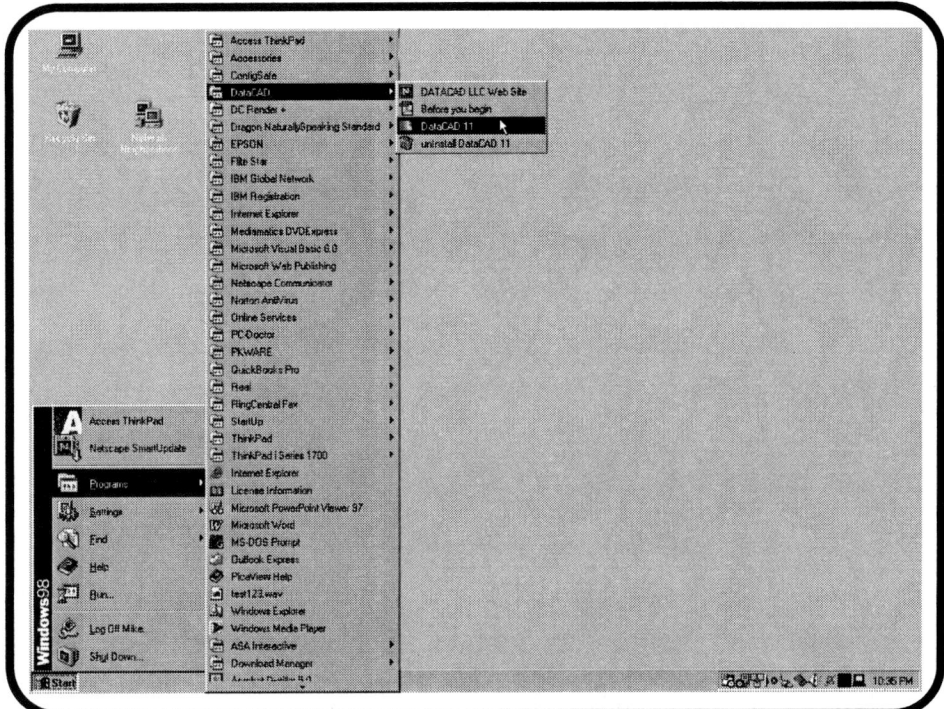

Figure 1-1, Starting DataCAD.

The exact appearance of your display will vary depending on what software has been installed.

Chapter 1, Page 3

The DataCAD 11 Project Book

After the DataCAD program has been loaded, go to the **File** menu in the Menu Bar at the top of the screen, and pick **New**. The **Enter name of drawing file to create** dialog box will appear as shown in Figure 1-2.

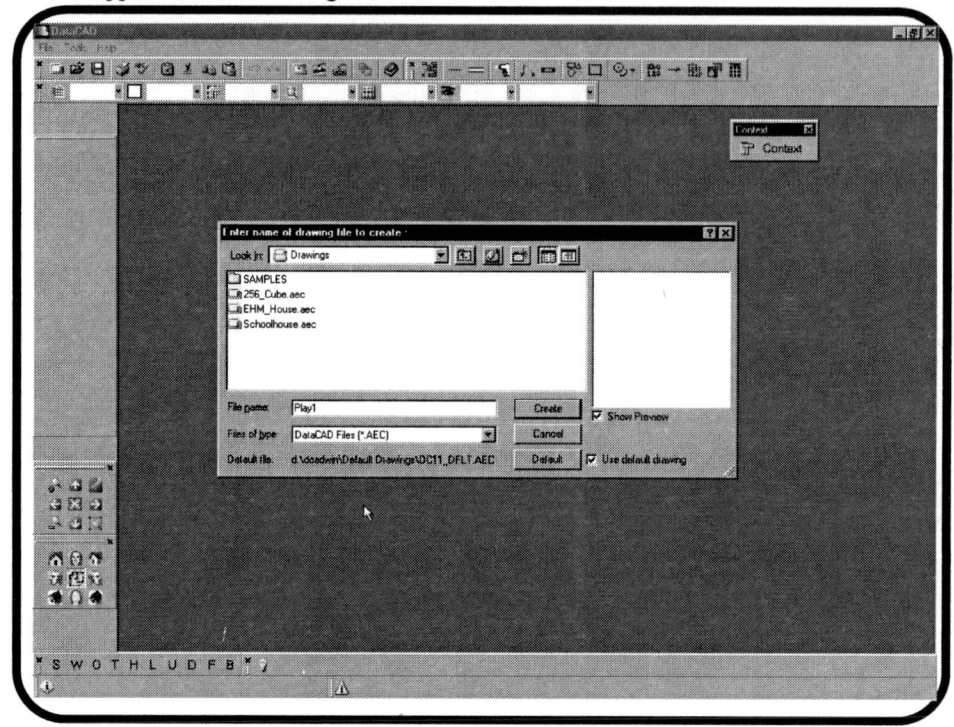

Figure 1-2, Enter name of drawing file to create.

Before you can start using DataCAD, you have to either select a file to work on, or enter a new drawing file name. The **Enter name of drawing file to create** dialog box also allows you to select the disk drive and directory path (the folder) where your drawing file will be stored.

> •• **NOTE:** If you are taking a class, ask your instructor for help in selecting or creating the proper folder for saving your files. *If you do not do this step properly, your drawings may be lost.*

3.__ Set the disk drive and directory path as desired, and then enter **Play1** in the **File name:** text box, and then click on the **Open** button.

Chapter 1, Page 4

Starting a New Drawing

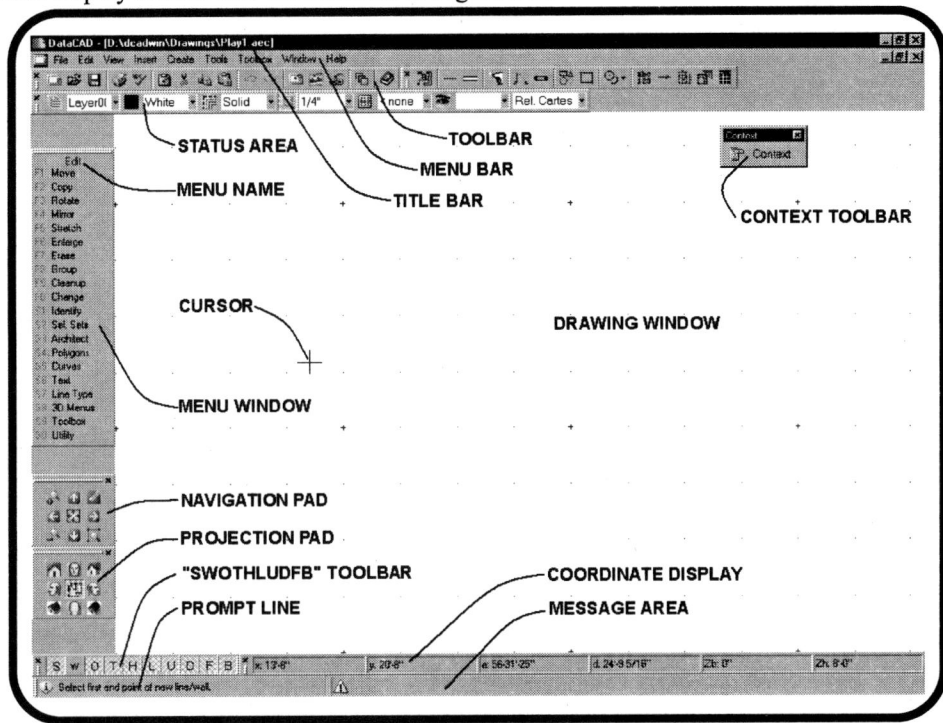

Your display should now look similar to Figure 1-3.

Figure 1-3, The DataCAD display.

•• **NOTE:** The display resolution, colors, and font style have been adjusted so that the figures in this book will print clearly. Your display may appear slightly different from the figures in this book.

Before you start making lines, you need to know a little about the DataCAD user interface. The horizontal line across the top of the display is the window **TITLE BAR**. This is common to all Windows programs and shows the name of the program and the name of the currently active file.

The next line below the **TITLE BAR** is the **MENU BAR**. The **MENU BAR** is also common to most Windows programs. It contains a list of words that provide access to pop down menu options. When you position the cursor over a word in the **MENU BAR** and click the *left* mouse button *once*, a pop down menu will open to reveal a list of options. An option is selected from the list by moving the cursor over the desired option and then clicking the *left* mouse button *once*.

Chapter 1, Page 5

The DataCAD 11 Project Book

The column on the left side of the display is the **Menu Window**. This is where the DataCAD menu options are shown. Above the menu buttons is the name of the current menu (in Figure 1-4 it's the **Edit** menu; in Figure 1-5 it's the **Utility** menu). Options may be selected by pressing the function key shown beside each option. F1 is the function key marked F1 on your keyboard. S1 means hold the *shift key* down and press the key marked F1.

The line on the bottom of the display is called the **Prompt Line**. This is where the computer tells you what it is expecting you to do next. This line will change as you use the program.

If you move the mouse around, the **Drawing Cursor** (which looks like a + sign) will move around on the display and the relative X and Y coordinates (shown near the bottom of the display) will change. As mentioned earlier, the bottom line of the display shows the prompt line.

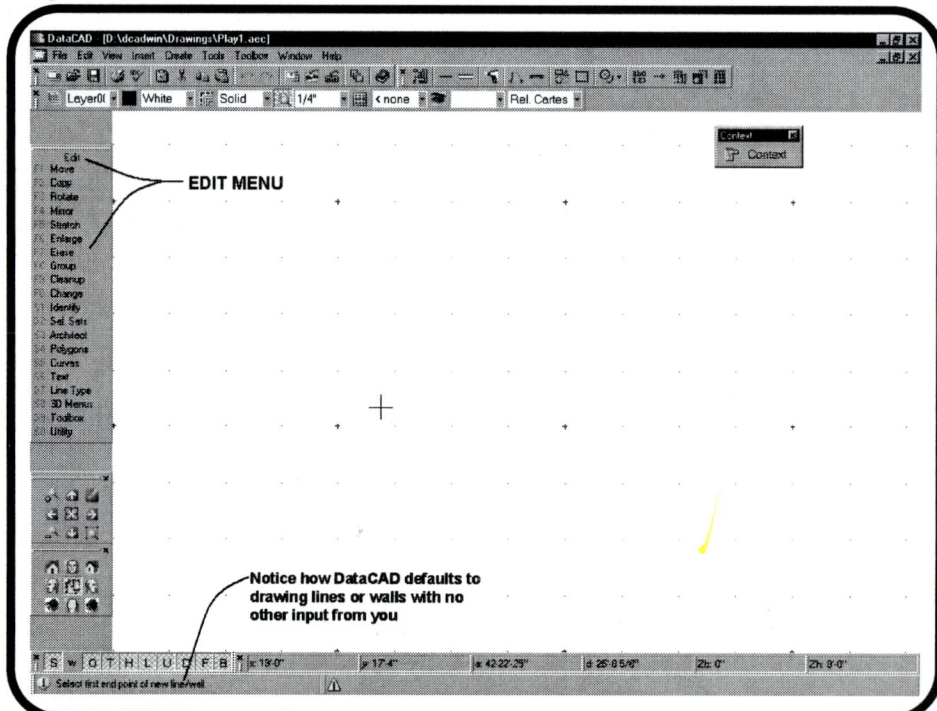

Figure 1-4, The Edit menu.

Starting a New Drawing

There are *two* main menus in DataCAD: the **Edit** menu and the **Utility** menu. Each time you select an option from a menu, the contents of the **Menu Window** will change. When you press the *right* mouse button, the system will back up one menu in the menu list. If the **Utility** menu is active, pressing the *right* mouse button will cause the **Edit** menu to appear. If the **Edit** menu is active, pressing the *right* mouse button will cause the **Utility** menu to appear. Try it.

4.__ Press the *right* mouse button *once* (or press [SHIFT] + [F10])

This brings up the **Utility** menu (as shown in Figure 1-5). The two main menus in DataCAD are the **Edit** menu and the **Utility** menu. You can jump between them either by pressing the *right* mouse button, or by selecting them from the menu.

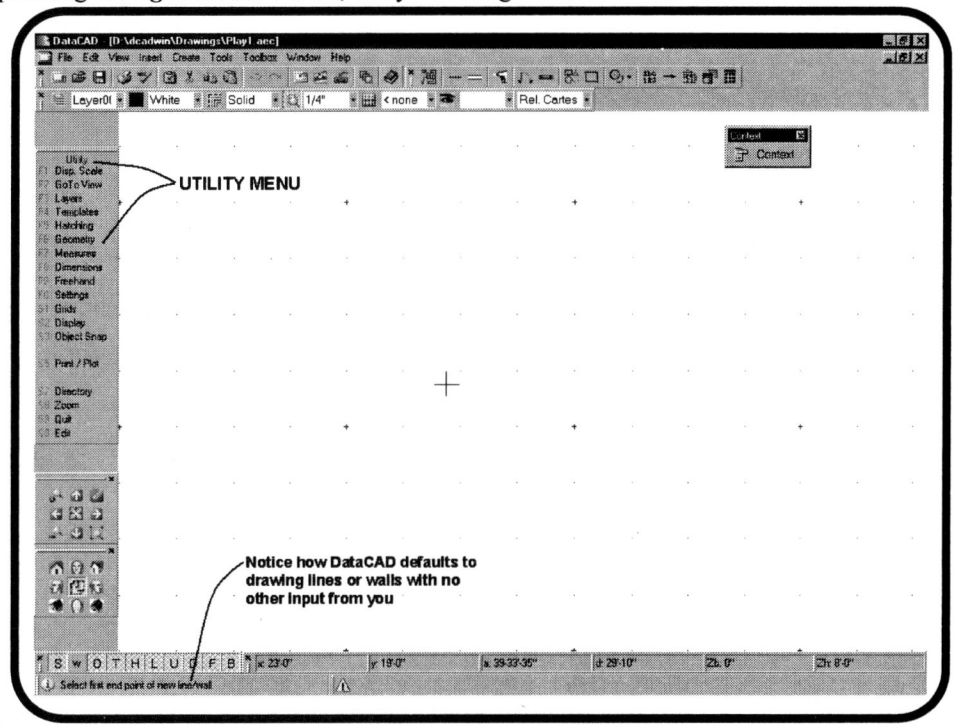

Figure 1-5, The Utility menu.

Chapter 1, Page 7

The DataCAD 11 Project Book

Selecting Menu Options

There are several ways to select a DataCAD function. You can press the function key (or the shifted function key) shown on the *left edge* of the menu item you want to select. Or, you can move the drawing cursor over the desired selection and press the *left* mouse button *once*. (Note that the drawing cursor changes from a + to an arrow when it is moved out of the **Drawing Window**.)

Another way to make a menu selection is by using a *DataCAD Shortcut*. There are some keys that you can press that will immediately jump to a menu. These will be discussed as they are needed. There is a reference page in the back of this book that lists the *DataCAD Shortcuts*. Try a couple of simple ones now:

- Hold down the [SHIFT] key and press the [+] sign. The cursor will change to fill the screen from top to bottom and left to right. Hold down the [SHIFT] key and press the [+] sign again and the cursor will change back to its original size.

- Press the **C** or **c** key and the **Copy** menu will appear as the current menu on the left side of the screen. Press the *right* mouse button and you will back up one menu, to where you were before pressing the **C** key.

It is also possible to select a DataCAD function or menu from the **TOOLBAR** across the top of the display. When you move the cursor over a **TOOLBAR** *icon*, an information line near the bottom of the display will tell you the function of that tool. A **tool tip** will briefly appear near the cursor when you move it over a **TOOLBAR** *icon*.

All of these methods have the same result. As you become familiar with the different functions, you will develop your own favorite way to navigate around the software.

> •• **NOTE:** The first several tools on the left side of the **TOOLBAR** are used to change the contents of the **TOOLBAR**. Because of system configuration differences, your display may not always match the figures shown in this book. The figures in this book were captured from an 1024 by 768 resolution display.

Chapter 1, Page 8

Selecting Menu Options

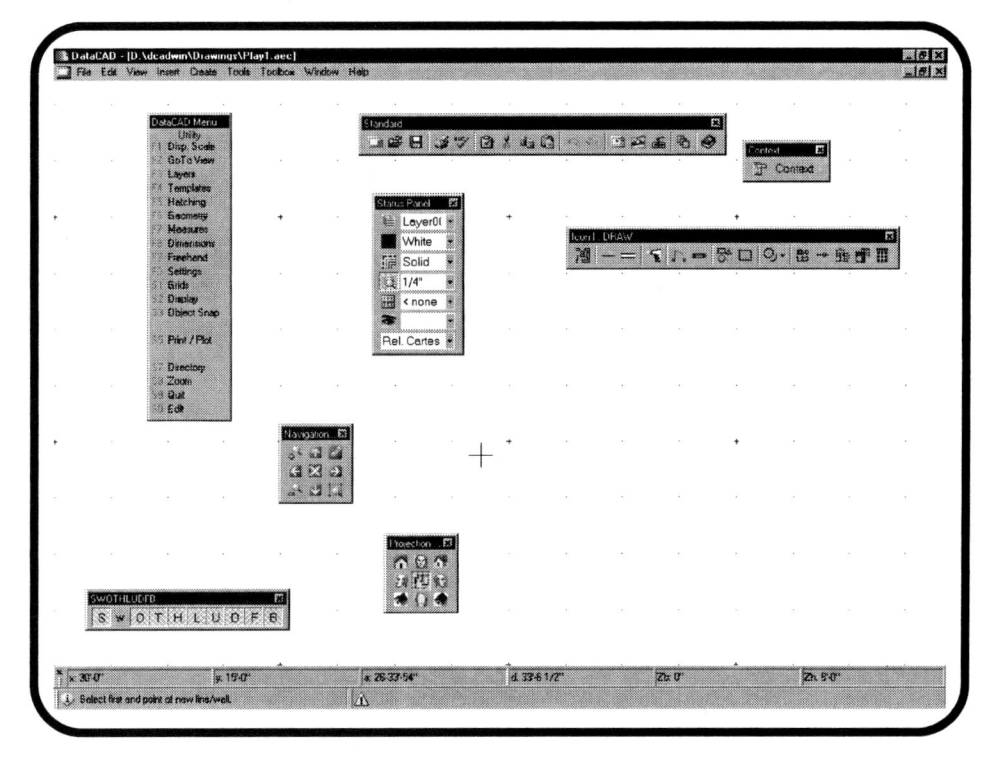

Figure 1-6, The Tool Bars and Menus Undocked.

There are several features of the DataCAD display that you should know about before you start drawing (see Figure 1-3). At the top of the display is a bar full of optoins known as the **Status Area**. The **Status Area** provides information about current system settings.

> •• **NOTE:** If the resolution of your computer monitor is high enough, the **Status Area** may be located at the bottom left corner of your display. At the 1024 by 768 resolution that the figures in this book were created at, the **Status Area** will only fit across the top of the display, as shown in Figure 1-3.

> •• **NOTE:** In Figure 1-6 you will see that DataCAD 11 actually allows you to "undock" most of the menus and tool bars. You can leave them floating as in Figure 1-6, or relocate them to suit your own personal work habits. *Double-clicking* on the top of any undocked menu will send the menu back to its original location.

Chapter 1, Page 9

The DataCAD 11 Project Book

Underneath the **Status Area** is what looks like the strange word: **SWOTHLUDFB**. These letters show the current status of some system toggles (toggles are like on-off switches). If the letter is lower case, the switch is off. If it is upper case, the switch is on. You will learn what each letter stands for as you need it to complete the exercises in this book.

Just above the bottom left corner of screen is the **Projection Pad**, which provides a quick way to switch between 3D views (see Figure 1-3). The symbol in the center of the 9 buttons (looks like a building floor plan) represents the plan, or "orthographic" view that we will be using to most everything in this book. Just above the **Projection Pad** is the **Navigation Pad**. The buttons in the **Navigation Pad** are used to move around your drawing and the **Drawing Window**. These pads will be discussed in more detail as they are needed.

Just to the right of the **Prompt Line** is a **Message Area**. As you make selections or changes, the **Message Area** will keep you informed of the action that has taken place. Above the **Message Area** is the **Coordinate Display**. As you move the drawing cursor around the display, the **Coordinate Display** will show the current X and Y coordinates of the cursor. It will also show the current angle and displacement of the drawing cursor, but only X and Y values will be used in the first drawing exercises in this book.

> •• **NOTE:** If the **Navigation Pad** or **Projection Pad** do not appear on your display, the resolution or the menu font has to be adjusted. Ask your instructor, tech support person, or dealer for help if these features are not shown on your display.

Turning On Walls

Your first job will be to create some double line walls for *Ferd's Playhouse*. The DataCAD program can automatically draw double lines to show the thickness of walls (but only if you ask for them). The **w** in **SwOTHLUDFB** shows if multiple line walls or single lines will be drawn. If the **w** is lower case, single lines will be drawn. If the **W** is upper case, multiple line walls will be drawn. Another way to select walls is by pressing the ▣ key (note that the ▣ key looks like the parallel lines of a wall).

> •• **NOTE:** If the **Edit** menu is not on the display, click the *right* mouse button or select **Edit** from the **Utility** menu.

Chapter 1, Page 10

Turning On Walls

There are *four* possible ways to select the **Architect** menu option in the next step. The first is to press [SHIFT] +[F3] when the **Edit** menu is displayed. The second is to move the drawing cursor over the menu until **Architect** is highlighted and click the *left* mouse button *once*. The third is to move the cursor over the **Architect** menu tool on the toolbar and click the *left* mouse button *once*. The fourth (and fastest) is to press the lower case letter **a** on the keyboard.

A complete listing of DataCAD quick (or shortcut) keys is found in **Appendix A** of this book. You might want to make a copy of this reference page and keep it handy whenever you are working with DataCAD. Before long, you will drawing with one hand, selecting commands with the other, and amazing your friends with your speed.

5.__ Press the **a** key to select the **Architect** menu option.

The DataCAD 11 Project Book

The **Architect** menu is shown in Figure 1-7. Note that the **Walls, 2 Line Walls,** and **Outside** buttons are depressed in Figure 1-7. The function key indicator is a different color when the key has been depressed. The buttons that have colored text are known as toggles.

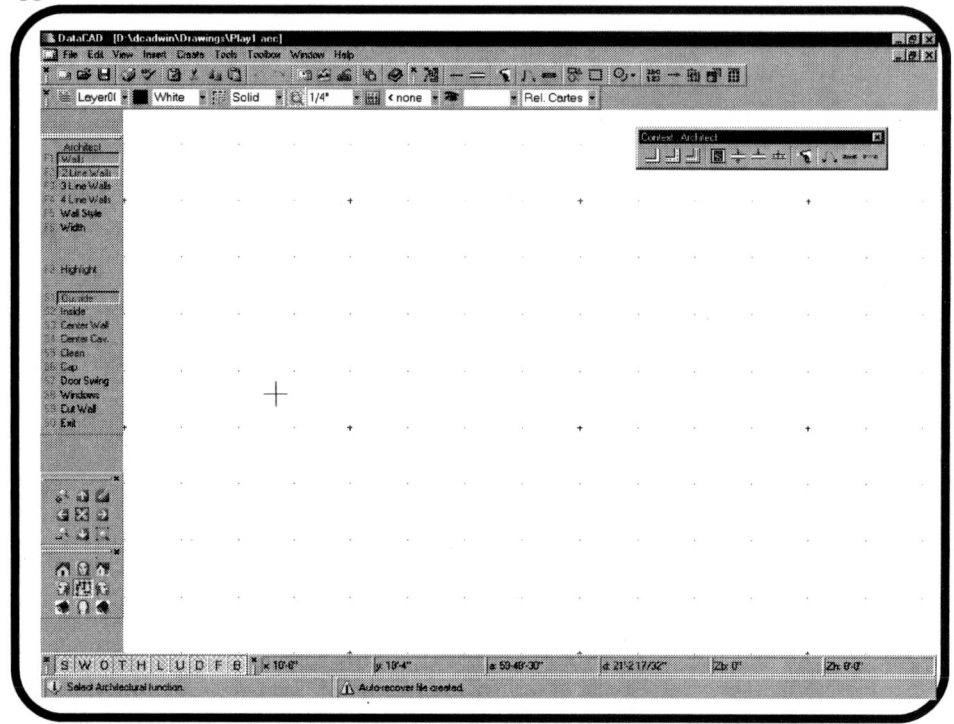

Figure 1-7, The Architct menu.

Toggles are like switches that are either on or off. When a toggle button is *depressed,* the function is turned *on*. This indicates that double line walls are now on. Also note that the **W** in **SWOTHLUDFB** is upper case in Figure 1-7. In the next step, you will turn on walls in your system.

6.__ When the **Architect** menu is displayed, **press function key** [F1] . (Or you can move the drawing cursor over the **Walls** button and click the *left* mouse button *once*.)

 •• **NOTE:** If the **Walls** button is not depressed, repeat the previous step. Since the **Walls** option is a toggle, each time it is selected it will be switched on or off.

Turning On Walls

Now that you have told the computer to draw double line walls, you have to tell it whether you will be specifying walls by outside or inside. In this example you need to use the **Outside** option.

7. ___ Toggle the **2 Line Walls** option **on**.

8. ___ Toggle the **Outside** option (in the **Architect** menu) **on** so that the **Outside** button in the **Architect** menu is depressed (the color of the function key should also change when the option is selected, see Figure 1-7).

9. ___ From the **Architect** menu, select the **Width** option.

The prompt near the bottom of the display should now read *Enter wall width* (see Figure 1-8). Just to the right of the prompt, the current width will be shown and highlighted. If you enter values from the keyboard at this point, the current width will be replaced with the value entered.

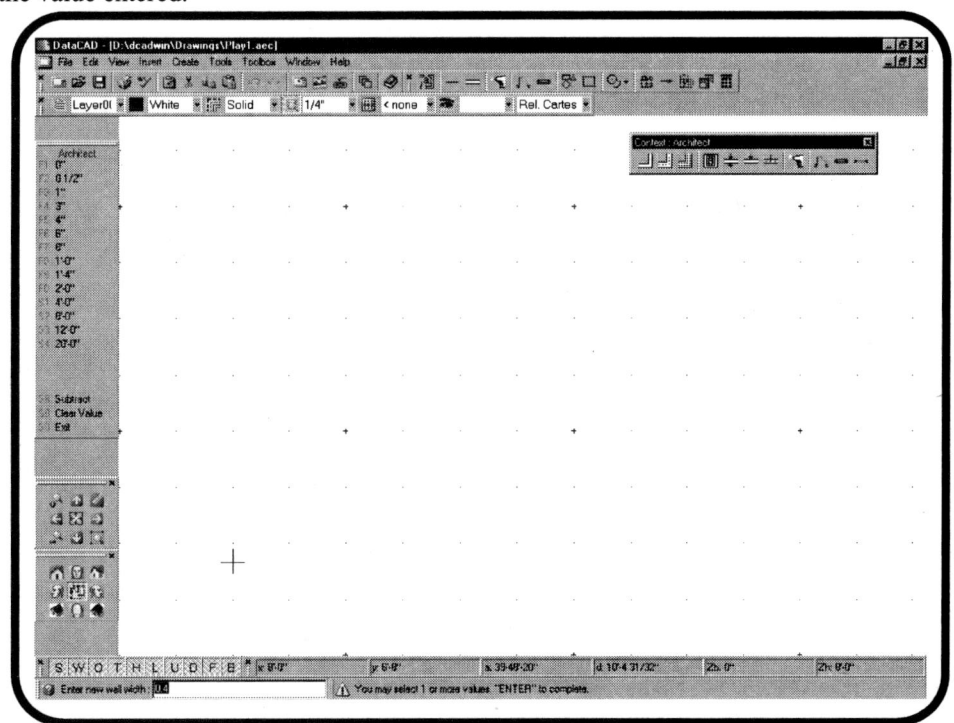

Figure 1-8, Entering wall width.

The desired width for the playhouse walls is 4 inches.

Chapter 1, Page 13

The DataCAD 11 Project Book

> •• **NOTE:** Before you enter wall width, you have to know how values are specified in DataCAD. Since feet, inches, and fractions of an inch are the most common measurements used in the USA, DataCAD makes it easy to enter this type of value by separating each part of the measurement with a period or a forward slash. For example, six feet three inches is entered as 6.3. Two feet 4 and 1/2 inches is entered 2.4.1/2. Four inches is entered as 0.4, or just .4 (no 0). So 2.5 would be 2 feet five inches, not 2.5'. And 1/4" could be entered as 0.0.1.4, or just ..1.4, or ..1/4. The periods and slashes are just place holders, and can be used interchangably. Likewise you can see that zeros can be used, or not.

10.__ At the prompt *Enter wall width*, type **0.4** (or just .4) and then press the **Enter** key.

11.__ Exit the **Architect** menu either by selecting **Exit** from the menu, or by clicking the *right* mouse button.

Drawing and Erasing Lines

Before you actually start drawing a floor plan, it is a good idea to practice drawing and erasing a few lines.

12.__ Click the *right* mouse button until the prompt line reads *Select first end point of new line/wall*. Move the drawing cursor somewhere in the drawing window and click the *left* mouse button *once*.

13.__ Move the drawing cursor to a new location and click the *left* mouse button *once*.

> •• **NOTE:** Because both the **Walls** and the **Outside** options are turned on, the computer wants to know which side of the line to make the *inside* of the wall. For now, it doesn't matter which side you choose.

14.__ When the prompt *Select a point to define the inside of the wall* appears, move the drawing cursor to one side of the line and click the *left* mouse button *once*.

This completes the first double line wall segment and also anchors the starting end of the next line.

15.__ Move the drawing cursor to a new location and click the *left* mouse button *once*.

You can continue drawing connected lines in this fashion. If you want to start a new series of lines, click the *right* mouse button *once*. This will end the current string of lines. If you click near the starting point of the first line, the system will close the series of walls. You should practice drawing lines until you feel that you have control of the process.

16.__ Draw several lines on the display.

Drawing and Erasing Lines

•• **NOTE:** As you draw and edit geometry, your display may not show exactly what is in the drawing until you refresh the display by pressing the ^ (**Esc**) key.

Next you need to know how to erase lines. (Actually lines are just one kind of drawing *entity* and erasing works the same for all types of entities.) There are two basic methods.

The standard Windows method:

17.___ Hold down the [CTRL] key and press the **Z** key.

Each time you do this the last entity or group that you drew will be erased. (If you draw several lines without clicking the *right* mouse button, these lines will be part of a **Group**. Rectangles and symbols also form groups, but more about these later.)

Pressing [CTRL] + **Y** will redraw the last thing you erased.

The DataCAD method:

18.___ The **Erase** menu can be activated either by selecting the **Erase** option from the **Edit** menu, by pressing the **e** [or the **E**] key, or by selecting **Edit, Erase** from the **MENU BAR**. You will then see 7 options: **Entity, Group, Area, Fence, Mask, Layer Search** and **Partial** (see Figure 1-9). You'll learn more about these options later, but for now you should pick the **Entity** option. Now, in the **Drawing Window**, place the cursor on or close to the entity you want to erase, the press the *left* mouse button. That entity will be erased.

One more way to erase several entities at one time is via the **Area** option in the **Erase** menu.

19.___ Press the the **e** [or the **E**] key.

The DataCAD 11 Project Book

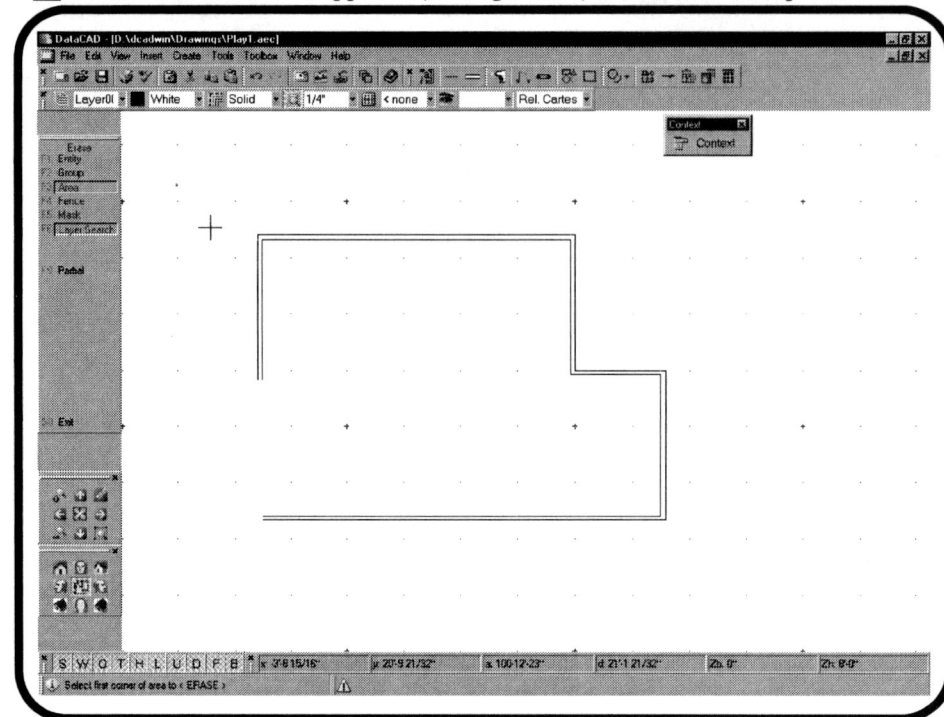

20.__ When the **Erase** menu appears, (see Figure 1-9) select the **Area** option.

Figure 1-9, The Erase menu.

21.__ Capture some lines in an area box by first clicking the *left* mouse button *once* at a point where you want the *upper left* corner of the area box to be, and then moving the mouse until you have captured the entities to be erased in a rectangle. Then click the *left* mouse button *once*.

Notice that only lines (or entities) completely inside of the area box will be erased.

22.__ Continue erasing until all entities have been removed from the display.

You can check to see that everything has been erased by refreshing the display. This is done by pressing the [ESC] key.

23.__ Press [ESC] to verify that the display is clear. If lines appear, erase them.

> •• **NOTE:** DataCAD 11 also has multiple undo and redo capabilities through the use of the **MENU BAR Edit** option. To use this feature, select **Edit** from the **MENU BAR**; then select either **Undo** or **Redo**.

Creating Rectangular Polygons

Creating Rectangular Polygons

There is one more menu to look at before you start creating the floor plan for Ferd's Playhouse. Since the house will be rectangular, you will tell the computer to draw a rectangle instead of straight lines. You could of course draw four lines to form a rectangle, but that would be more work than needed.

24.___ Click the *right* mouse button until the **Edit** menu is on the display [or press the [:] key].

25.___ Select the **Polygons** option from the **Edit** menu.

The **Polygons** menu is shown in Figure 1-10. You will use the **Rectangle** option, so it has to be selected.

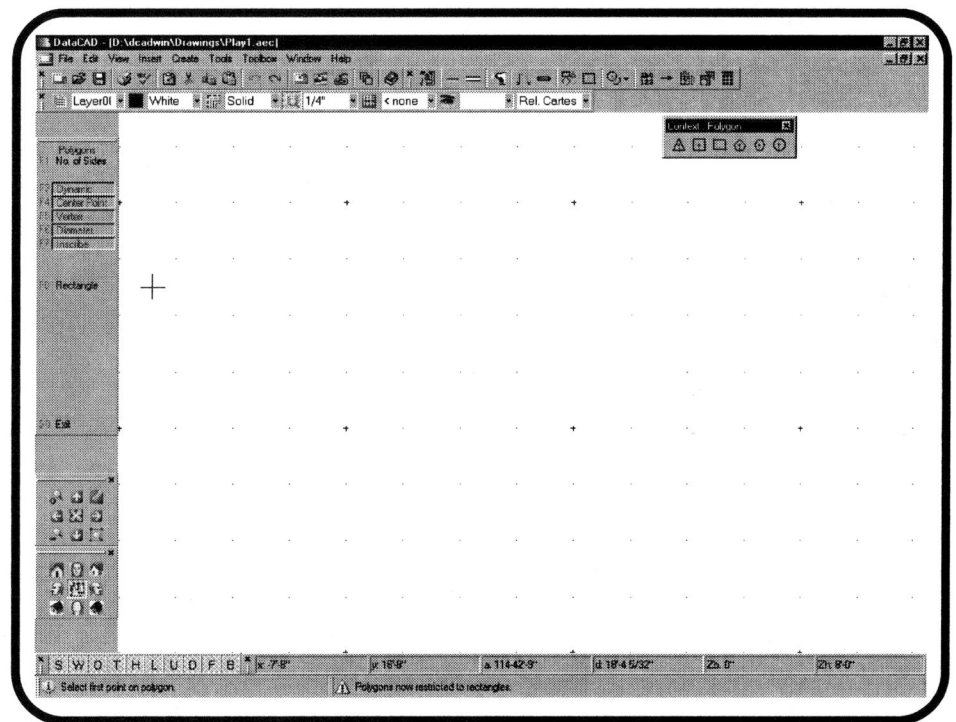

Figure 1-10, The Polygons menu,

26.___ Select the **Rectangle** option from the **Polygons** menu either by pressing the **F10** function key, or by moving the drawing cursor until the **Rectangle** button is highlighted and then clicking the *left* mouse button *once*.

Chapter 1, Page 17

The DataCAD 11 Project Book

•• **NOTE:** Rectangles are created by first moving the drawing cursor to the *lower left* corner of the rectangle, *clicking* the *left* mouse button *once*, and then moving the cursor to the *upper right* corner and *clicking* the *left* mouse button *once* again. By watching the coordinate readout near the bottom of the display, you can make the rectangle exactly the size you want. Ferd's Playhouse is 12 feet wide by 10 feet high. Since DataCAD is accurate out to several decimal points, there is one setting that you should check, which will make it easier to draw like this. *Click* the *right* mouse button until the **Utility** menu is displayed. Select the **Grids** option by *clicking* on it with the *left* mouse button. Make sure the **Grid Snap** option is on (it will be green, and the button will appear depressed). Now *click* on **Grid Size**, then **Set Snap**, and finally *click* on the **1"** option. *Click* the *right* mouse button three times and you'll be right back to the **Utility** menu, where you started. Now your cursor will jump, or "snap", at 1" increments as you move the cursor.

••• **WARNING**••• Some systems have very sensitive mouse buttons. If you press too hard or too long on the mouse button, the system might think you pressed the button more than once. When the instructions say *click* the button, just give it a soft quick tap.

Creating Rectangular Polygons

27.__ Move the drawing cursor near the *lower left* corner of the drawing area and *click* the *left* mouse button *once*.

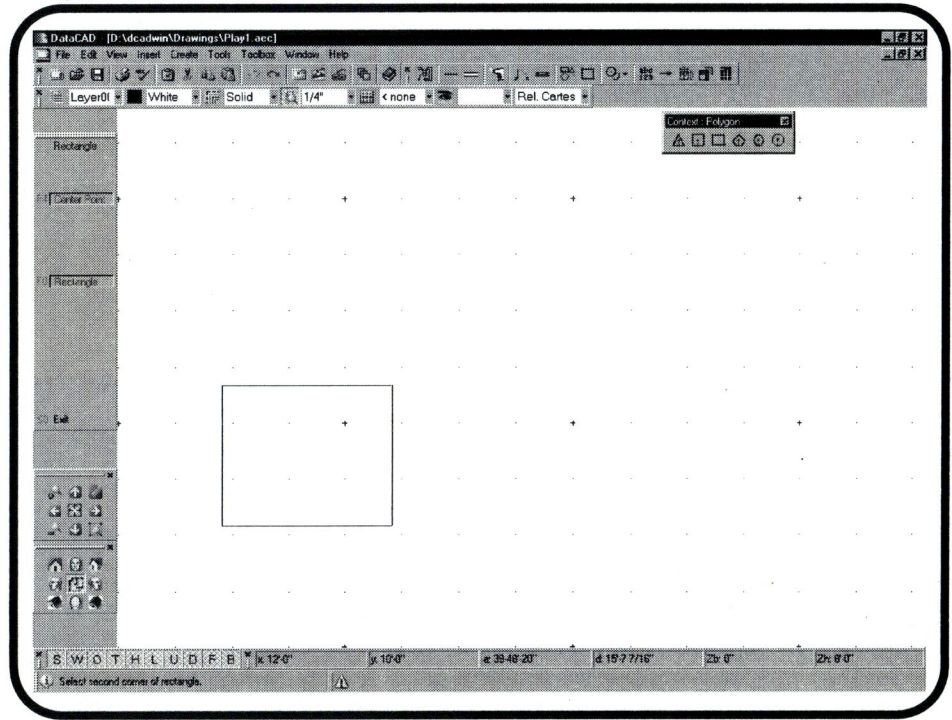

Figure 1-11, Creating a rectangle.

28.__ Move the drawing cursor until the coordinates show **X=12'-0"** and **Y=10'-0"**. Click the *left* mouse button *once* when these numbers are shown (as shown in Figure 1-11). The rectangle is drawn.

As soon as you tell the computer where the corners of the rectangle are, it wants to know where to locate the inside of the wall. (This is because you asked for double line walls with the **Outside** option turned **on**.) Since the outside dimensions of the Playhouse must be 12 by 10, the other side of the walls need to be placed inside the rectangle.

29.__ When the prompt *Select a point to define the inside of the wall* appears, move the drawing cursor somewhere *inside* the rectangle and click the *left* mouse button *once*.

The DataCAD 11 Project Book

Notice that not only did the computer draw a rectangle with double lines, it also automatically cleaned out the corner intersections as shown in Figure 1-12. This is much faster than if you had to carefully draw the same thing with pencil and paper.

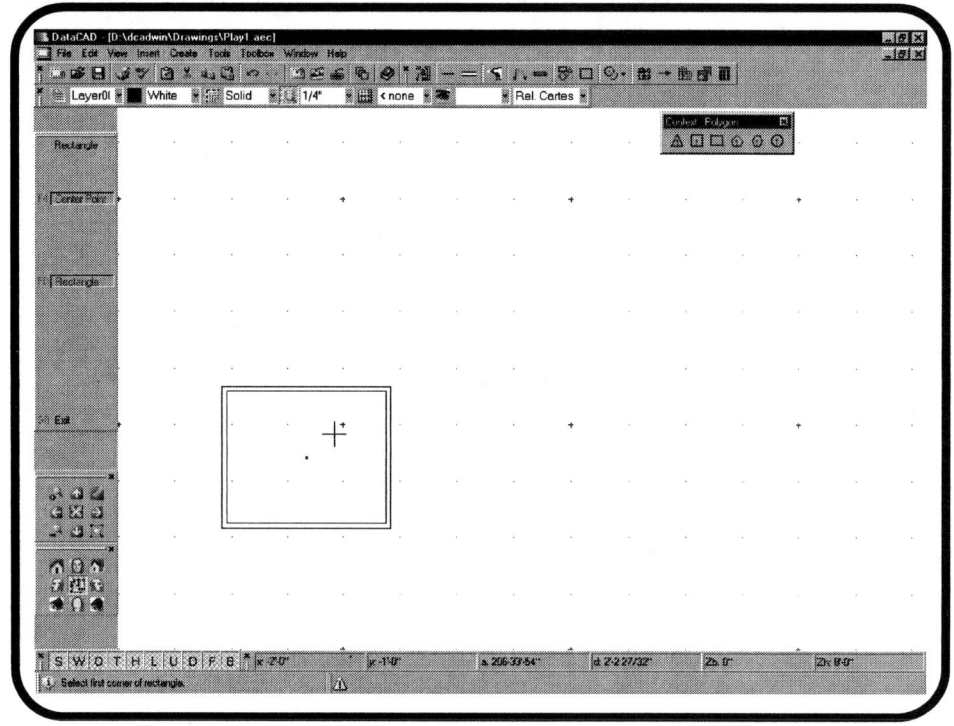

Figure 1-12, Creating a rectangle.

•• NOTE: If you have to stop the drawing and get back to it another time, don't worry. Just click the *right* mouse button until the **Utility** menu is on the display. Then select **Quit** from the **Utility** menu. Next select the **Yes** option. The next time you start DataCAD, select the desired file name from the **Enter name of drawing file to open:** dialog box. (If you do quit this project early, make a note of what step you are on so that you will know where you stopped.)

Chapter 1, Page 20

Adjusting the Display

Adjusting the Display

The rectangle for Ferd's Playhouse does not fit the display very well in Figure 1-12. The display is like a window to your drawing world, just like the window in a room is a window to the world outside. You can move closer or farther from the window, or you can move left or right to change how much of the world you see. Pressing the **PageUp** or **PageDown** keys on the keyboard has an effect similar to moving closer to or farther from the window in a room. The *arrow keys* on the keyboard move the drawing window up, down, left, and right.

30.__ Change your drawing window by pressing the **PageUp** and **PageDown** keys several times. Press the *arrow keys* to move the drawing window around.

> •• **NOTE:** Selecting the *arrow* buttons on the **Navigation Pad** has the same effect as pressing the *arrow keys*. Try it.

The DataCAD software provides you with some additional ways to adjust the drawing window. If you want to center the window on a spot on the drawing, just move the drawing cursor to that spot and press the key marked **Home** [HOME]. There are other ways to adjust the drawing window using a couple of **Zoom** options.

31.__ Press the [/] key (do not confuse it with the [\] key) to get the **Zoom** menu on the display. Press function key [F2] (**F2, ReCalc**) to make the computer calculate how to make the drawing fit the window.

32.__ Or, *click* on the button in the center of the **Navigation Pad** (the one that has four red arrow within a box).

33.__ Or, in the **Menu Bar** at the top of your screen, *click* on **View**, and **Zoom to Extents**.

These **Extents** options will automatically change the display to fit the current drawing into your drawing window. If you want to zoom in on a small part of the drawing, you can also do this in several ways.

34.__ Press the [/] key to get the **Zoom** menu on the display. Note that the prompt line for the **Zoom** menu says *Select first corner of the Zoom window*.

Chapter 1, Page 21

The DataCAD 11 Project Book

•• **NOTE:** If the **Free Zoom** option is **on**, the system will use whatever display scale it takes to make the most use of the screen. If **FreeZoom** is **off**, the system will make the drawing as large as possible with a standard display scale. I prefer to keep **FreeZoom on**. You should experiment with both options and decide which you prefer.

When you are in the **Zoom** menu, if you move the drawing cursor somewhere and click the *left* mouse button *once*, and then move the cursor, it will drag an area box out on the display. The next click of the *left* mouse button will make a new window which shows what was in this area box. Try it:

35.__ Press the **Page Up** key several times to make the plan smaller on the display. Go to the **Zoom** menu then *click* the *left* mouse button at each of two corners of a zoom window and try to zoom on an area similar to that shown in Figure 1-13.

As with most options in DataCAD, there are several ways to do the same thing. Use whichever ones you find fastest and most convenient. Here are two more options for getting to the **Zoom** menu so that you can select an area box for zooming:

36.__ *Click* on the button in the lower right corner of the **Navigation Pad** (the one that looks like a magnifying glass with a box around it). Then select your area box to zoom into.

Adding a Door

37.___ Or, in the **Menu Bar** at the top of your screen, *click* on **View**, and **Zoom**. Then select your area box to zoom into

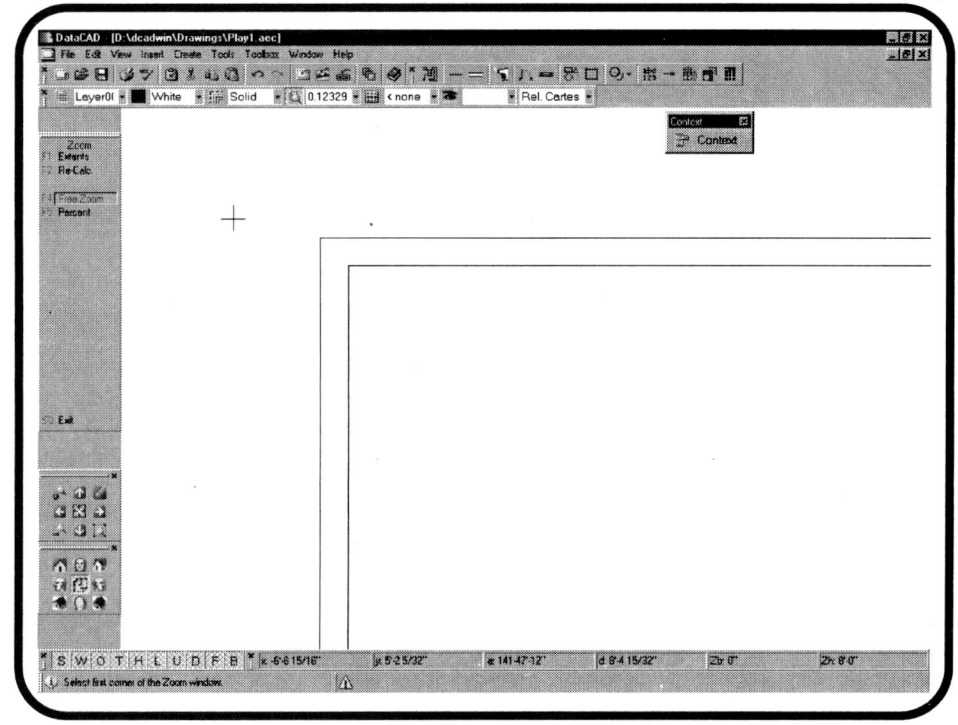

Figure 1-13, Zooming in on an area.

38.___ To get back to a full display, just select **Extents** from the **Zoom** menu (remember all the options for getting there), or the center button in the **Navigation Pad**.

Adding a Door

Now it is time to add a door so that Ferd can get into his Playhouse.

39.___ Press the **a** (*not* the A) key to jump to the **Architect** menu.

40.___ Select the **Door Swing** option from the **Architect** menu.

41.___ Select the **Door Style** option from the **Door Swing** menu.

42.___ Toggle the **Single** option **on**; then *click* the *right* mouse button *once* to return to the **Door Swing** menu.

Chapter 1, Page 23

The DataCAD 11 Project Book

The door swing menu is shown in Figure 1-14. Before you locate the door, it would be handy if the coordinate 0,0 location could be reset so you could use the coordinate readout to help locate the door. There is a quick key press that resets the 0,0 location. The quick key for resetting the coordinates is the unshifted ~ key. (Note that ` and ~ share one key, and that ' and " share another. Because it is difficult to tell the difference between the ` and the ' when they are printed, this book will refer to the unshifted ~ key whenever the ` key is required.)

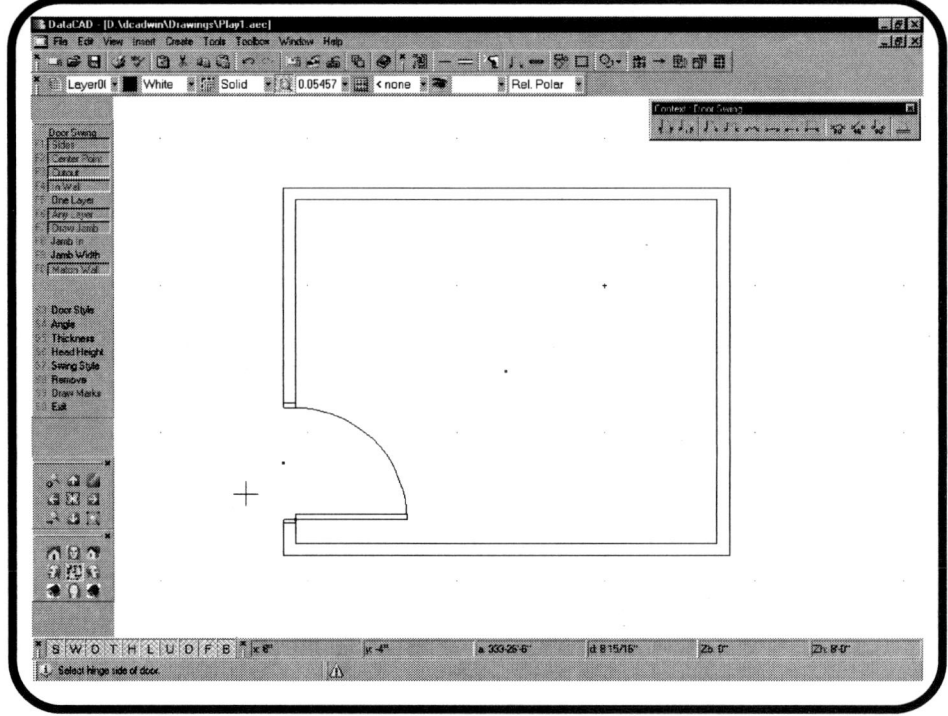

Figure 1-14, The DoorSwng menu.

•• **NOTE:** When a step calls for the *object snap button*, either press the **n** [or **N**] key, or you may use the *middle* mouse button if you have one and it is set up to do so. Using the middle mouse button for snapping is one of the fundamental time savers in DataCAD, so if yours does not work as such, see your instructor to get it set up.

43.___ Press the unshifted ~ key. Then move the drawing cursor to the *lower left* corner of the outside of the Playhouse and *click* the *object snap button* once.

Chapter 1, Page 24

Adding a Door

> •• **NOTE:** The *object snap button* causes the cursor to jump to the end of the nearest line. This makes it easier to locate something exactly at the end of a line rather than wherever the cursor happens to be. Object snap can also be set to jump to the middle of a line, but that will be covered in a later project.

44.__ Move the mouse around and verify that the X=0, Y=0 coordinate is at the lower left corner of the Playhouse. If it is not, you should repeat the previous step. If you can not make object snap work, ask your teacher for help.

45.__ When the prompt *Select hinge side of door.* is on the display, move the drawing cursor to a point **1' 0"** above the *lower left* corner of the house (watch the coordinate readout) and click the *left* mouse button *once*.

> •• **NOTE:** The cursor does not have to actually touch the thing you are snapping to; it just has to be close.

This anchors the hinge side of the door. Now you have to tell the computer where to put the other side of the door.

46.__ At the prompt *Select strike side of door.* move the drawing cursor **3'0"** above the hinge (watch the coordinate readout) and click the *left* mouse button *once*.

Once you tell the computer where to put the hinge and strike sides of the door, it will ask you which way you want the door to swing. You will have the door swing into the house.

47.__ At the prompt *Select direction of door swing.* move the drawing cursor somewhere *inside* of the house and click the *left* mouse button *once*.

The computer cuts the walls for the door, but now it wants to know where the outside of the wall is so that it can add the details correctly.

48.__ At the prompt *Select any point on the outside of the wall.* move the drawing cursor somewhere *outside* of the house (but on the same side of the house where the door is) and click the *left* mouse button *once*.

The computer now draws the whole door, complete with hinge, jambs, and swing arc. Use the **Zoom** option to zoom in close and check the details of what the computer just did for you. (Remember that the quick key for **Zoom** is the ▢ key, or use the **Zoom to area** button in the lower right corner of the **Navigation Pad**.) When you are done examining the door, use the **Extents** button in the center of the **Navigation Pad** button to get the whole Playhouse back on the display.

Chapter 1, Page 25

The DataCAD 11 Project Book

Adding Windows

Next, a couple of windows will be added to the Playhouse. There is a general housekeeping item to take of first. If you have been working for a while, you may have noticed that the system occasionally seems to stop and not let you do anything for a short time.

What is happening is that DataCAD is designed to save a backup copy of your drawing every so many minutes. This is so that if the power goes off, you will only lose what you have done since the last backup. It can be a bother if the save delay time is too short. On the other hand, if the save delay time is too long, you could lose a lot of important work. Here is how to adjust the save delay time.

49.__ Click the *right* mouse button until the **Utility** menu is on the display [or press [SHIFT] + [:]].

50.__ Select **Settings** from the **Utility** menu.

51.__ Select **Save Delay** from the **Settings** menu.

52.__ Enter a file save interval of **10** minutes.

> •• **NOTE:** You can set the save delay to whatever you want. Remember, if the power goes off, or if your system crashes, you will lose everything you have done since the last save. Do not set the save delay interval to be *too* long.

Now it is time to add some windows to the playhouse plan. The **Windows** menu is found in the **Architect** menu. You can jump directly to the **Architect** menu by pressing the **a** [not the A] key.

53.__ Press the **a** [not the A] key to jump to the **Architect** menu.

54.__ From the **Architect** menu select the **Windows** option.

Adding Windows

The **Windows** menu is shown in Figure 1-15. You create windows about the same way that doors are created. You locate each side of the window using the drawing cursor, and then you tell the computer where the outside of the wall is so that it knows how to draw the window details.

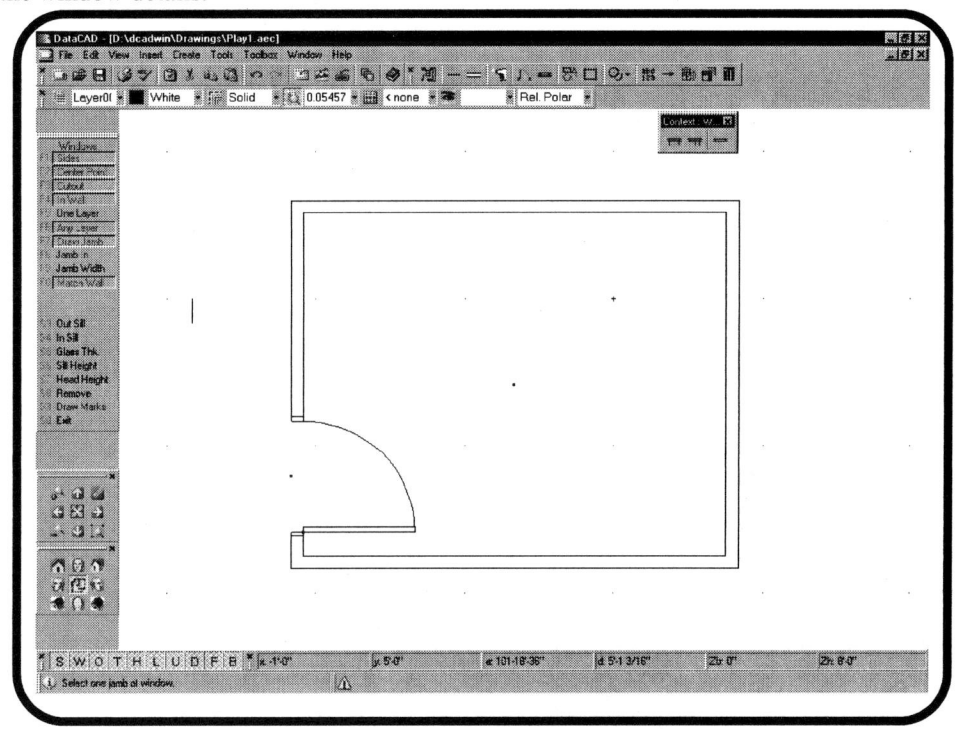

Figure 1-15, The Windows menu.

Chapter 1, Page 27

The DataCAD 11 Project Book

55.___ Insert two windows the sizes and locations shown in Figure 1-16. You will probably want to reset the coordinate 0,0 using the unshifted ~ key as explained earlier, (but with 0,0 at the upper left corner of the house) to help locate each window properly.

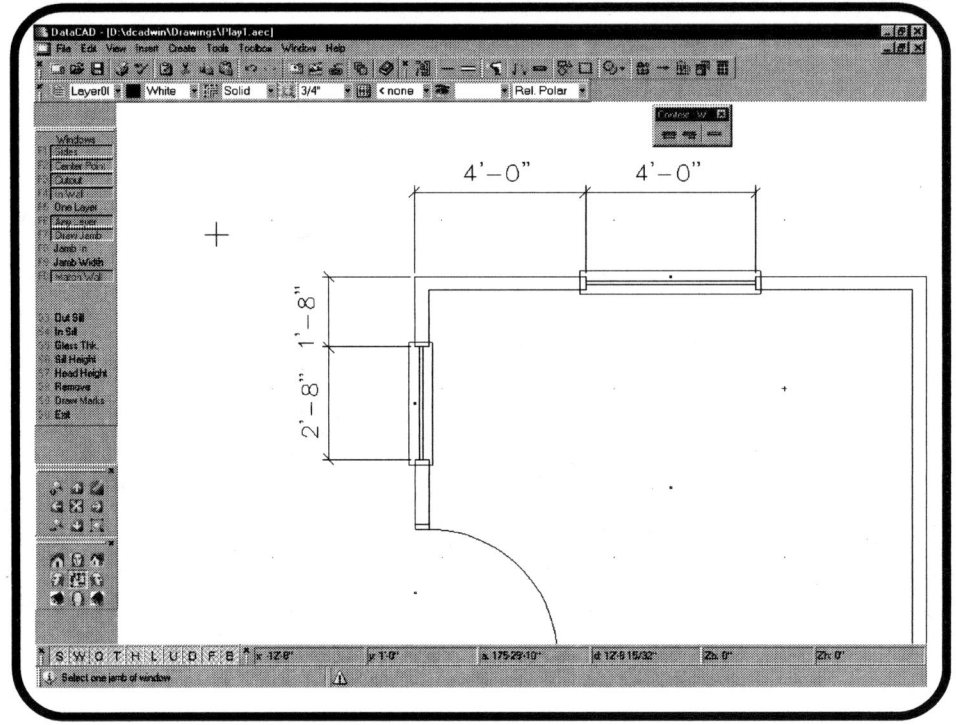

Figure 1-16, Adding windows.

This is all that is going to be done with the Playhouse in this project. Next time some furniture and text will be added to the Playhouse drawing. Before you quit this session, you have to save your drawing on the disk. The computer will do this automatically for you if you properly end the session.

Quitting DataCAD

Quitting DataCAD

56.__ From the **MENU BAR**, select **File, Exit**.

57.__ At the prompt *Save Changes to File Play1.aec?* select **Yes**.

> •••**WARNING**••• Never remove a disk from the disk drive or turn the computer off when the drive busy light is on. This could damage the disk (and possibly the drive) and you will lose all of your data files.

Review and Important Points

- When starting a new drawing: *first*, set the disk drive and directory path, *second*, enter the new drawing name.

- Whenever you use the *cursor* to position entities, keep the snap turned ON and set as large as possible.

- Remember that there are several different ways to select menu options. Either press the appropriate function key or move the cursor over the desired option and click the left mouse button.

- You can jump directly to a menu by either using a keyboard shortcut (see Appendix A) or clicking on a **TOOLBAR** tool.

- To find out what a **TOOLBAR** tool does, position the cursor over the icon and read the pop up *tool tip*, or read the **MESSAGE AREA** near the bottom of the display.

- Save your drawing often by pressing **Ctrl + S** (this means hold the **Ctrl** key down, and while holding the **Ctrl** key down, press the **S** key once).

- If you make a mistake, select **File, Open**, answer **No** to the *Save changes* prompt, and reload the last saved version of your drawing and start from there. You will almost always fix a mistake faster by redrawing then by trying to edit your way out of a problem.

- A DataCAD wall contains more than just two parallel lines. It has a Z-height and includes lines joining the Z-base and Z-height lines. If a wall loses integrity (through a mistake), it must be welded back together or replaced. Otherwise you may not be able to insert a door or window opening into a wall that has not been properly repaired.

- Use the CTRL + **X** keys to erase that last thing you drew, and the CTRL + **Y** keys to redraw the last thing you erased. You can also use the **Edit, Undo** or **Edit, Redo** options to erase or replace.

- The **Erase** menu (press **E** or **e**) provides a number of options for erasing entities.

- Use the **Navigation Pad** to quickly zoom or move around on the display.

Chapter 1, Page 29

The DataCAD 11 Project Book

- Walls, doors, and windows are found in the **Architect** menu (press **a** [not A].
- Use the unshifted ~ key and then click near the end of a line to reset the coordinate read-out to 0,0 at that point.
- Select **File, Exit** to exit DataCAD. Do *not* select **No** unless you want to cancel all changes made to the drawing since the last save.

Quitting DataCAD

Practice Drawings

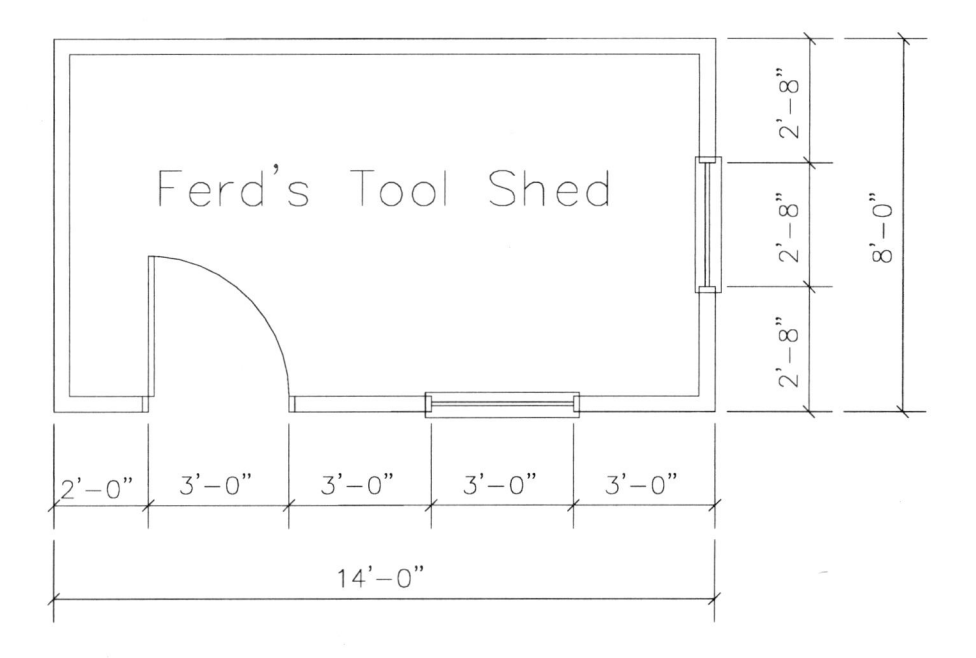

Assignment P1-A, Tool Shed

- Use the dimensions shown above and create Ferd's tool shed.
- Do not include dimensions in your drawing.
- The pictorial is for reference only.

-
-

Assignment P1-B, You design it!

- Create your own playhouse or tool shed.

Chapter 1, Page 31

The DataCAD 11 Project Book

Notes

CHAPTER 2

Adding Furniture and Text

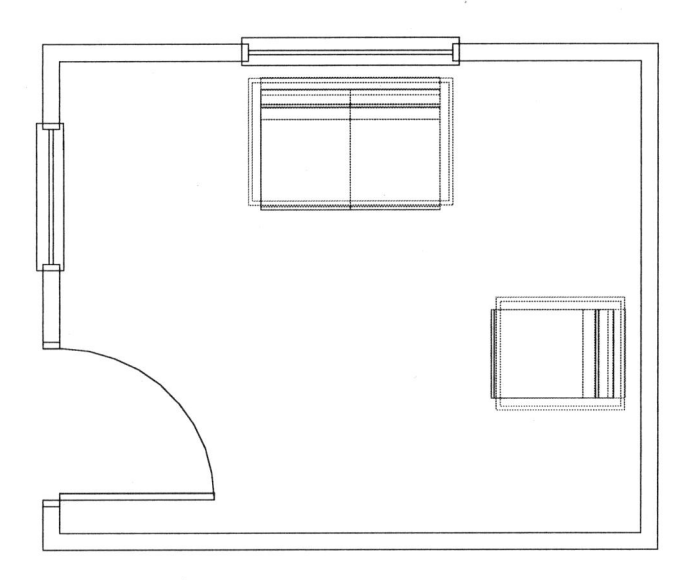

Ferd's Playhouse
Drawn by: Ferd Snodfield

Project Goals

When you have completed this project, you will be able to:

1. Open and modify a DataCAD drawing.
2. Add furniture to a drawing.
3. Add text to a drawing.
4. Create a new drawing layer.
5. Move entities between layers.

The DataCAD 11 Project Book

Prerequisite: Chapter 1

Loading a File

This project requires that you have completed Chapter 1 and have saved a drawing file named **Play1** on a floppy disk in drive a: (or on a drive with a directory path you remember).

1.__ Turn on your computer and start the DataCAD software.

- **• NOTE:** The computer will display the file names of the drawings on the currently active disk drive and directory path in the **Open DataCAD** files dialog box.

2.__ Open the file named **Play1** created in Chapter 1.

To preserve the original **Play1** file, a new file will be created with the name **Play2**.

- **• NOTE:** From this point on in this book, when a step asks you to *enter* a value or text, it means to type the value or text in the appropriate text box, and then press the **Enter** key.

3.__ From the **MENU BAR**, select **File, Save As**, and *enter* **Play2** as the file name.

Adding Furniture (Using Symbols)

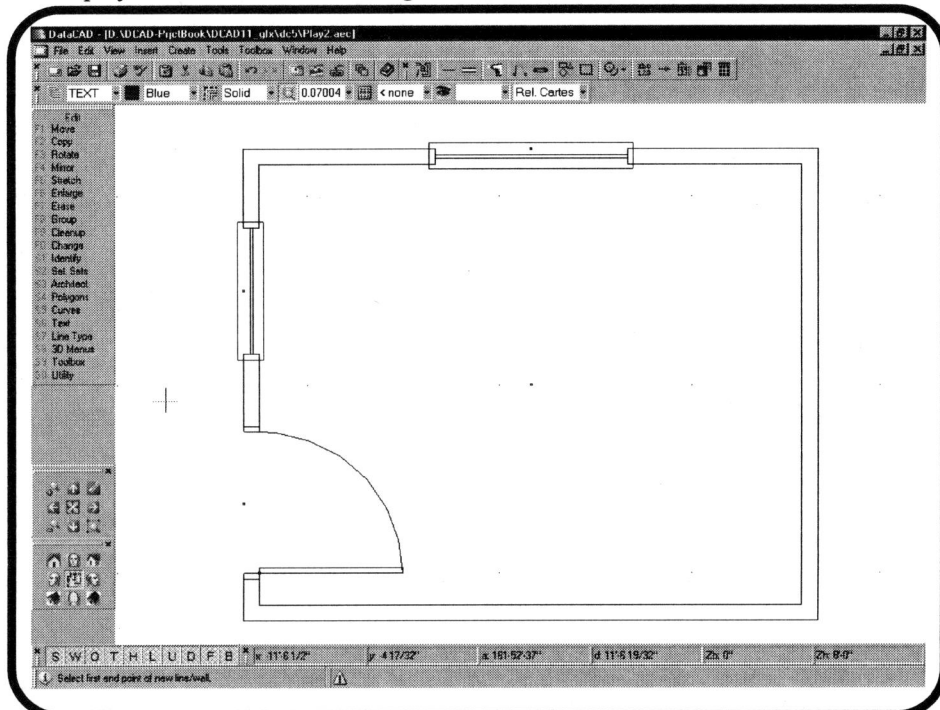

Your display should look similar to Figure 2-1.

Figure 2-1, The Play2 drawing,

Adding Furniture (Using Symbols)

In this project, you will first add some furniture for Ferd to sit on. There is a whole collection of furniture (and other goodies) stored on the hard drive. The furniture is available through the **Symbol Browser** menu.

There are three ways to get into the **Symbol Browser** menu. One is to select the **Symbol Browser** option from the **View** menu in the **Menu Bar**. Or, you can select a toolbar icon. The other (faster) way is to use a *shortcut key*.

4.__ Hold down the [CTRL] key and press the **T** [or the **t**] key to enter the **Symbol Browser** menu.

The DataCAD 11 Project Book

This opens the DataCAD **Symbol Browser** dialog box as shown at ther right side of the screen in Figure 2-2.

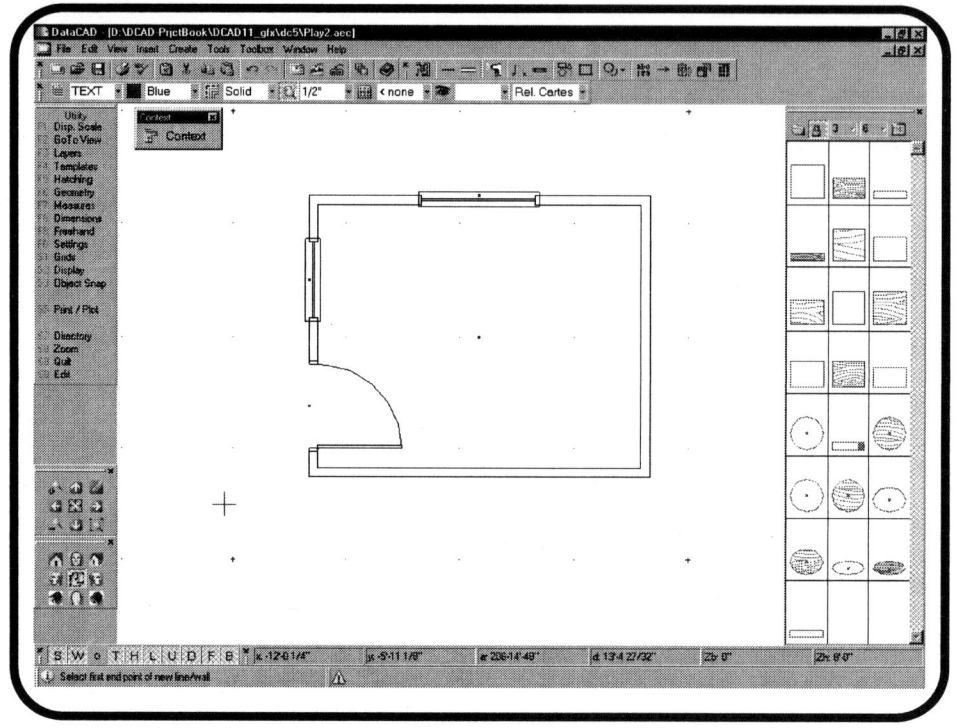

Figure 2-2, The DataCAD Symbol Browser.

- •• **NOTE:** The **Symbol Browser** will default to the display of the last group of symbols selected. Your **Symbol Browser** dialog might look different than the one in Figure 2.2.

5.__ To locate the folder with the symbols you need, *click* on the icon that looks like an open folder at the upper left corner of the **Symbol Browser**, and *click* on the text that says **Folder**. A dialog box like the one in Figure 2-3 will be displayed.

- •• **NOTE:** If DataCAD was installed using the default settings, symbols should be found with the path **C:\DATACAD\SYMBOLS**.

Chapter 2, Page 36

Adding Furniture (Using Symbols)

6. ___ One of the sub-folders in the **Symbols** folder is called **Living Room**, and one of the sub-folders in that folder is called **3D Furniture**. *Click* on that **3D Furniture** folder and the **Symbol Browser** should show you plan views of all the 3D sofas and chairs in that folder, as shown in Figure 2-3.

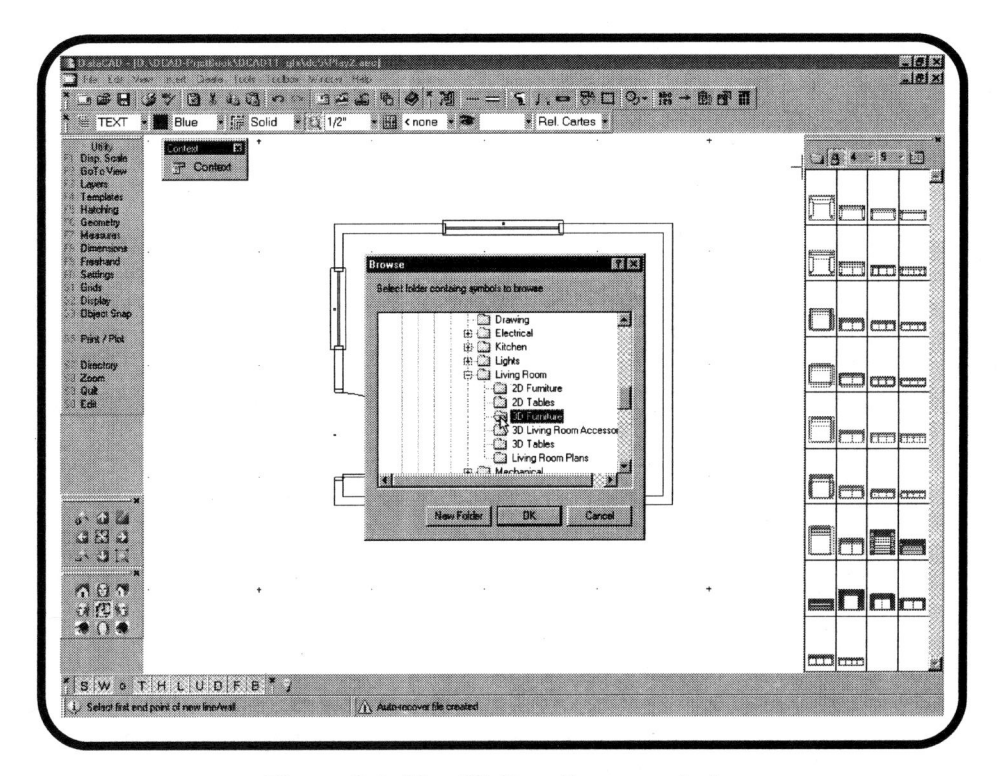

Figure 2-3, The 3D Furniture symbols.

Notice that as you move the drawing cursor over the different symbols, the symbol name (and in fact its whole path) appears in two places: just above the prompt line (where the Coordinate Display is displayed), and inside a yellow "fly-out" tag just below your cursor.

Let's give Ferd a love seat and a lounge chair to sit on.

7. ___ Move the drawing cursor down to the 7th row from the top, and then to the second column from the left. The name of the symbol should say **LOVESEAT**. *Click* the *left* button *once*.

The DataCAD 11 Project Book

8. __ Position the **LOVESEAT** as shown in Figure 2-4 and click the *left* mouse button *once*. Notice that there is still an outline of a love seat attached to your cursor, even after placing the furniture. In this way you can keep placing more symbols in the drawing without having to pick the symbol each time. To finish placing symbols, *click* the *right* mouse button once..

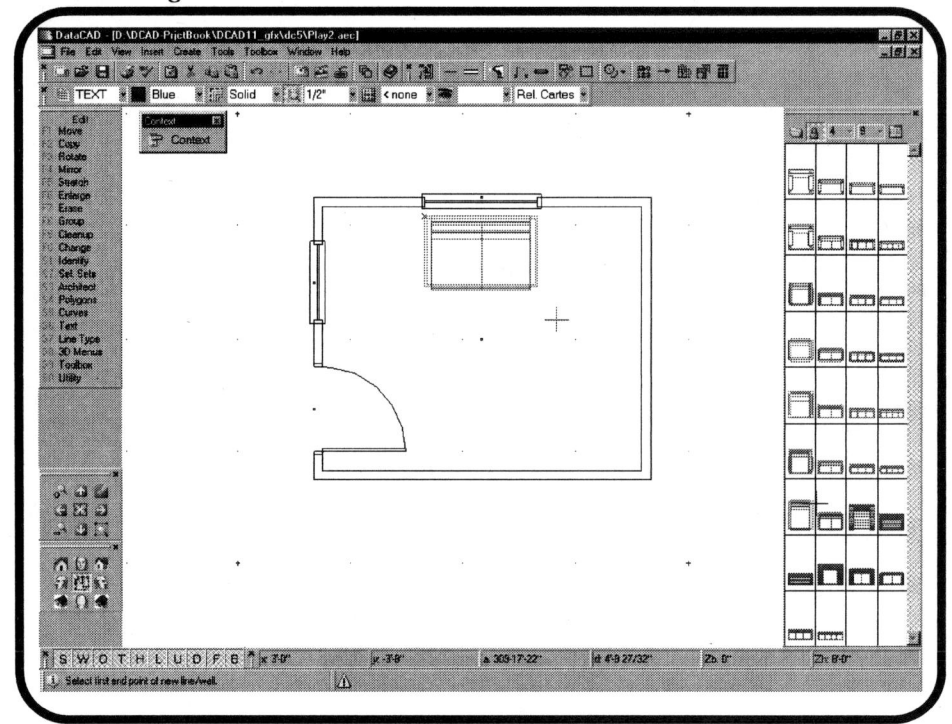

Figure 2-4, Adding a love seat.

- • **NOTE:** Notice that when you click on a symbol, the drawing cursor becomes an outline of the symbol. Unless you do something to change the angle of rotation, the symbol will be transferred to your drawing with the same orientation that it has on the **Symbol Browser**. When you add the lounge chair in the next steps, you will have to change this angle. There are two ways to do this: dynamically, or by typing in the angle.

Adding Furniture (Using Symbols)

9. ___ To insert a symbol and rotate it dynamically, select the symbol first, then pick the **Dyn. Rotate** (Dynamic Rotate) option from the **Ins. Symbol** menu that now appears at the *left* side of your screen (see Figure 2-5). Dynamic Rotate allows you to place the symbol in the drawing, then rotate it around to any angle before placing it.

The second method is to insert a symbol with a specific angle of rotation, and that's what we'll do now.

10. ___ Move the drawing cursor down to the 7th row from the top, and in the first column on the left. The name of the symbol should say **LOUNGE**. *Click* the *left* button *once* to pick it.

11. ___ A new menu appears at the *left* side of your screen, called **Ins. Symbol**, and there you will see another option called **Dyn. Rotate** (see Figure 2-5). If it is turned off (the button is not depressed, and the text is black), then all you have to do is click on it. If it's not already on, you'll have to turn it on then turn it off before you can input an angle. Strange, I know.

12. ___ After turning off **Dyn. Rotate** another menu will appear with 13 different angles. Note also that the prompt line is asking you to "Enter angle of rotation." Now you can enter an angle either by typing the value on the prompt line, or you can select a value from the menu window. Observe that each time you select a new value from the menu window, it will be added to the value shown on the prompt line. The angle is not actually entered until you either press the **Enter** key, or *click* the *right* mouse button *once*. The safest way to enter a value is to actually type it, followed by pressing the enter key.

13. ___ Enter an angle of **270** and press the **Enter** key, or click the *right* mouse button *once*.

Chapter 2, Page 39

The DataCAD 11 Project Book

14.__ Position it as shown in Figure 2-5. (Note: You may have to adjust the display with the arrow keys or the **Navigation Pad** buttons to move the right wall of the Playhouse into the display.)

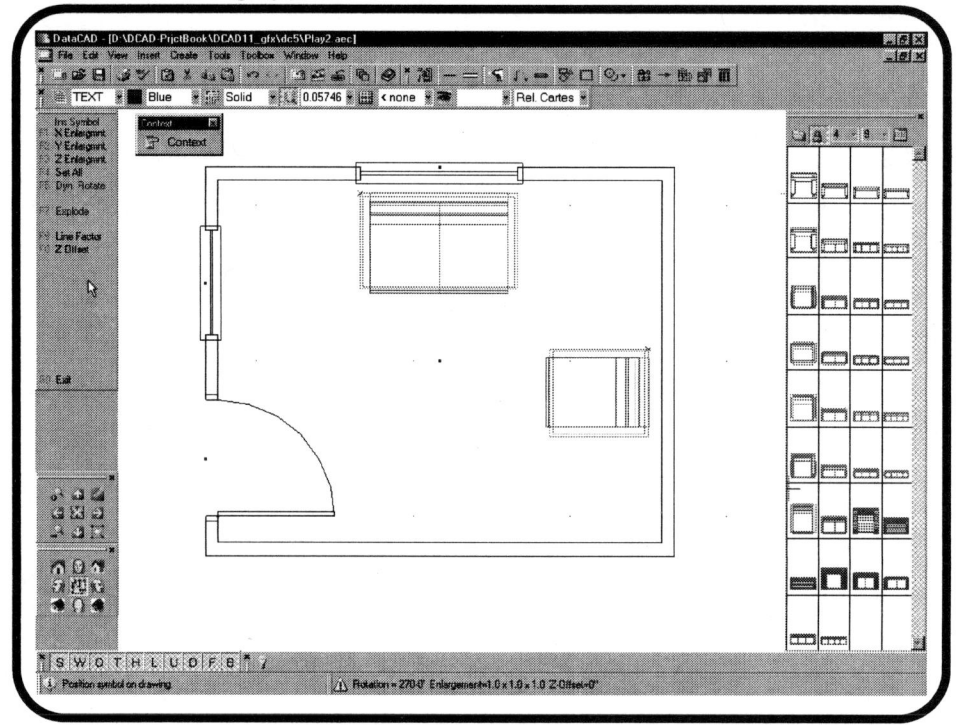

Figure 2-5, Adding a lounge chair.

15.__ Click the *left* mouse button to place the lounge chair on the drawing. To finish placing symbols, *click* the *right* mouse button once.

Since you are done adding furniture, you can turn **off** the **Symbol Browser** menu.

16.__ Click on the **X** in the upper right corner of the **Symbol Browser**, or press [CTRL] + **T** again.

> •• **NOTE:** Incidentally, the reason that the shortcut uses the letter "T" is that the primary way to get to symbols in DataCAD prior to version 11 was to access them through a "Template". This is still available in DataCAD 11, but we won't be using it here.

Chapter 2, Page 40

Creating and Naimg Layers

••• **WARNING**••• Have you saved your drawing lately? (Either select **File, Save** from the **MENU BAR**, or press **Ctrl+S** to update the current drawing file.)

As you can see, the computer does a lot of work for you. Just by selecting a symbol, you can have the computer draw a lot of lines with one click of the mouse button.

Creating and Naming Layers

Next, you will learn how to add text to a drawing. The text will be placed on a different drawing layer to keep things better organized. Having parts of a drawing on different layers is kind of like having a stack of sheets of clear glass with a different part of the drawing on each sheet. It is possible to turn any drawing layer on or off. Turning a layer off is the same as removing a sheet of glass from the stack. Turning a layer on is the same as inserting a sheet of glass. The **Layers** menu is used to control drawing layers.

In this drawing you will keep the floor plan of the Playhouse on one layer and some text on a different layer. This is done so that when you make a three dimensional view of the Playhouse in the next project, you can turn the text layer off so that it won't mess up the picture.

17.__ Press **Ctrl+L** (or move the cursor over the icon just to the left of the layer name box in the **Status Area** and click the *left* mouse button).

Chapter 2, Page 41

The DataCAD 11 Project Book

This opens the **Layer Manager** window as shown in Figure 2-6.

Figure 2-6, The Layer Manager window.

One new layer will be added, and then the layer names will be changed.

18.___ Select **New** from the **Layer Manager** window.

19.___ Enter **1** in the **Create New Layer(s)** dialog box; then click the **Ok** button to add one new layer.

20.___ Double click on the **Layer002** name in the **Layer Manager**..

21.___ Enter **TEXT** as the new name.

22.___ Double click on the **Layer001** name in the **Layer Manager**.

23.___ Enter the name **PLAN.**

Creating and Naming Layers

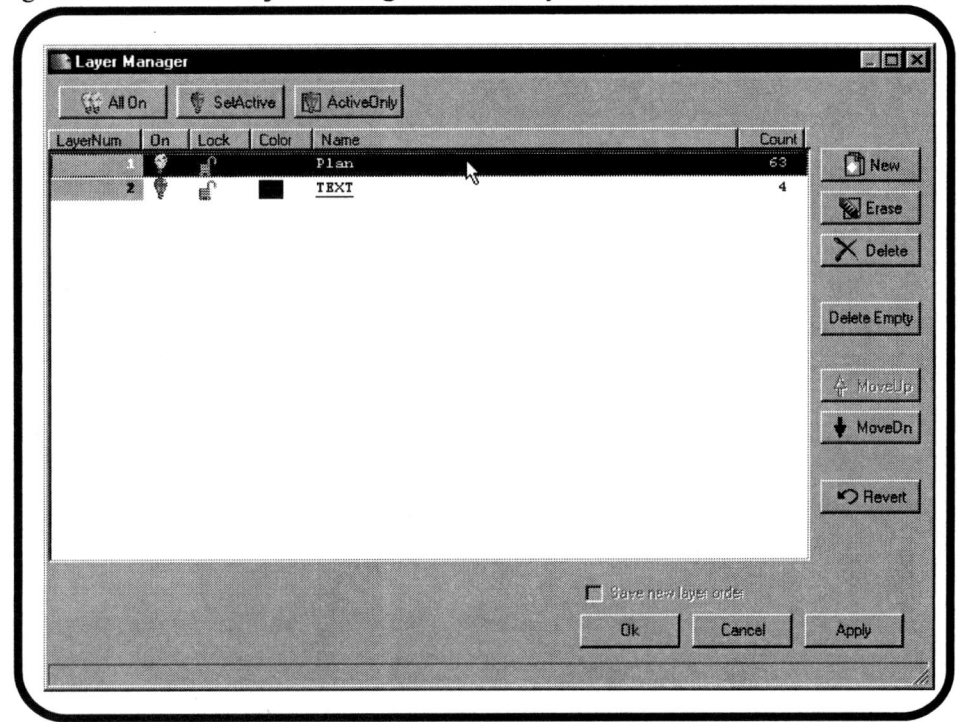

Figure 2-7 shows the **Layer Manager** after the layers have been renamed.

Figure 2-7, The Layer Manager Window.

There are a few features of the **Layer Manager** window that you should know about.

- The **ON** column shows a light bulb that is one of three colors; *green* for ON and active, *white* for OFF, and *yellow* for ON.

- The **Lock** option prevents any changes to a locked layer. It is toggled on or off by clicking on the lock for each layer.

- The **Count** column shows the number of entities on the layer.

- The three buttons near the top of the **Layer Manager** dialog box can be used to toggle all layers on, set a layer to be active, or turn all layers off except for the active layer.

Chapter 2, Page 43

The DataCAD 11 Project Book

- The seven buttons on the right side of the **Layer Manager** dialog box can be used to add **New** layers, **Erase** the entities on a layer (without removing the layer from the drawing), **Delete** a layer from the drawing, **Move** a layer up or down the layer list, **Delete Empty** layers (layers that have no entities in them), or **Revert** to the previous settings. (**Revert** essentially cancels any changes made while the **Layer Manager** has been open.)

- A layer can be moved in the list by positioning the cursor over the layer name, clicking and holding the left mouse button down, and dragging the layer to a new position.

- Clicking on the word **Name** at the top of the list of layer names will sort the layers in either ascending or descending alphabetical order. Unless the *Save new layer order* box is checked, the layer order changes will be temporary. The same is true of the **On**, **Lock**, **Color** and **Count** headings. Try it.

24.__ Make the color of the **PLAN** layer **White**, and the **TEXT** layer **Blue** by clicking on the **Color** box for a layer and then selecting a color from the **Color Palette** dialog box.

25.__ Click **Ok** to close the **Layer Manager**.

Shortcuts

There are two more DataCAD shortcuts that can help you when working with layers: the **K** (or **k**) key, which will cycle through the 15 standard DataCAD colors, so that you can quickly change colors while drawing; and the Tab key, which will cycle through all the layers that are turned on, making it easy to move between layers. (And by holding down the [SHIFT] key while pressing any of those keys, you can cycle backward through the colors and layers. This is true of most of DataCAD's shortcuts.)

26.__ Press the **Tab** key. Note the **LAYER NAME** in the **Status Area**.

27.__ Press the **Tab** key again. Note the **LAYER NAME** in the **Status Area** has changed.

28.__ Click on the color box next to the **COLOR NAME** in the **Status Area**. This opens the **Color Palette** dialog box and allows you to change the color of the currently active layer. Or press the **K** (or **k**) key as noted above.

Adding Text

•• **NOTE:** If you change the drawing color for an active layer, only entities created after the change will have the new color. Also note that if the drawing window background color has been set to white, and the drawing **Color** has also been set to white, the lines on the display will appear black.

Adding Text

Now that you have created a drawing layer for the text, you can add a title and your name to the drawing.

29.__ Press the **Tab** key until the active layer reminder shows that the **TEXT** layer is active.

30.__ Press the [:] key to return to the **Edit** menu, and then select **Text** from the **Edit** menu [or press [ALT] +[T] to jump directly to the **Text** menu].

The DataCAD 11 Project Book

Figure 2-8 shows the **Text** menu. There are quite a few options, but for now let's just deal with the size of the text. Notice that when you enter the **Text** menu, the cursor changes shape.

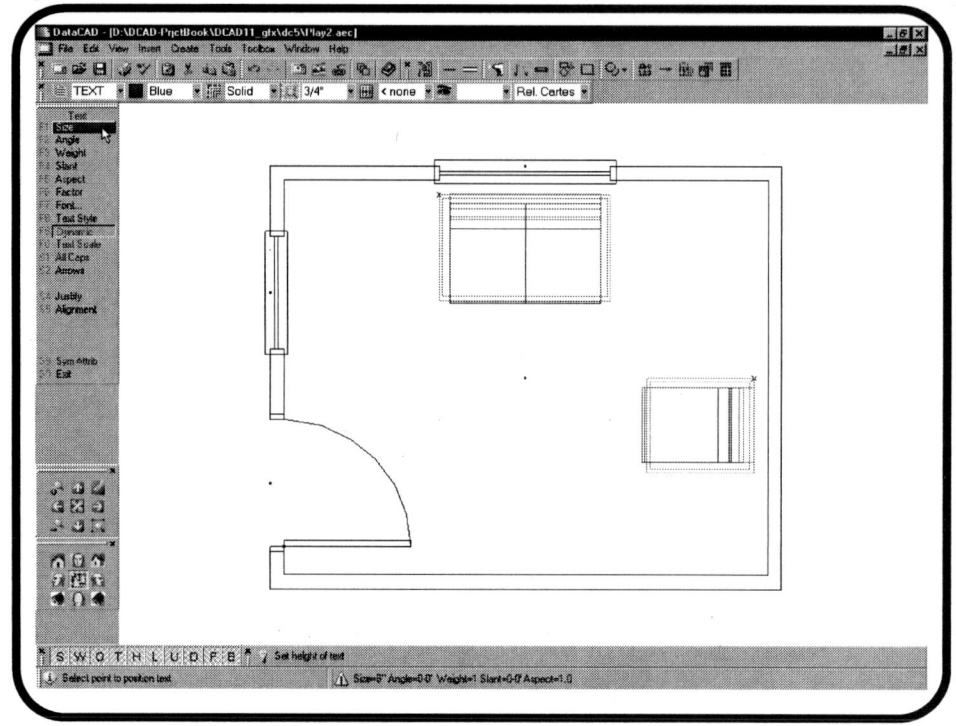

Figure 2-8, The Text menu.

When the **Text** menu is active, the size of the cursor shows the current text size.

31.___ Select **Size** from the **Text** menu.

> •• **NOTE:** There are two ways to enter a text size of 1'0". You can use the keyboard to enter 1.0 (or simply **1**), or you can select 1'0" from the **Size** menu. (If you select it from the **Size** menu, you will have to press either the **Enter** key, or the *right* mouse button to return to the **Text** menu.) If you select the size from the menu with the mouse, be careful. Each time you press the *left* mouse button, another size increment will be added. If you press too hard or long on the mouse button, you will get the wrong size. It is better to enter the size from the keyboard.

32.___ Enter a size of **1'0"**.

Chapter 2, Page 46

Adding Text

The size text to use depends on the size of the drawing you are working on. Unlike drawing on paper (where you have to draw at a reduced scale to get a house to fit on the paper), you always draw with full scale dimensions with CAD. When you get around to setting up the CAD drawing for the plotter, you will worry about paper sizes and drawing scale. The size paper your drawing will be plotted on will influence what size text to use. For this drawing, a size of 1'0" for big text and 6" for small text will work out well.

> •• **NOTE:** When you have some free time, look up the **Text Scale** option in the DataCAD online manual. This is something that you simply must learn to use for text in the future. It will make your life much simpler.

Now that you have set the text size, move the mouse and notice what has happened to the cursor. It has changed from the drawing cursor to the text cursor, and the text cursor shows the current text size. Let's add a title to this drawing.

33.__ Move the text cursor a little ways below the Playhouse.

34.__ Click the *left* mouse button *once*.

> •• **NOTE:** Substitute your name for Ferd in the next step.

35.__ Enter **Ferd's Playhouse** from the keyboard (you may want to substitute your name for Ferd).

36.__ Click the *right* mouse button *once* to exit text entry mode.

> •• **NOTE:** You can enter several lines of text at once by pressing **Enter** at the end of each line, and continuing to type.

Next you will add your name to the drawing. The text size for your name will be half the size of the title text. Before you change text size there is something you have to know about: how to specify feet and inches in DataCAD.

> •• **NOTE:** To specify inches in DataCAD, place a period in front of the number. For example, 4 inches would be entered as 0.4 (or just .4). If you wanted to enter 2'8," you would enter 2.8. To enter six inches, you would enter either 0.6 (again, the 0 is optional). Now let's change the text size and add your name to the drawing.

37.__ Select **Size** from the **Text** menu and enter **0.6** (or just .6) to enter a size of 6 inches.

38.__ Move the text cursor below the title and click the *left* mouse button *once*.

39.__ Enter **Drawn by:** *your name*.

40.__ Press the *right* mouse button *once* to exit text entry mode.

The DataCAD 11 Project Book

Moving Text

It's not very often that you get text placed exactly where you want it on the first try. One way to fix misplaced text it to erase it and enter it again. A better way is to move it around until it is in the right spot.

In Figure 2-9 you can see that the text is not nicely centered under the Playhouse. The following steps explain how to move text entities.

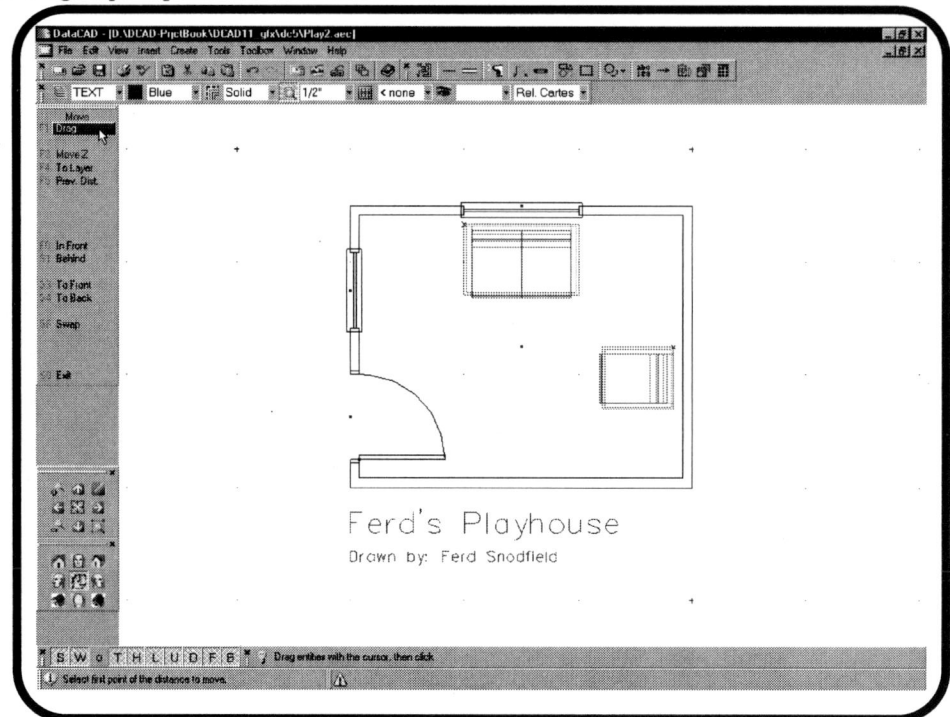

Figure 2-9, Poorly placed text.

41.___ Jump to the **Move** menu by selecting **Move** from the **Edit** menu, or by pressing the **m** [or the **M**] key.

42.___ Select **Drag** from the **Move** menu.

Moving Text

43.__ Toggle the **Entity** option **on** from the **Drag** menu (see Figure 2-10). (The **Entity** button will be depressed in the menu when the **Entity** option is toggled **on**.)

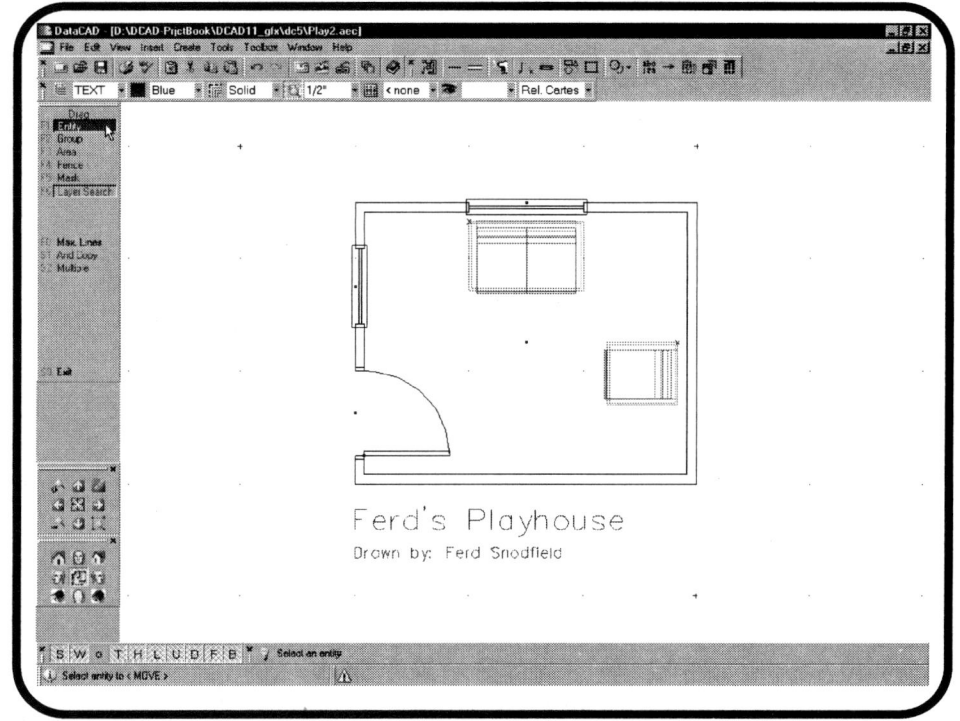

Figure 2-10, Preparing to move text.

44.__ Select the text entity to be moved by clicking the *left* mouse button *once* when the drawing cursor is positioned near the *lower left corner* of the first letter in the text entity.

> •• **NOTE:** As soon as you have picked the entity to move, the prompt line will ask you to *Select a point to drag from*. The exact point you pick is not critical. In this case it might be helpful to pick a point near the center of the text you are moving.

45.__ Select the point to move from by positioning the drawing cursor at the desired point and clicking the *left* mouse button *once*.

The DataCAD 11 Project Book

•• **NOTE:** If **Ortho** mode is toggled **on** in your system (the **O** in **SwOTHLUDFB** will be upper case), it might be hard to place the text exactly where you want it. If you want to toggle **Ortho** mode on or off, press the **o** [or the **O**] key.

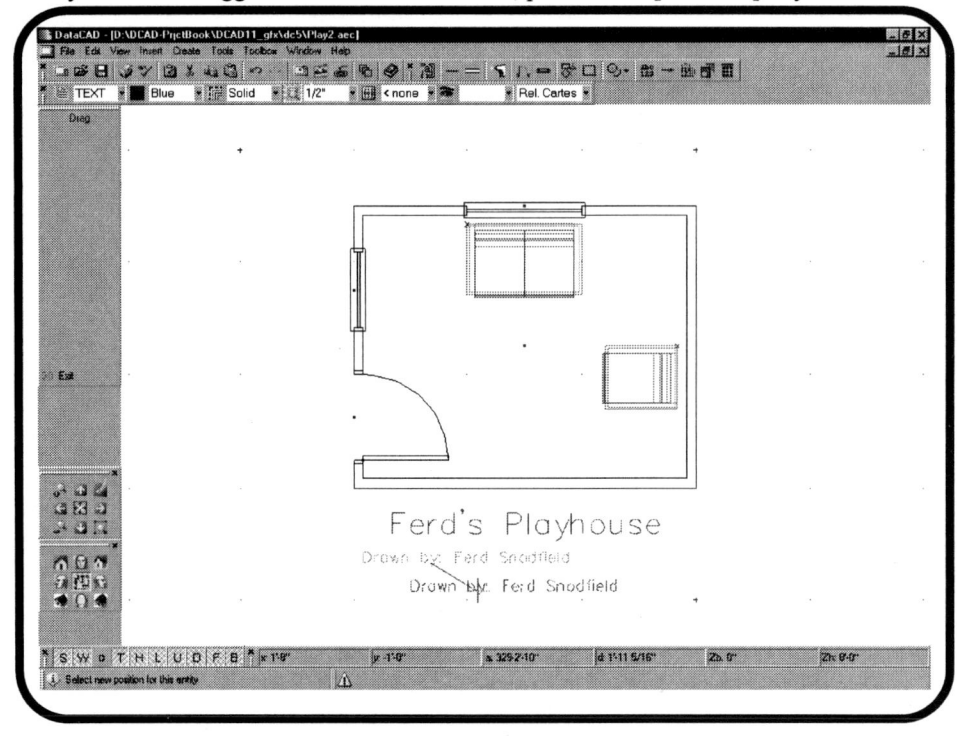

Figure 2-11, Moving text.

46.___ Move the to the desired location and click the *left* mouse button *once*. Move your text entities until they are properly positioned (see Figure 2-11).

•• **NOTE:** If you want to move more than one entity at a time, use the **Area** option in the **Drag** menu.

The **Move**, **Drag** option can be used with any drawing entity, not just text. There are some other **Move** and **Drag** menu options, but they will be left for a later project.

••• **WARNING**••• Have you saved your drawing lately?

Moving Text

This completes the Playhouse plan with furniture and text. You have to save the changes made to the Playhouse during this session or they will be lost. If you want to save a drawing without leaving the DataCAD software, you can use the **File, Save** option. If you use the **File, Exit** option the system will prompt you as to whether you want to save the drawing or not.

47.___ From the **MENU BAR**, select **File, Exit**.

48.___ Select **Yes** to respond to the *Save changes to File?* prompt.

In the next project, you will create some exciting three dimensional views of Ferd's Playhouse. You will even be able to take a 3D "walk through" of your computer house model.

The DataCAD 11 Project Book

Review and Important Points

- To edit an existing drawing, select it from the DataCAD opening menu. (If it does not appear, adjust the directory path in the **Look in** text box found in the **Enter name of drawing file to open** dialog box.)

- Symbols are found in the **Symbols** directory by using the **Symbol Browser**. In most installations, the complete directory path will be **C:\DATACAD\SYMBOLS**.

- The name of a symbol will appears in two places when the cursor is moved over the symbol in the **Symbol Browser**: just above the prompt line; and inside a yellow "fly-out" tag just below your cursor.

- To set a fixed rotation angle for a symbol, select **Dyn. Rotate** twice from the **Ins. Symbol** menu, then enter the desired angle.

- Positive angles are counterclockwise in DataCAD.

- Use the **Layers** menu (press **l** or **L**) to add or rename layers, or to turn layers on or off.

- The quick key for cycling through the layer list is the **Tab** key. Pressing [SHIFT] +**Tab** will cycle through the list in reverse order

- The quick key for cycling through the color list to select an active color is **k**. Pressing **K** will cycle through the list in reverse order.

- Use the **Text** menu ([ALT] +**T**) to set text size and other properties and to add text.

- You can edit text that is already placed by selecting **Edit, Change, Text, Contents**.

- Use the **Move, Drag** function to move an entity, area, or group to a new location.

Chapter 2, Page 52

Moving Text

Practice Drawings

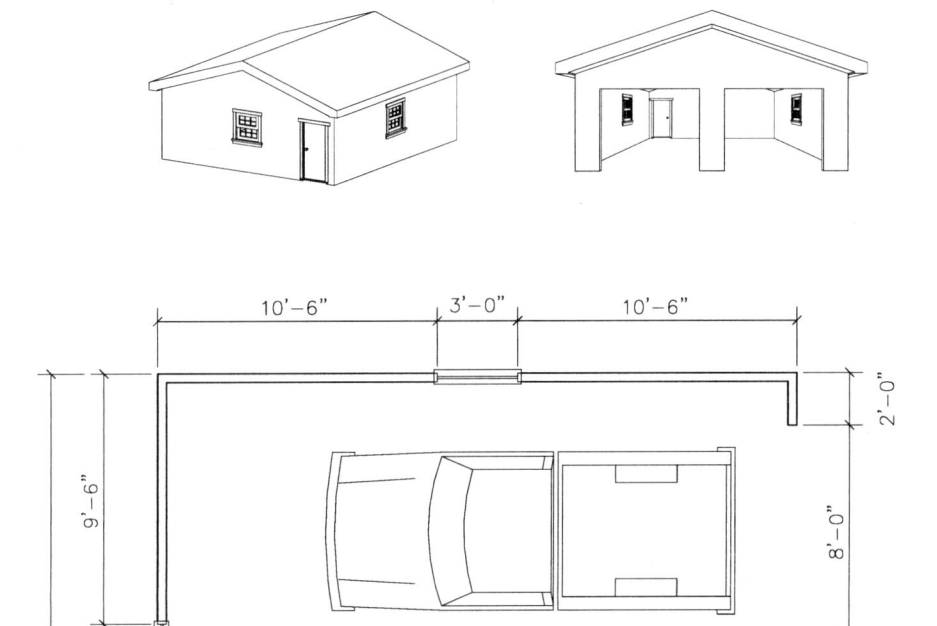

Assignment P2-A, Garage

- Use the dimensions shown above and create Ferd's garage.
- Use the **Cut Wall** option in the **Architect** menu to add the garage doors.

The DataCAD 11 Project Book

Practice Drawings
Assignment P2-2, You design it!

- Create your own custom garage design.

CHAPTER 3

3D Drawing

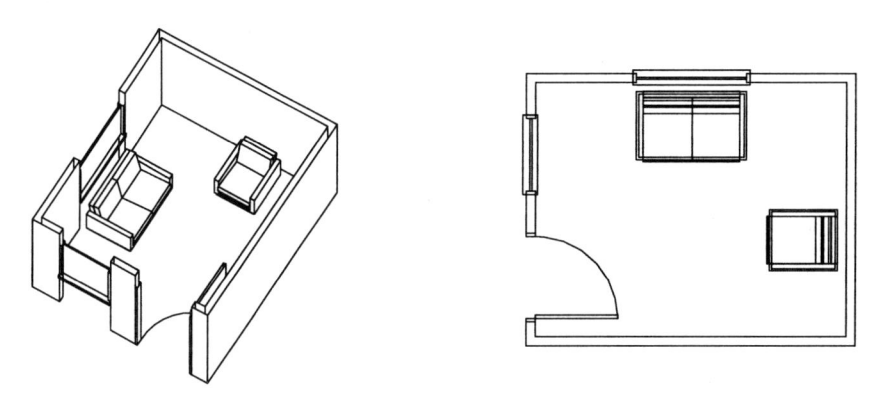

Ferd's Playhouse
Drawn by: Ferd Snodfield

Project Goals

When you have completed this project, you will be able to:

1. Create isometric views.
2. Create perspective views.
3. Remove hidden lines from pictorial views.
4. Combine pictorial and plan views for presentation drawings.
5. Backup a drawing file with the **Save** option.

The DataCAD 11 Project Book

Prerequisite: Chapter 2

Introduction

In the first two projects in this book, you used the DataCAD software as a two dimensional system. While you were creating the plan for Ferd's Playhouse, the software was creating a three-dimensional computer model of your plan. In this project you will see how to create different pictorial views of this model.

1.__ Start the **DataCAD** software and **Open** your **Play2** drawing file.

To preserve the Play2 version of the design, a new file will be created. This process allows the designer to create a new version of a design without losing the original.

2.__ From the **MENU BAR**, select **File, Save As**, and *enter* **Play3** as the new file name.

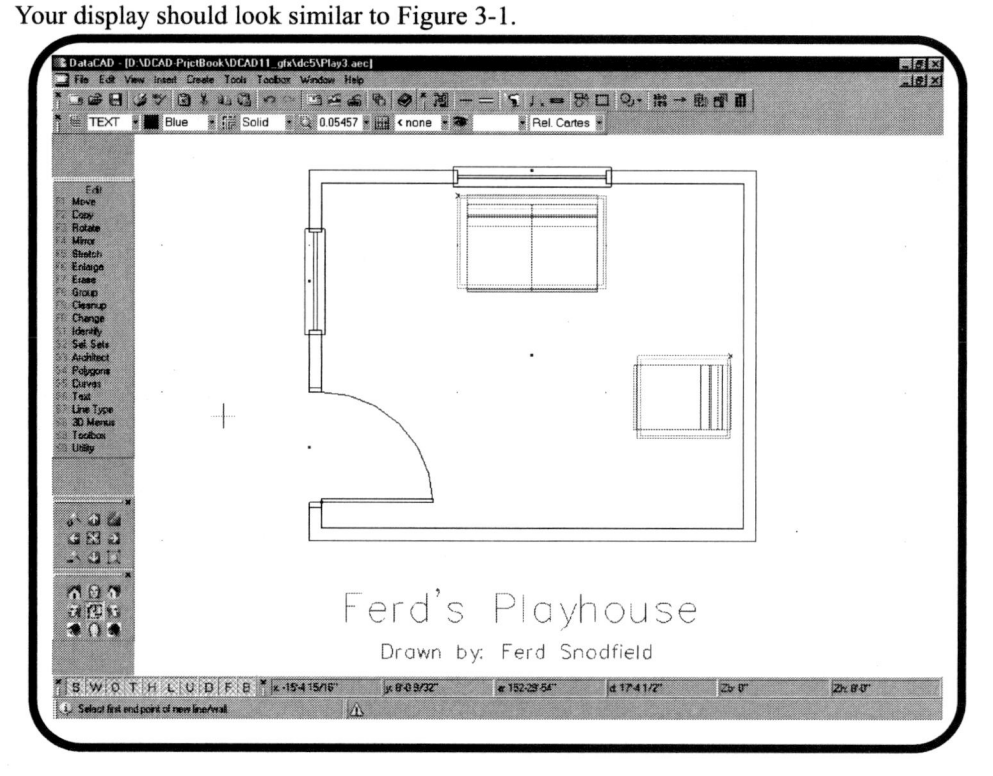

Your display should look similar to Figure 3-1.

Figure 3-1, The playhouse drawing.

You may recall that in Chapter 2 you created a new drawing layer for the text. Since the text will not look very good in a pictorial view, the text layer will be turned off.

3. __ Press the **I** [or the **L**] key to enter the **Layers** menu.

4. __ Select the **On/Off** option from the **Layers** menu.

5. __ Select **TEXT** from the **On/Off** menu to turn the text layer **off**.

Figure 3-2 shows the **Layers, On/Off** menu after the text layer has been turned **off**. Notice that the **TEXT** button is not depressed. This shows that the layer is off. Also notice that the information line on your display shows what action the system has completed. Also notice that the text has been removed from the display.

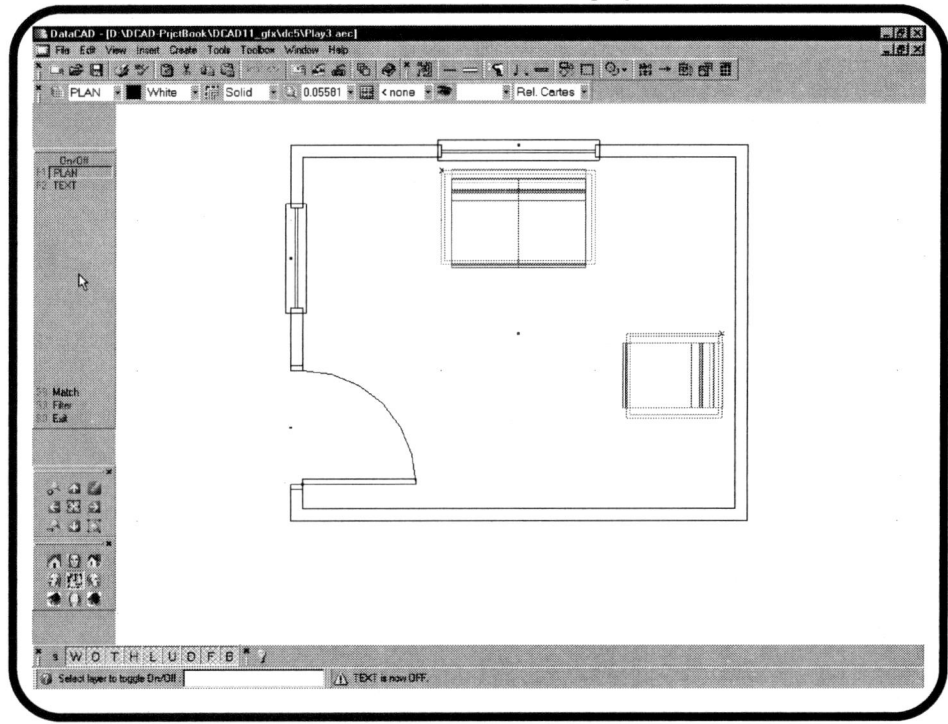

Figure 3-2, The Layers, On/Off menu.

You should note that the text has not been removed from the drawing file. It is still there. Whenever you want the text to appear on the display, just turn the text layer back on.

The DataCAD 11 Project Book

Creating 3D Views

Now that the text has been set aside, you can concentrate on creating 3D views of the Playhouse. There is a special menu that provides the 3D views options.

6.__ Enter the **3D Views** menu either by selecting **3D Menus** from the **Edit** menu and then selecting the **3D Views** option from the **3D Menus** menu, or by pressing the **y** [*not* the **Y**] key.

The **3D Views** menu is shown in Figure 3-3. There are four different view types on the **3D Views** menu: **Plan** (currently active in Figure 3-3), **Parallel**, **Perspective**, and **Oblique**. First an isometric view (which is a special case of parallel projection) will be created.

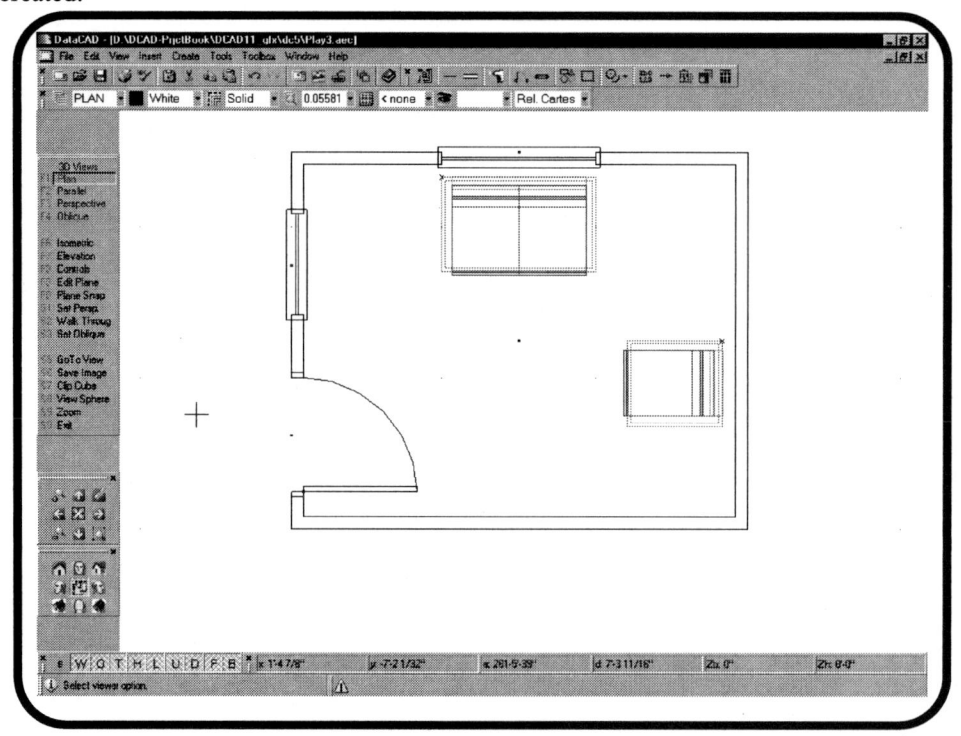

Figure 3-3, The 3D Views menu.

7.__ Select **Isometric** from the **3D Views** menu.

Creating 3D Views

As soon as you select the **Isometric** option, your display will look similar to Figure 3-4. Notice that when you created the double line walls, the computer automatically created walls that were eight feet tall. When you added the doors and windows, the computer cut a hole in the walls and inserted the door and placed the windows at the proper elevation. Also notice that the love seat and lounge chair that you added to the plan were constructed in 3D.

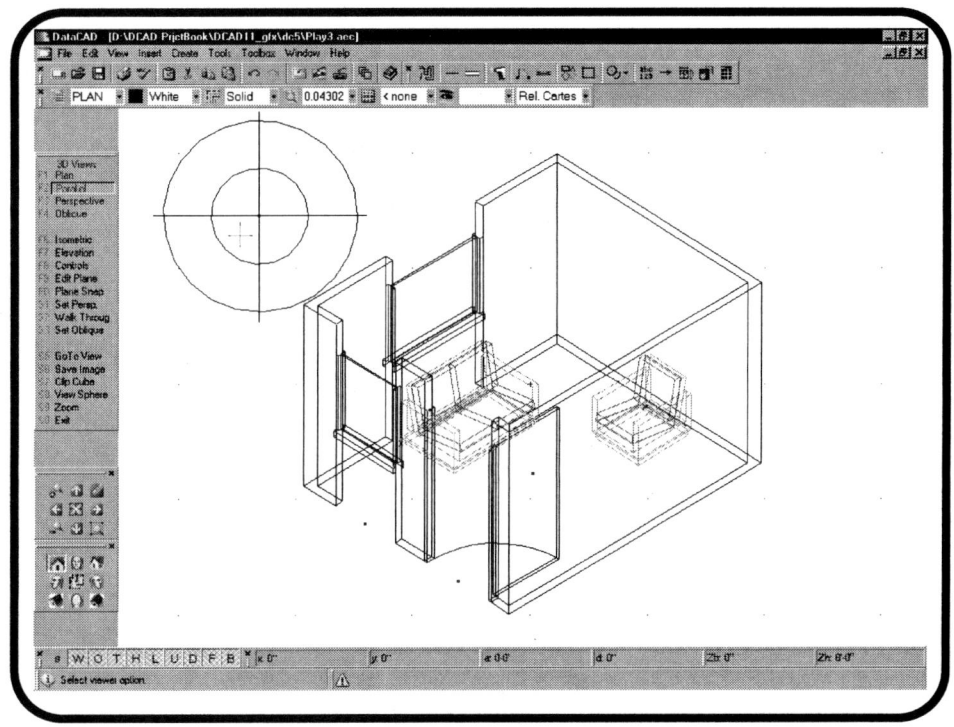

Figure 3-4, An isometric view.

To make an isometric drawing, such as that shown in Figure 3-4, using traditional pencil and paper techniques would take an experienced technical drawing expert quite a while to create. The DataCAD software provides tremendous power for manipulating 3D computer models.

Notice the double circle with the cross hairs shown in Figure 3-4. This represents a *viewing sphere* which gives us even more power to create different pictorial views. Let's first create a new parallel projection view, and then discuss how the viewing sphere works.

Chapter 3, Page 59

The DataCAD 11 Project Book

Using the Viewing Sphere

8.__ Move the drawing cursor to the point near the cross hairs in the viewing sphere as shown in Figure 3-4 (or Figure 3-5) and click the *left* mouse button *once*.

Notice in Figure 3-5 how the view of the Playhouse on your display now has changed from Figure 3-4. The new view is from a different side of the house looking at a different angle.

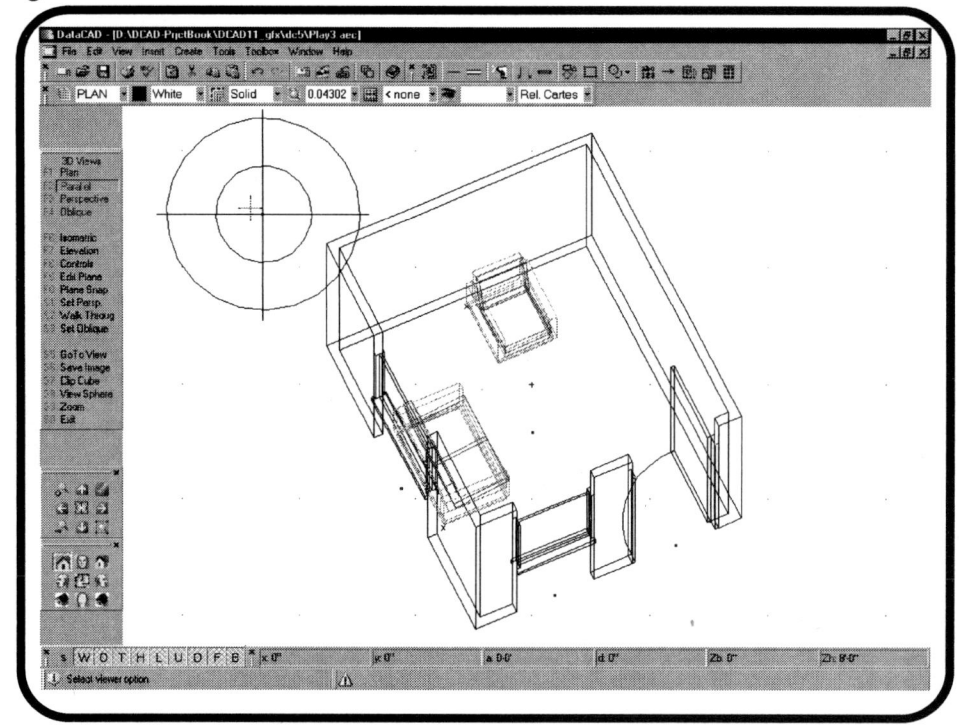

Figure 3-5, A new parallel projection view.

Here is how the viewing sphere works. Imagine that the double circle is actually a flattened out ball or globe. The point where the lines cross is the "north pole" of the globe. The smaller of the two circles is the "equator" of the globe and the large circle is the "south pole" (squished out to a circle because the screen is flat). To create a new view of the Playhouse, you first position the drawing cursor in one of the four quadrants (the pie shaped pieces between the lines) of the globe and click the left mouse button. Each quadrant will produce a view from a different side of the object.

Using the Viewing Sphere

The closer the drawing cursor is to the north pole of the viewing sphere when you click the mouse button, the more of a bird's eye view you will get. The closer the cursor is to the equator, the more of a people's eye view you will get. If you go below the equator (which means the cursor is positioned somewhere between the two circles), you will get a worm's eye view. Try a few views. (If the view goes off center or off the display, just use the **Zoom view** button on the **Navigation Pad** (the one in the center with four red arrow within a box) or the **Zoom** option to get it back).

The pictorial view on your display is what is called a *wire frame* view. Notice that the objects in the view look like they are made from a bunch of wires, and you can see between all of the wires. It would look a little more realistic if the hidden lines were removed. The amazing DataCAD software can actually do that job for you. First, you need to get a nice view of the Playhouse on the display.

9.__ Create a view similar to Figure 3-6 by positioning the drawing cursor near the point marked **1** in Figure 3-6, and then clicking the *left* mouse button *once*.

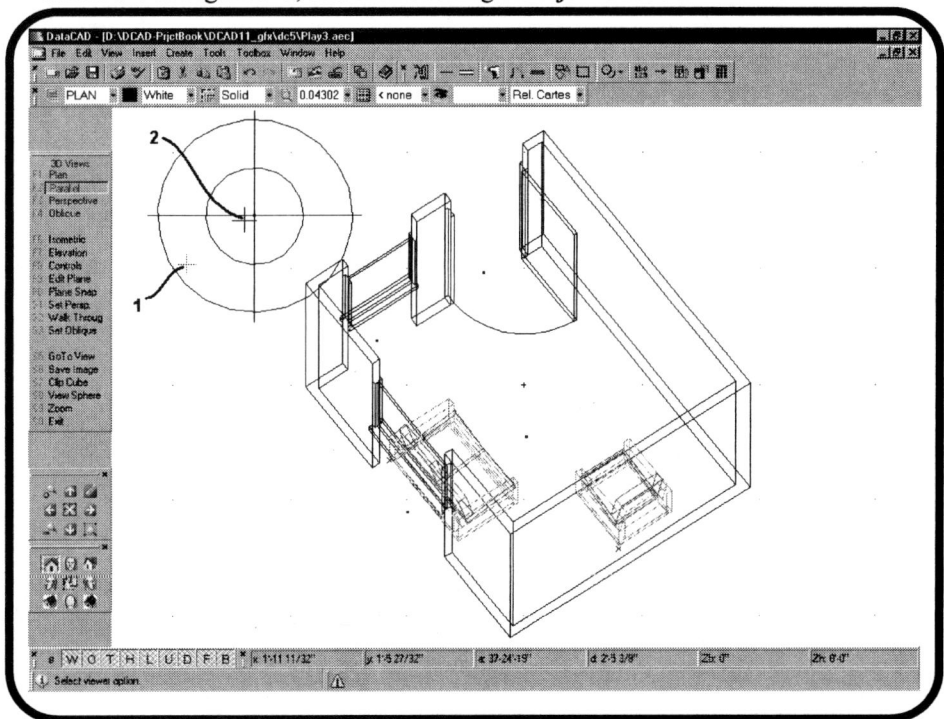

Figure 3-6, A new parallel projection view.

Chapter 3, Page 61

The DataCAD 11 Project Book

Hidden Line Removal

There is a menu that provides you with the hidden line removal options.

10.__ Press the *right* mouse button until the **3D Menus** menu appears, and then select the **Hide** option, or press the **Y** [not the **y**] key to jump to the **Hide** menu.

The **Hide** menu is shown in Figure 3-7. The computer can do some calculations to determine which lines (or parts of lines) will be hidden in the current view. It will then display the view with the hidden lines removed.

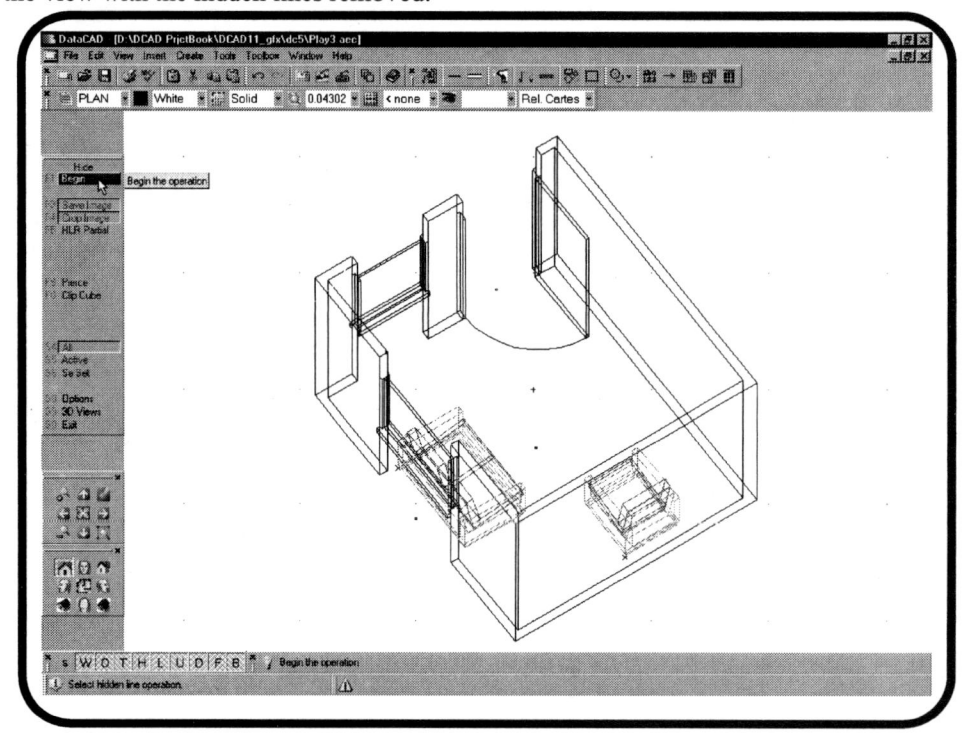

Figure 3-7, The Hide menu.

Hiding the lines in a view does not change the 3D model. As soon as you change views, the hidden line version will disappear from the display. There is an option you can use which will save a 2D copy of the view with the hidden lines removed on a new drawing layer. This is the **Save Image** option which is found on the **Hide** menu.

•• **NOTE:** Unless the **Save Image** option is **on**, the hidden line view will be lost when the displayed view is changed.

Chapter 3, Page 62

Hidden Line Removal

11.__ Turn **on** the **Save Image** option in the **Hide** menu.

Now you can have the system remove the hidden lines from the view. If you have an old slow system, this process can take quite a bit of time. The time will depend on how many objects there are in the model and how fast a processor chip your computer has. The information line near the bottom of the display will let you know what is happening.

12.__ Select **Begin** from the **Hide** menu.

In this example, you can see that a worm's eye view has been created. Before saving the image, a new bird's eye view will be created.

13.__ Click the right mouse button to return to the **3D Menus** menu and then select **3D Views** (or press the **y** [*not* the **Y**] key).

14.__ Click in the viewing sphere near the point marked **2** in Figure 3-6.

15.__ Press the *right* mouse button until the **3D Menus** menu appears, and then select the **Hide** option, or press the **Y** [not the **y**] key to jump to the **Hide** menu.

16.__ Select **Begin** from the **Hide** menu.

As soon as the hidden line removal process is complete, the **Save Image** menu will appear and provide some options for dealing with the new image. This image will be placed on a new drawing layer.

17.__ After the computer has finished hiding lines, select the **New Layer** option from the **Save Image** menu.

18.__ At the prompt *Enter name of new layer to be added*. enter **PICTURE1**

19.__ Select the new layer **ON** option.

> •• **NOTE:** What you just created is not a view. It is a new drawing layer which contains a 2D rendered image of the 3D model. Also, all of the lines in this new layer are in a **Group**. The lines on this new layer can be treated just like the lines on any other layer. Next, the plan view layer and the pictorial image layer will be arranged into a presentation drawing which shows both the plan view and the pictorial image at the same time.

20.__ Press the **y** [not the **Y**] key to jump to the **3D Views** menu. (You can also press the *right* mouse button until the **3D Views** menu appears.)

21.__ Select the **Plan** option from the **3D Views** menu.

Chapter 3, Page 63

The DataCAD 11 Project Book

When you have selected the **Plan** view and have both the **PLAN** layer and the **PICTURE1** layer on the display at the same time, the display will probably look very confusing (as shown in Figure 3-8). This problem can be solved with the **Move** function used in an earlier project. First, you have to get back into the 2D drawing mode.

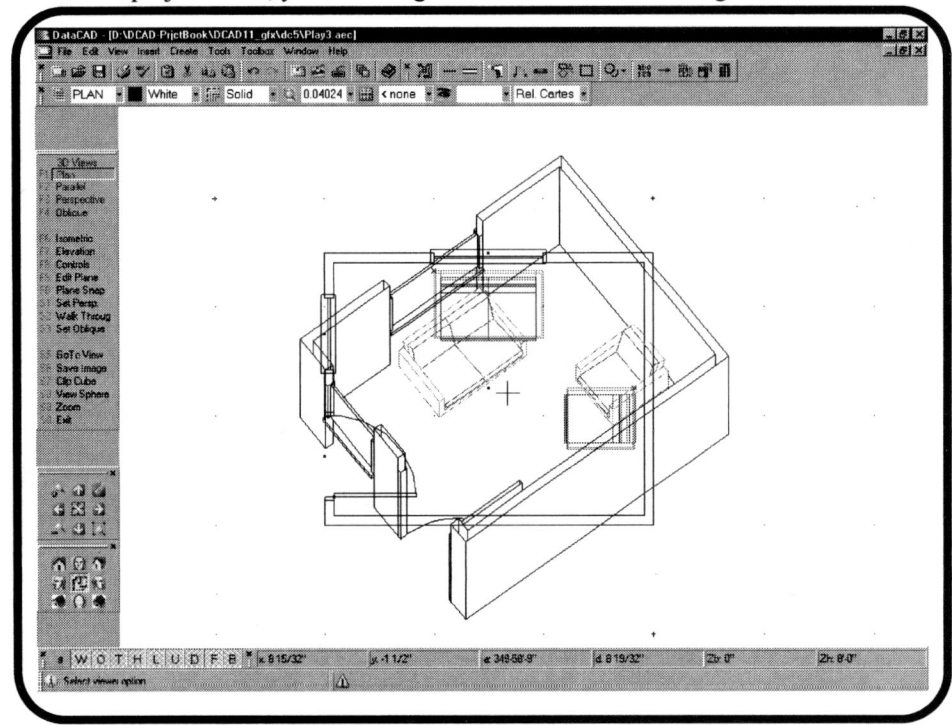

Figure 3-8, The PLAN and PICTURE1 layers.

22. __ Exit the **3D Views** menu (by selecting the **Exit** option, or by clicking the *right* mouse button).

23. __ Select the **2D Menus** option from the **3D Menus** menu (or press the ; key to jump back to the **2D Edit** menu).

Moving the Pictorial Image

Now you will move the pictorial view away from the plan view.

24. __ Press the **m** [*not* the M] key to enter the **Move** menu (or select the **Move** option from the **Edit** menu).

25. __ Select the **Drag** option from the **Move** menu.

Moving the Pictorial Image

The **Drag** menu is shown in Figure 3-9. There are two options on the **Drag** menu which can be used to move the pictorial view, either **Area** or **Group**. Since all of the lines in the pictorial view are on the **PICTURE1** layer, the **Area** option will be used. This requires that only the active layer be searched.

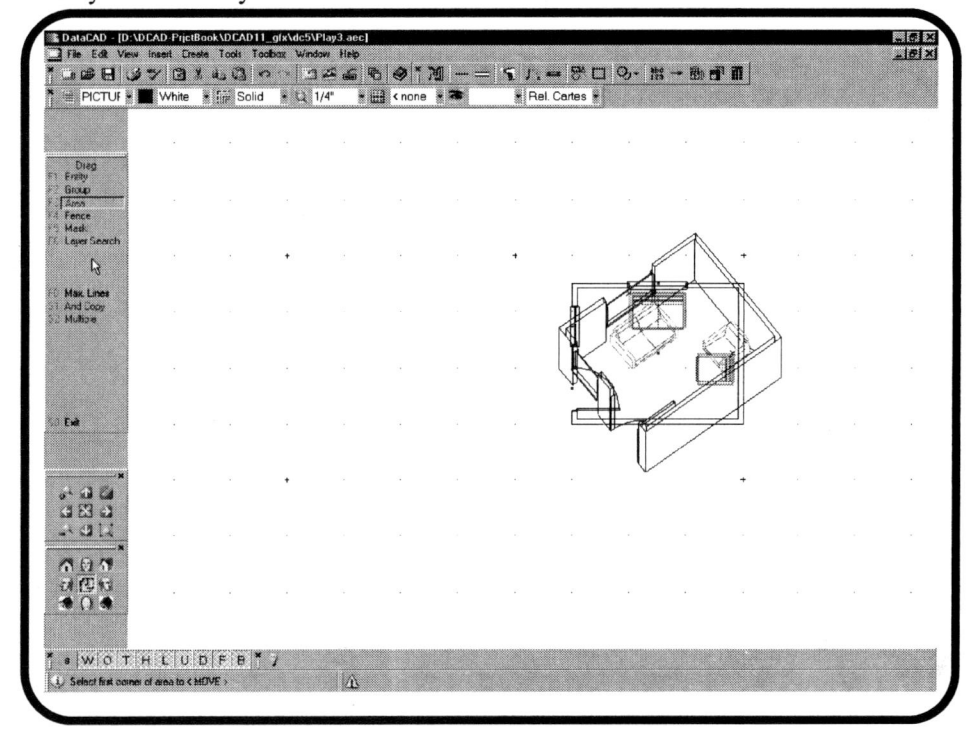

Figure 3-9, The Drag menu.

26.___ Turn the **Area** option **on** in the **Drag** menu.

27.___ Turn the **Layer Search** option **off** in the **Drag** menu.

28.___ Press the **Tab** key to make the **PICTURE1** layer active.

29.___ Use the **Navigation Pad** (or the **PageUp/PageDown**) buttons to make your display look similar to Figure 3-9.

The DataCAD 11 Project Book

Now you are ready to move the pictorial view. At the prompt *Select first corner of area to move*. click once above and to the left of the pictorial view see Figure (3-10).

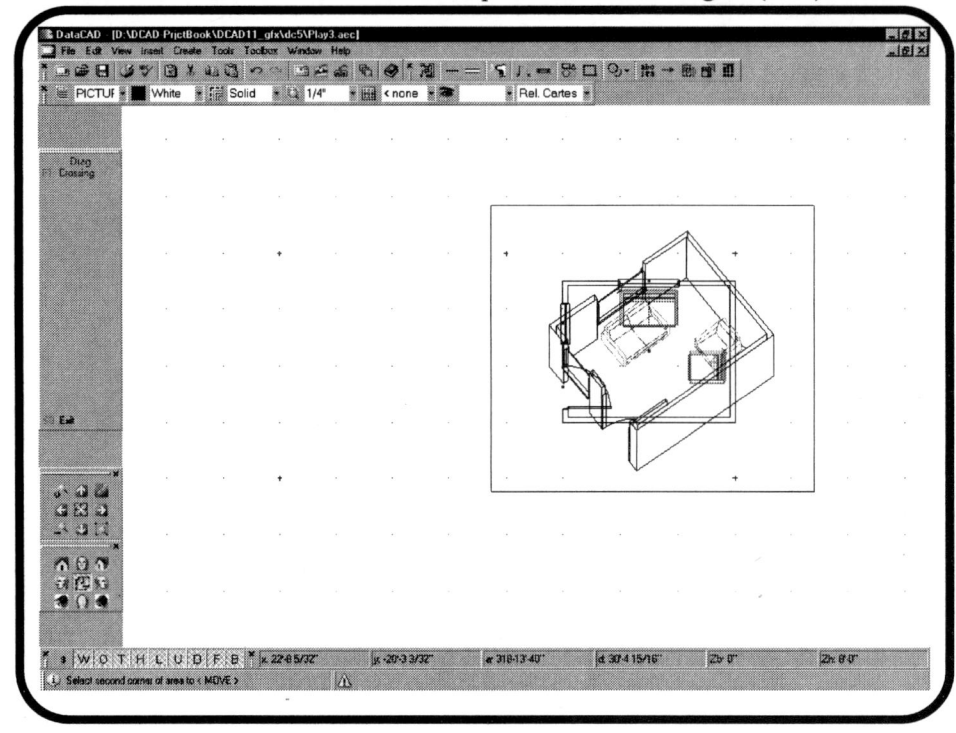

Figure 3-10, Capturing an area.

30.__ At the prompt *Select second corner of area to <MOVE>*. click once below and to the right of the pictorial view.

31.__ Look to see that the computer has made the pictorial view dashed, and then pick a spot on the pictorial view to move *from*.

Chapter 3, Page 66

Changing Enlargement Factors

32.___ Drag the group outline box to a new location (as shown in Figure 3-11) and click the *left* mouse button *once*.

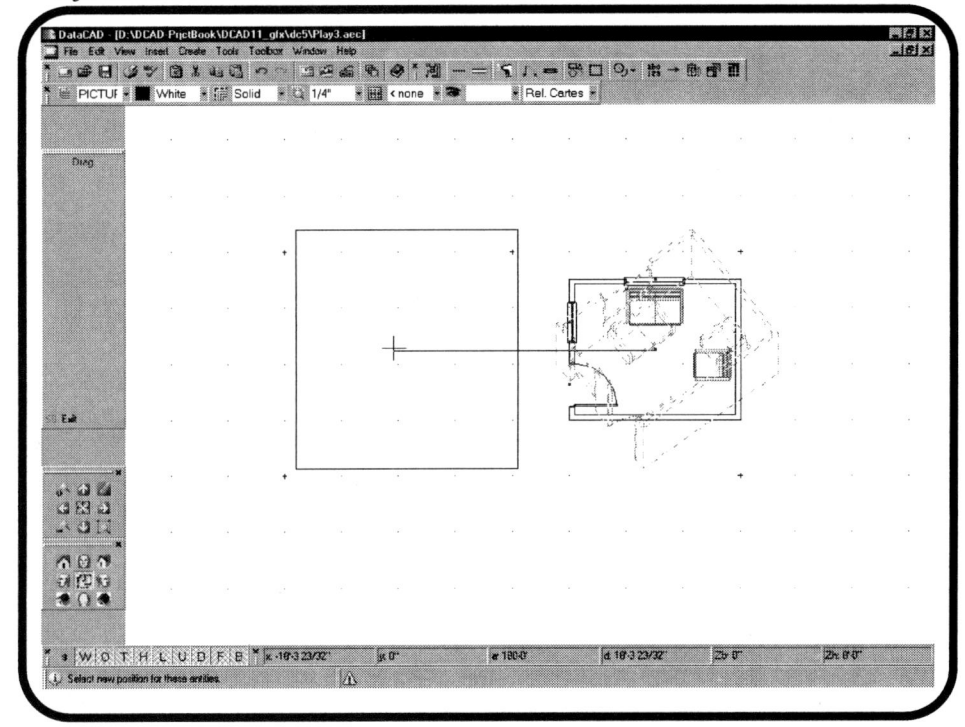

Figure 3-11, Moving a view.

33.___ Press the **Esc** key to redraw the display.

This is much better. It would even look a little better if the pictorial was a little smaller. This provides a chance to explore another DataCAD function, the **Enlarge** option.

Changing Enlargement Factors

34.___ Click the *right* mouse button until the **Edit** menu is on the display.

35.___ Select the **Enlarge** option from the **Edit** menu (or press [ALT] + [E] to jump to the **Enlarge** function).

36.___ When the system prompts you to enter the center of the enlargement, move the drawing cursor to the *center* of the pictorial view and click the *left* mouse button *once*.

Chapter 3, Page 67

The DataCAD 11 Project Book

The **Enlarge** menu (as shown in Figure 3-12) should now be on your display. Notice that the **Enlarge** menu has **Area** and **Layer Search** options just like the **Move** menu. The **Area** option should be turned **on** for the next step.

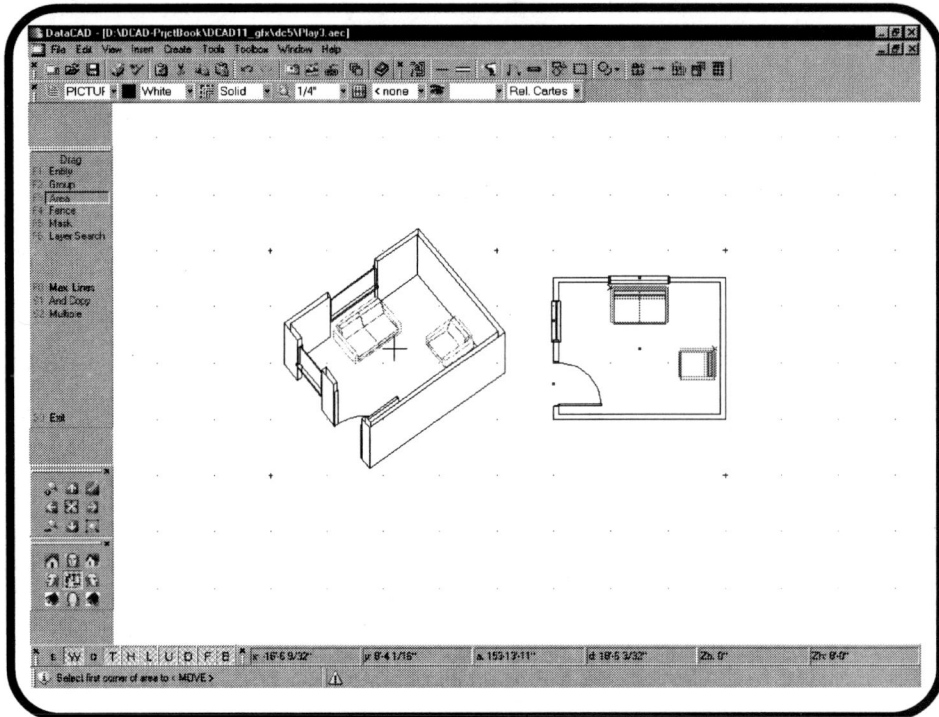

Figure 3-12, The Enlarge menu.

The **Enlarge** function allows for both making things bigger or for making things smaller. The amount of enlargement depends on the enlargement factor. This is set with the **Enlargement** option on the **Enlarge** menu. Enlargement factors smaller than one will reduce the size of objects.

37.__ Select the **Enlargement** option from the **Enlarge** menu.

38.__ Select the **Set All** option and enter **0.75** (or simply **.75**).

39.__ Press the *right* mouse button to return to the **Enlarge** menu. Toggle the **Area** option **on** in the **Enlarge** menu.

40.__ Capture the pictorial view in an area box.

Changing Enlargement Factors

•• **NOTE:** If you make a mistake while enlarging, or want to change the enlargement factor, and want to go back to the way things were, select **Edit, Undo Enlarge** from the **MENU BAR**.

41.__ After the enlargement is drawn, press the **Esc** key to redraw the display. Use the **Move** option to re-locate the pictorial if necessary.

To complete the presentation drawing, the **TEXT** layer should be turned back on and the text moved to a location centered below the two views.

42.__ Press the **I** [or the **L**] key to jump to the **Layers** menu.

43.__ Select the **On/Off** option from the **Layers** menu.

44.__ Turn the **TEXT** layer **on**.

Your display should now look similar to Figure 3-13. Next, you will move the text around.

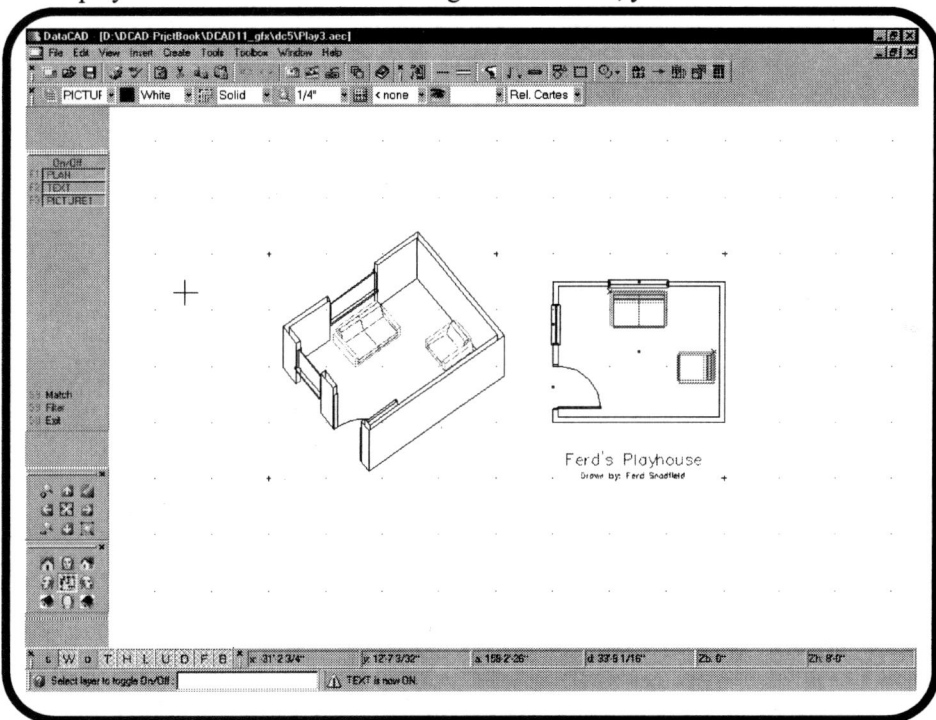

Figure 3-13, Arranging the text.

45.__ Click the *right* mouse button to exit the **On/Off** menu.

Chapter 3, Page 69

The DataCAD 11 Project Book

46.__ Jump to the **Move** menu by pressing the **m** [*not* the M] key.

47.__ Select the **Drag** option from the **Move** menu.

The two lines of text were created as two different entities. Both of them can be moved together by using the **Area** option and capturing all of the text in an **Area** box.

48.__ Select the **Area** option from the **Drag** menu.

49.__ Press the **TAB** key until the **TEXT** layer is active.

50.__ Check to be sure that the **Layer Search** option is **off**.

51.__ Create an **Area** box around the text by clicking the *left* drawing cursor at the upper left and lower right corners of the desired area. (If both lines of text are not selected, press the *right* mouse button and try a bigger area capture box.)

52.__ Pick a point to drag from and drag the text to a location similar to that shown in Figure 3-14.

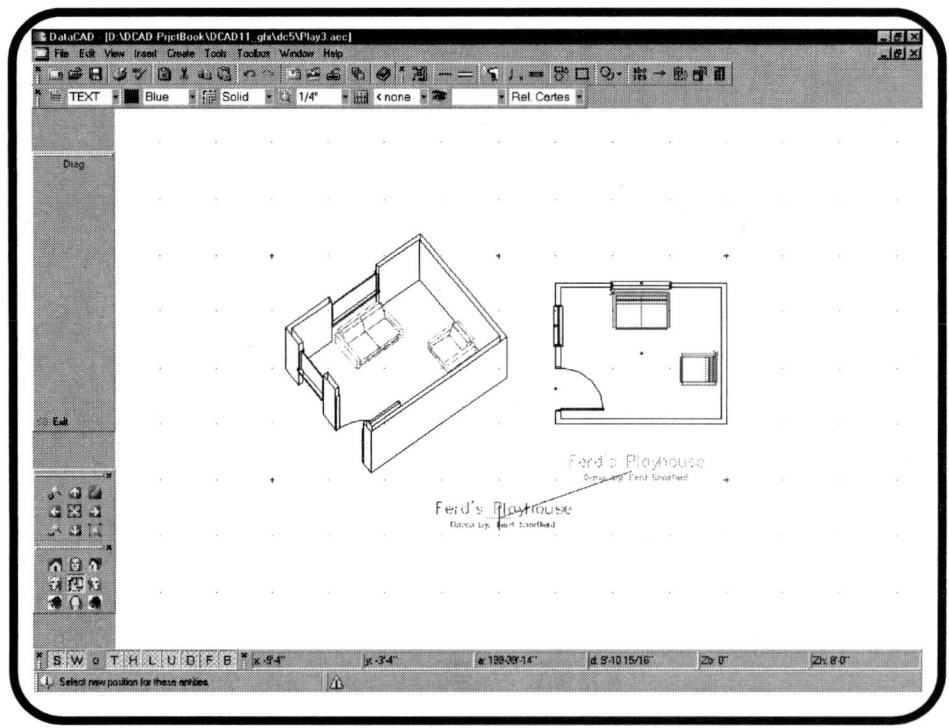

Figure 3-14, Dragging text.

Chapter 3, Page 70

Saving the Drawing

After changing the display window with the **Zoom, Re-Calc.** option, your display should look similar to Figure 3-15.

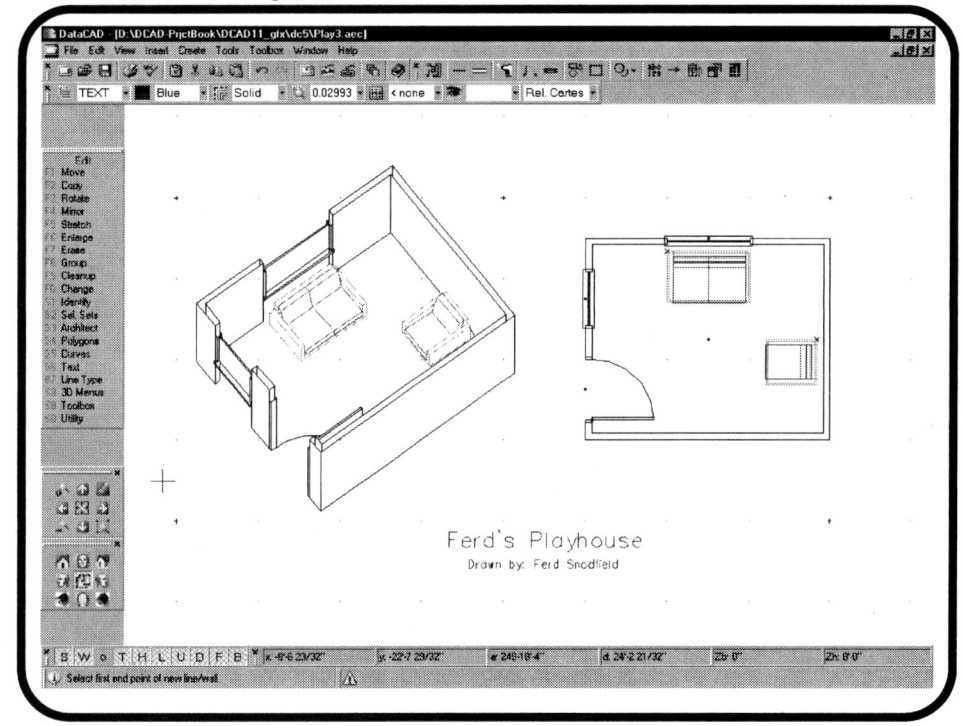

Figure 3-15, The completed drawing.

Saving the Drawing

Next, some more 3D Views options will be explored, but first an updated version of the drawing should be saved.

53.___ Select **File, Save** from the **Menu Bar** (or press **Ctrl + S**).

> •• **NOTE:** If later on in this session you make some changes that include some errors that you would like to get rid of, all you have to do is select **File, Close** and respond **No** to the *Save Changes to File?* prompt. The system will cancel all changes made to the file since the last time the drawing was saved. You can then reopen the last saved version of the file.

The DataCAD 11 Project Book

Creating Perspective Views

Earlier in this chapter you explored creating different parallel projection views. These are views where all parallel lines appear parallel no matter which way you are looking at the objects. This is not quite how things look in real life.

If you stand beside a straight section of railroad tracks or beside a straight long highway and look toward the horizon, the tracks or edges of the road appear to come together at a vanishing point. This is called perspective. Making perspective drawings by hand is very difficult.

If you tried to draw the railroad tracks in perspective, for example, the spaces between the railroad ties would change as the tracks got farther from your point of view. Measuring these changing distances while making a perspective drawing by hand is difficult and time consuming. DataCAD to the rescue!

The DataCAD software can automatically create perspective drawings for you almost as quickly and easily as it draws parallel projection views. Before creating some perspective views, you should turn off the TEXT and PICTURE1 drawing layers.

54.__ Press the **l** [or the **L**] key to jump to the **Layers** menu.

55.__ Select the **On/Off** option from the **Layers** menu and turn the **TEXT** and **PICTURE1** layers **off**.

56.__ Click the *right* mouse button to exit the **On/Off** menu.

57.__ Press the **y** [*not* the Y] key to jump to the **3D Views** menu.

To create a perspective view you have to tell the computer where you want to place your eye, and what direction to look. Unless you tell it differently, the computer will automatically put your eye 5'0" above the ground level. This results in a nice "human eye" view of the objects. Before the perspective viewing direction is specified, the **Ortho** cursor control mode should be toggled **off**. This provides a little more control in locating the viewing direction.

58.__ Press the **o** [or the **O**] key to toggle **Ortho** mode **off**.

> •• **NOTE:** The **Status Area** should show **SWoTHLUDFB**, (not **SWOTH-LUDFB**).

59.__ Select **Set Persp.** from the **3D Views** menu.

Chapter 3, Page 72

Creating Perspective Views

60. __ Use the **Page Up** or **Page Down** and **Navigation Pad** buttons to make your display look similar to Figure 3-16.

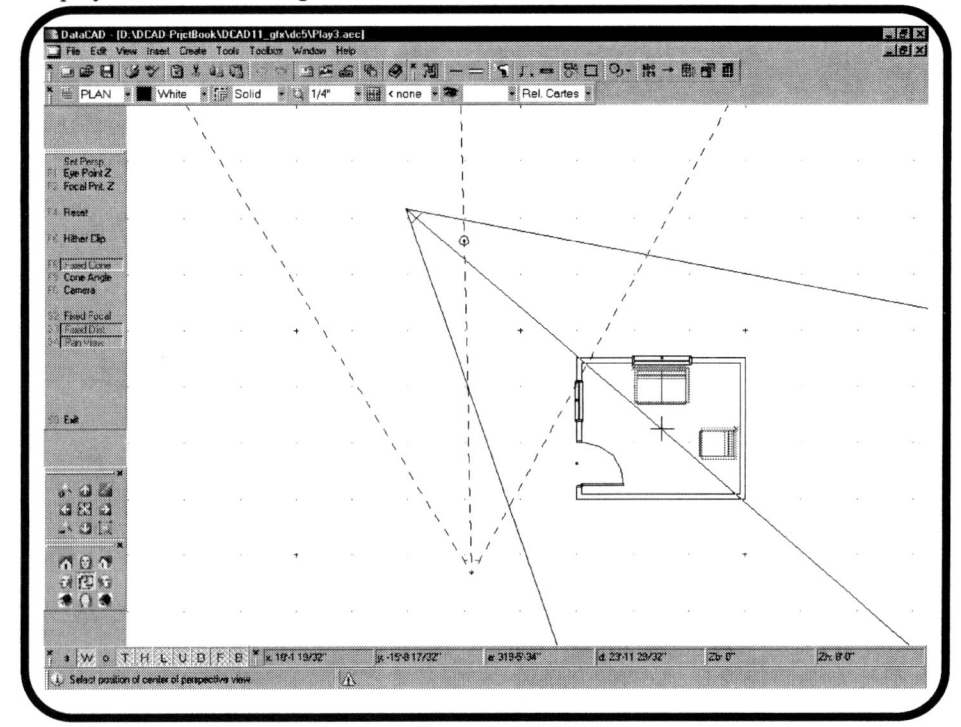

Figure 3-16, Creating a perspective view.

61. __ In the **Set Persp.** menu, toggle the **Fixed Cone**, **Pan View**, and **Fixed Dist.**, options **on**.

62. __ In the **Set Persp.** menu, toggle the **Fixed Focal** option **off**.

63. __ Follow the prompts and locate the **eye** position as shown in Figure 3-16, and at the prompt *Select position of center of perspective view*, click in the **center** of the house.

Chapter 3, Page 73

The DataCAD 11 Project Book

As soon as you have located the eye and center points, the computer will create a perspective view for you. This should look similar to Figure 3-17.

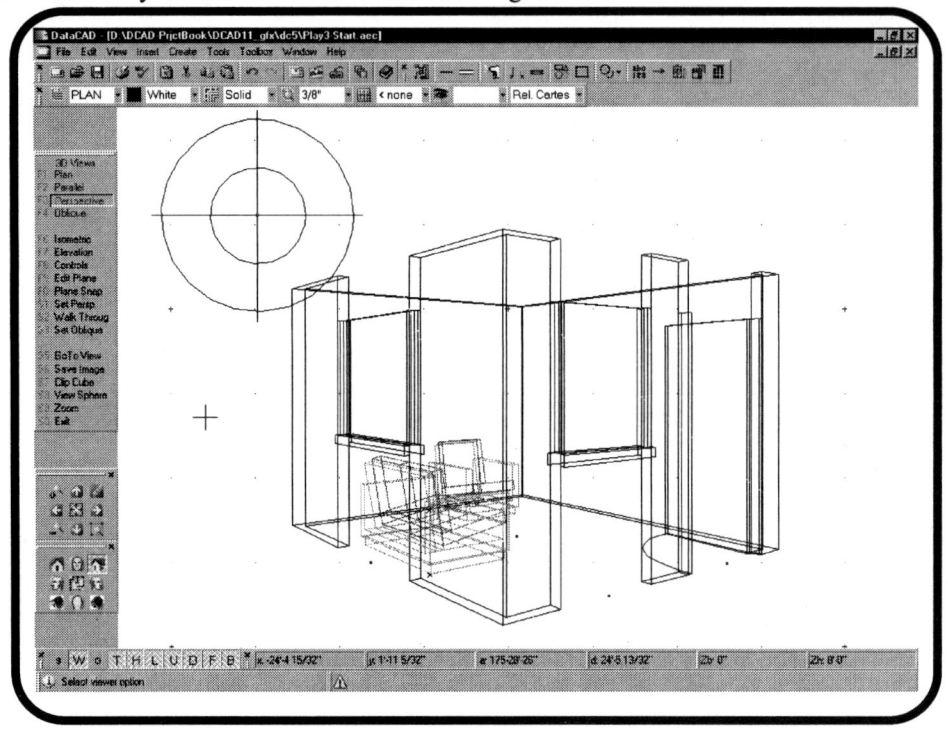

Figure 3-17, A perspective view.

The exact appearance of the perspective view will depend on where you place the eye position and view center. This view doesn't look too good because of all the lines that fall on top of each other in this view. The previous steps set the center of the view near the center of the house. The next steps will demonstrate panning the view at a fixed focal distance.

64.___ Select **Set Persp.** from the **3D Views** menu.

65.___ In the **Set Persp.** menu, toggle the **Fixed Dist.** option **off** and the **Fixed Focal** and **Pan View** options **ON**.

Chapter 3, Page 74

Creating Perspective Views

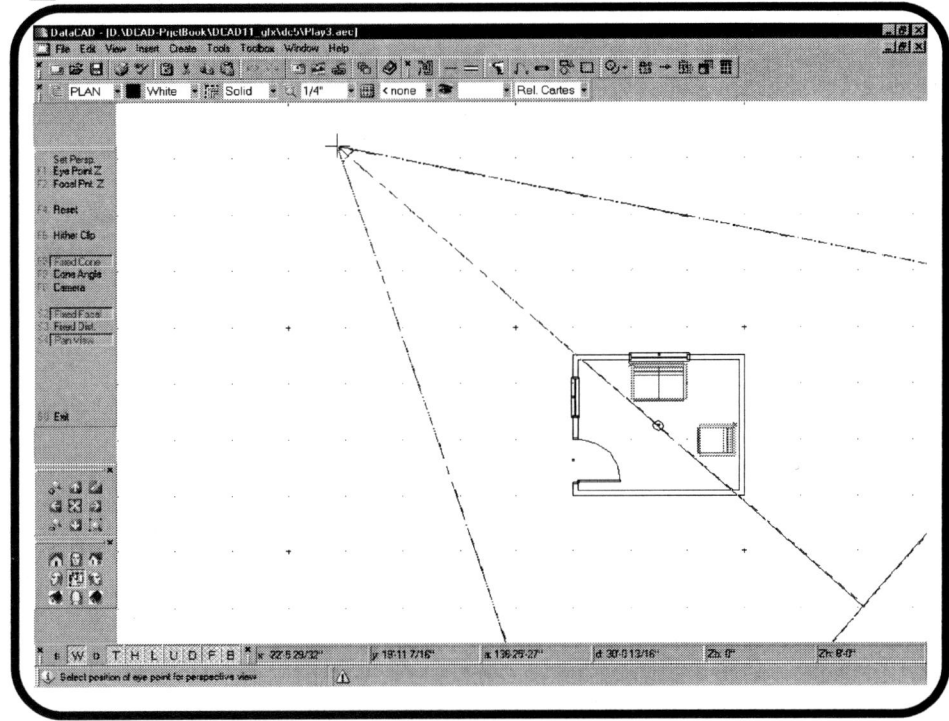

66.__ Set the eye position similar to that shown in Figure 3-18.

Figure 3-18, Creating a perspective view.

Next, the view will be changed by simulating a "walk-through" of the computer model.

67.__ Select **Walk Throug** from the **3D Views** menu.

68.__ Use the **Step Up, Step Left, Step Right,** etc. commands from the **Walk Throug** menu to adjust your perspective view.

Chapter 3, Page 75

The DataCAD 11 Project Book

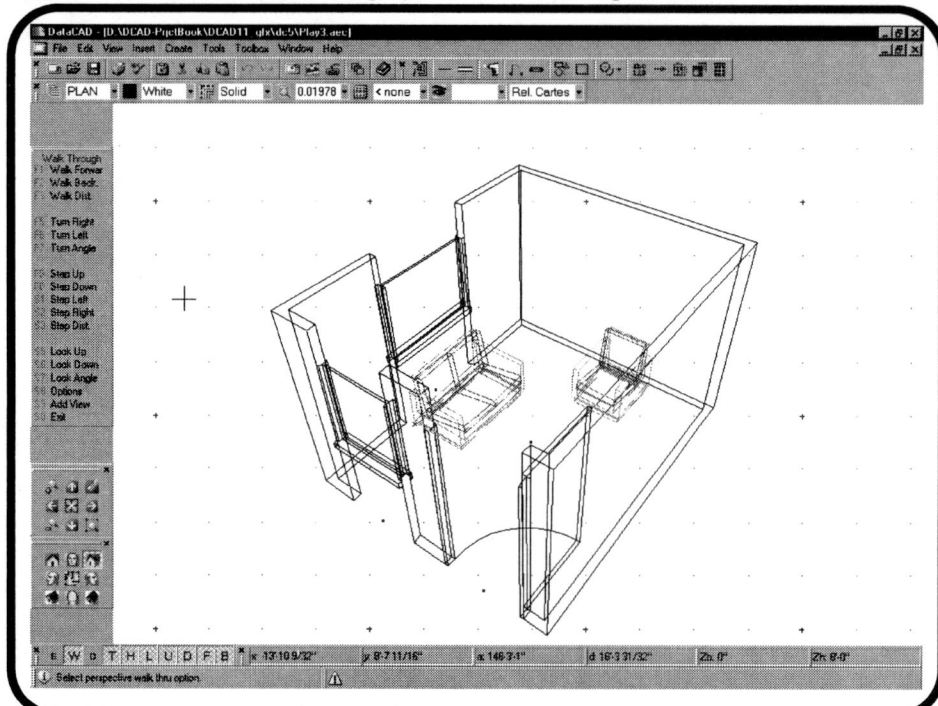

69.___ See if you can make your display look similar to Figure 3-19.

Figure 3-19, A perspective view.

•• **NOTE:** You can change the walk and step distance and turn angle by selecting those options from the **Walk Throug** menu. Also note that the navigation pad arrows function differently when you are in perspective mode. Remember that if a new view falls off of the display, you can use the **Re-Calc.** and **Zoom** functions to get it back.

It is possible to walk right through walls. If you click the *right* mouse button to exit the **Walk Throug** menu, you will see the viewing sphere displayed in the perspective view. The viewing sphere works here the same as is did in parallel projection. It is very easy to get lost in space using the viewing sphere with perspective views. If you do get lost in space, just select **Set Persp.** from the **3D Views** menu and start over.

Creating Perspective Views

Once you have a perspective view that you like, you can use the **Hide** function to remove the hidden lines. If you want to save an image that has the hidden lines removed, you have to go through the same process that was used earlier to create a new drawing layer containing the parallel projection view.

INFO STOP

The **Shader** button, directly below the **Hide** button on the **3D Menus** menu, allows for generating color shaded images of your design. Unfortunately, there are just too many options to discuss here. However, here are a couple of tips. In the **Shader** menu, select **Settings** to open up the **Shader Settings** dialog box. Toggle the **Right Click to Exit** option **on**. (This will allow you to return from a shaded image to the DataCAD display without saving a bit-mapped image file.) *Click* on the **Shade** button. When you are through viewing the image, *click* the right mouse button. If you have some time, experiment with the various settings and try turning different lights on and off. You can do a quick shaded renderings by selecting **View, Shade** from the **MENU BAR**.

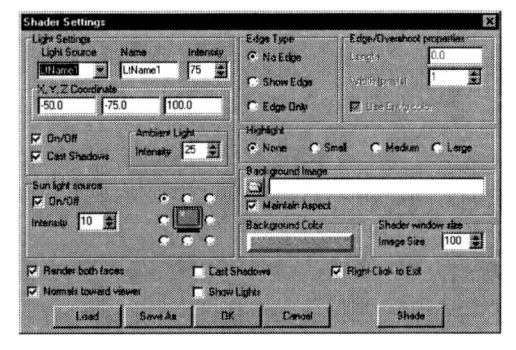

70.__ When you have finished experimenting with **3D Views,** you can select **File, Exit**.

In the next project you will return to the business of creating architectural plans; also, more of the powerful options available in DataCAD will be explored.

Chapter 3, Page 77

The DataCAD 11 Project Book

Review and Important Points

- Press **y** [not Y] to jump to the **3D Views** menu.

- Create an **Isometric** view and then use the *viewing sphere* to generate new parallel projections.

- If you want to save the results of the **Hide** option, you must toggle the **Save Image** option on before doing the hide.

- The image generated by the **Hide** function lies in the **Plan** view and is a 2D image of the 3D model. The 2D image can be moved or edited without affecting the 3D model.

- Use the **Enlarge** function to change the size of a pictorial image created with the **Hide** function.

- To save a drawing without leaving it, select **File, Save** from the **Menu Bar**, or press [CTRL] + [S].

- To create a perspective view, select **Set Persp.** from the **3D Views** menu. Then click the *left* mouse button first at the eye position, then at the center of the view.

- If you set the center of a perspective view away from the house, using the perspective viewing sphere may cause the view to fall off the display.

- Use **Walk Throug** from the **3D Views** menu (after you set a perspective view) to do a perspective tour through your design.

- Use **Hide** or the **Shader** to obtain a more realistic 3D view of your design.

Creating Perspective Views

Practice Drawings

Project Hints
Assignment P3-A, Perspective interior

- Use **3D Views, Set Persp.** to create several interior perspectives as shown above.
- Use the **Hide** function and save the rendering on a new drawing layer.
- Use the **Camera** option and select a **28mm** (wide angle) lens for viewing.

The DataCAD 11 Project Book

Practice Drawings

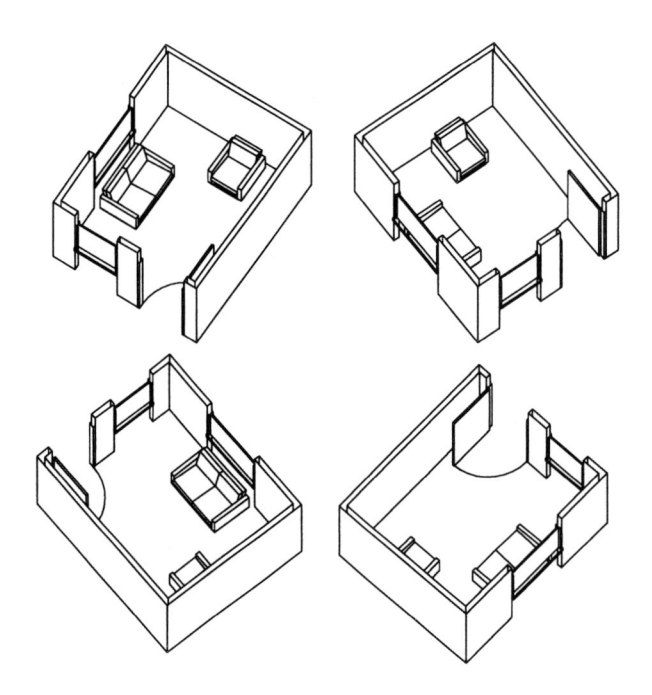

Project Hints
Assignment P3-B, Bird's-eye views

- Create birds-eye views from four different viewing directions as shown above.
- Use **Hide** and save each view on a different drawing layer.
- Use the **Move** function to arrange the views.

Chapter 3, Page 80

CHAPTER 4

Ferd's Motel Room, Phase 1

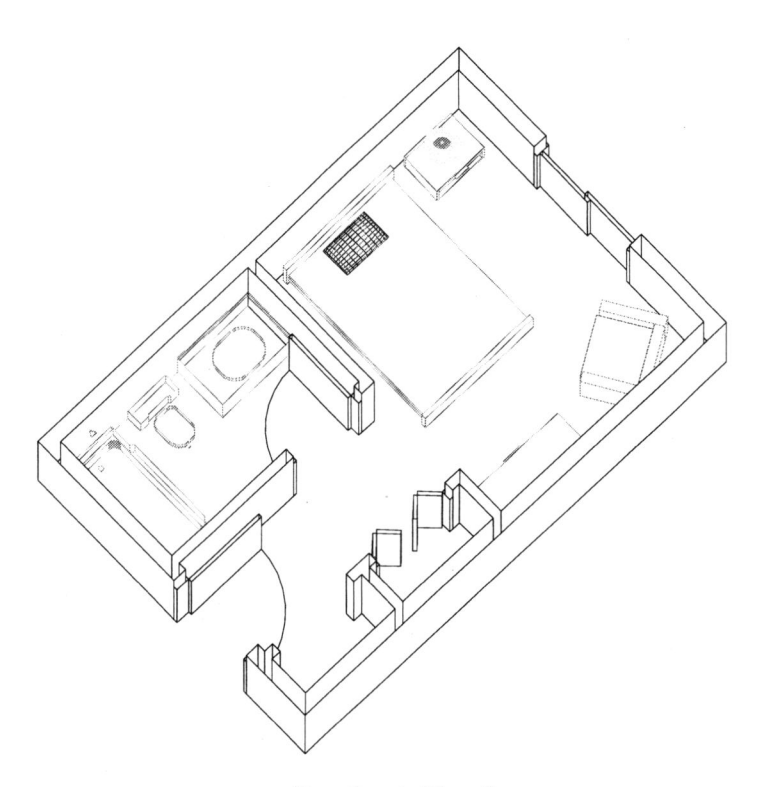

Project Goals

When you have completed this project, you will be able to:

1. Create and name drawing layers.
2. Change the width of walls.
3. Adjust the grid snap size.
4. Clean up wall intersections.
5. Toggle the **Big Cursor** on and off.

The DataCAD 11 Project Book

Prerequisite: DataCAD 11 interface experience

Introduction

In Chapter 1, Ferd's Playhouse was created. The playhouse was a simple rectangle with double line walls and did not require knowing anything about how to create intersections of walls, change wall width, or many other things that are required to create useful architectural designs. In the next several projects you will gain knowledge and experience that will allow useful architectural designs to be created using DataCAD.

1.__ Start the **DataCAD** software.

2.__ Select File, New. Enter the file name **Motel1** in the *File name:* box of the *Enter name of drawing file to create:* dialog box. Then select **Create**.

Creating and Naming Drawing Layers

In the earlier project, a new drawing layer was added after the drawing was partly finished. This time you will create and name the needed layers right at the beginning.

> **INFO STOP** If you were working in an architectural office involved in producing a lot of similar designs, you would create one or more *Default Drawings* to speed the process of starting a new file. Since you should have quite a bit of experience using DataCAD before committing to a particular *Default Drawing,* this exercise will not introduce the concept of using *Default Drawings.*

3.__ Press the **I** key [or the **L** key] to enter the **Layers** menu.

4.__ Select the **New Layer** option from the **Layers** menu and add **3** new layers.

5.__ Select the **Rename** option from the **Layers** menu and name the layers as follows: F1-**PLAN**, F2-**TEXT**, F3-**PLUMB**, F4-**FURN**.

6.__ When you're done, *click* the *right* mouse button once, or select **Exit**.

Creating and Naming Drawing Layers

•• **NOTE:** When the layers have been added and named, the display should look similar to Figure 4-1.

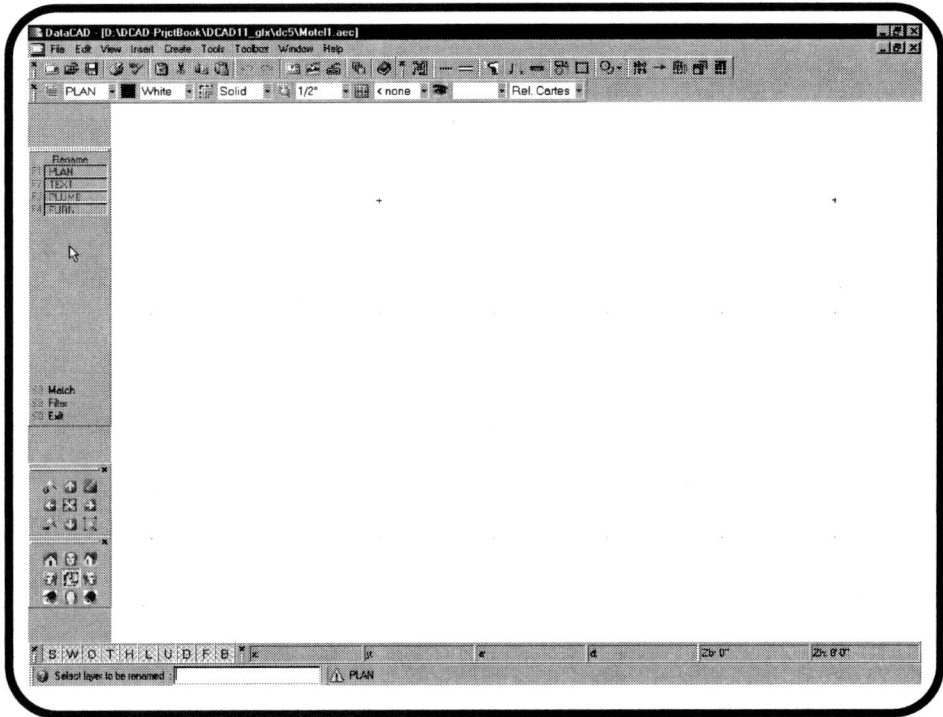

Figure 4-1, Naming new drawing layers.

It is useful to make each layer a different color. Layer colors can be set either from the **Layers** menu, or by using immediate mode commands.

7. ___ Press the **Tab** key to select a layer (note the layer name in the **Status Area**). Press the **k** [or the **K**] key to scroll forward [or backward] through the color list until the desired color has been selected.

8. ___ Repeat the previous step until a different color has been selected for each layer.

   •• **NOTE:** From here on in this book you will no longer be reminded that you can get back to the **Edit** and **Utility** menus by clicking the *right* mouse button [or by using the [;] or [:] shortcut keys].

9. ___ Select the **Settings** option from the **Utility** menu.

Chapter 4, Page 83

The DataCAD 11 Project Book

10.__ Select **Save Delay** from the **Settings** menu and set the auto save delay time to **15** minutes.

11.__ When you're done, *click* the *right* mouse button twice, or select **Exit** twice.

Creating Walls

In the earlier Ferd's Playhouse project, the default wall thickness was accepted. This time the outside walls of the motel room will be 6 inches thick and the inside walls 4 inches thick.

12.__ Press the **a** key [not the A key] to jump to the **Architect** menu.

13.__ Select the **Width** option from the **Architect** menu.

14.__ Enter a *width* of **6** inches (remember, 6 inches is entered **0.6** or .6).

15.__ Toggle **Walls, 2 Line Walls** and **Outside on** in the **Architect** menu.

Next, you will create a rectangle for the motel room. The room is to be exactly 18 feet by 12 feet (outside dimensions). There is a way to create a rectangle by entering the dimensions directly from the keyboard. This is more precise and easier than trying to control the drawing cursor.

INFO STOP
DataCAD provides several different methods of entering sizes from the keyboard. The method that will be used in this exercise is called *Relative Cartesian Coordinates*. This is just a way of giving the computer measurements in the X (which is left and right, or horizontal, on the display), and Y (which is up and down, or vertical, on the display) directions.

The first corner of the rectangle will be located using the drawing cursor, and then coordinate entry mode will be used to enter the second corner of the rectangle.

16.__ Select the **Polygons** option from the **Edit** menu.

17.__ Select the **Rectangle** option from the **Polygons** menu.

18.__ Check the **Status Area** to make sure that the **PLAN** layer is active. (If it is not, press the **Tab** key until it is.)

19.__ Click the *left* mouse button *once* somewhere near the *lower left* corner of the display to place the first corner of the rectangle.

Chapter 4, Page 84

Setting Snap Size

•• **NOTE:** In the next step, watch the prompt to be sure the system asks for the *relative* X distance. If not, press the *right* mouse button; then press the **Insert** key until *relative cartesian* coordinates are active, and then continue.

20.___ Press the **space bar** and enter a value of **18** for X and **12** for Y.

21.___ At the prompt *Select a point to define the Inside of the wall*. click the *left* mouse button *once* with the cursor *inside* of the rectangle.

This creates a rectangle for the motel room with walls exactly 6 inches thick. Let's check this out.

22.___ Jump to the **Zoom** menu by pressing the [/] key [or by selecting the **Zoom to area** button in the **Navigation Pad**].

23.___ Capture one corner of the rectangle in an area box to zoom-in.

24.___ Press the unshifted ~ key, and reset 0,0 by locating the drawing cursor on the *outside corner* of the motel walls and pressing the *object snap button* (the **n** [or **N**] key, or the *middle* button on the mouse).

25.___ Move the drawing cursor to the *inside* corner of the motel walls and **observe** the coordinate readout to check the wall thickness.

••• **WARNING**••• OK, I tricked you!

The drawing cursor jumps around and will not stop at the inside corner of the motel walls. If you watch the coordinate readout while you are moving the cursor, you will see that it is making 4 inch jumps each time it moves. This is called snapping, and the size of the jump (known as the snap size) can be changed or turned off.

Setting Snap Size

••• **WARNING**••• If you are drawing by watching the coordinate readout, then it is *not* a good idea to turn the snap off while drawing. If the cursor did not jump to pre-set locations, you would never get things to line up properly. But in time you will learn that it is often best, and more accurate, to draw by typing in the lengths and coordinates of the drawing entities.

To get the cursor to jump to the inside corner of the motel walls, the snap size must be reduced to 2 inches. One inch would also work of course, but it is always best to keep the snap set as large as makes sense.

26.___ Press the **Tab** key until the **PLUMB** layer is active.

Chapter 4, Page 85

The DataCAD 11 Project Book

27.__ Press the **s** key (*not* the S key) to jump to the **snap grid size** menu.

28.__ Enter values of 2 inches (**0.2**) for both X and Y.

Now if you try checking the wall corner coordinates again, you can verify that the walls are 6 inches thick.

> •• **NOTE:** When drawing by the coordinate readout, always keep the snap turned **on** while drawing, and keep the snap grid size as *large* as possible to make drawing easier.

Adding Plumbing

The next thing to do is add some plumbing fixtures to the bathroom. Then the bathroom walls will be built around the plumbing group. This is because the size of the plumbing fixtures is not yet known. There is a symbol folder that contains some standard plumbing fixtures that will be used.

29.__ Select the **Zoom extents** button from the **Navigation Pad** [or press the [/] key and select the **Extents** option from the **WindowIn** menu].

30.__ Hold down the [CTRL] key and press the **T** [or the **t**] key to enter the **Symbol Browser** menu.

Adding Plumbing

31.___ Set the disk path to **C:\DATACAD\Symbols\Plumbing\Residential** (see Figure 4-2), and *click* on **OK**.

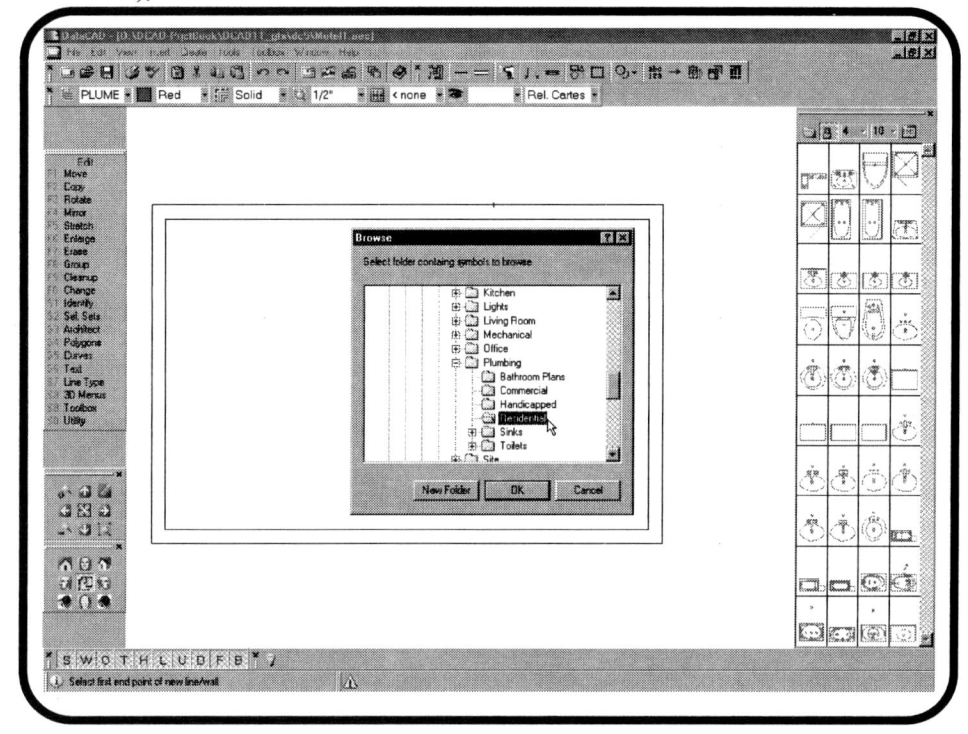

Figure 4-2, Setting the Symbol Browser disk path and file name.

Chapter 4, Page 87

The DataCAD 11 Project Book

The display should now look similar to Figure 4-3. (You may have to pan the drawing with the arrow keys to get the right side of the motel room to appear beside the Symbol Browser).

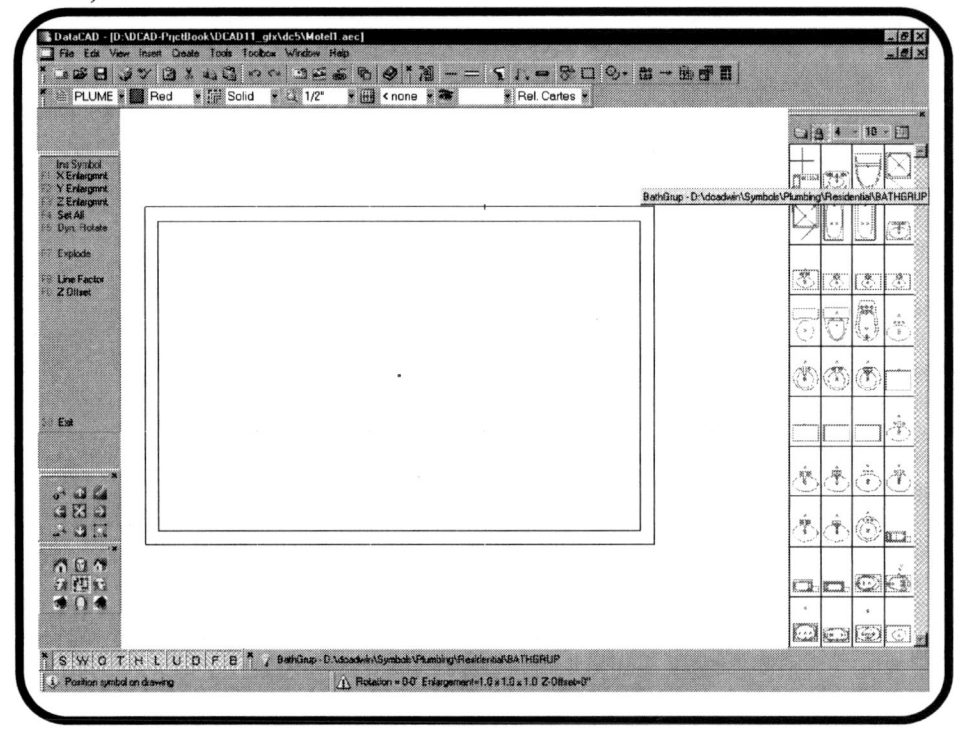

Figure 4-3, The residential plumbing symbols.

The bath group will be placed in the lower right corner of the motel room. The bath group symbol in the template is rotated opposite of what it should be to fit in the room. No problem. A rotation angle of 180 degrees will be specified before using the symbol.

32. ___ Select the **BATHGRUP** symbol from the Symbol Browser (it's in the upper left corner, as shown in Figure 4-3) by moving the drawing cursor to the symbol in the Symbol Browser and *click* the *left* mouse button *once*. The symbol will be attached to your cursor, and the **Ins. Symbol** menu will be displayed at the left side of your screen (see Figure 4-3).

33. ___ Select the **Dyn. Rotate** option from the **Ins. Symbol** menu (if it was already highlighted then a new menu with a list of angles will appear. If it was not, then you must select **Dyn. Rotate** a second time to enter the angle specification mode).

Adding Interior Walls

34.__ Enter an angle of **180**.

35.__ Locate the bath group symbol outline in the *inside lower right* corner of the motel room and click the *left* mouse button *once*.

36.__ Press the *right* mouse button *once*, or select the Exit button to finish working with that symbol.

37.__ To close the Symbol Browser *click* on the **X** at the upper right corner of the Symbol Browser, or hold down the [CTRL] key and press the **T** [or the **t**] key.

The display should now look similar to Figure 4-4.

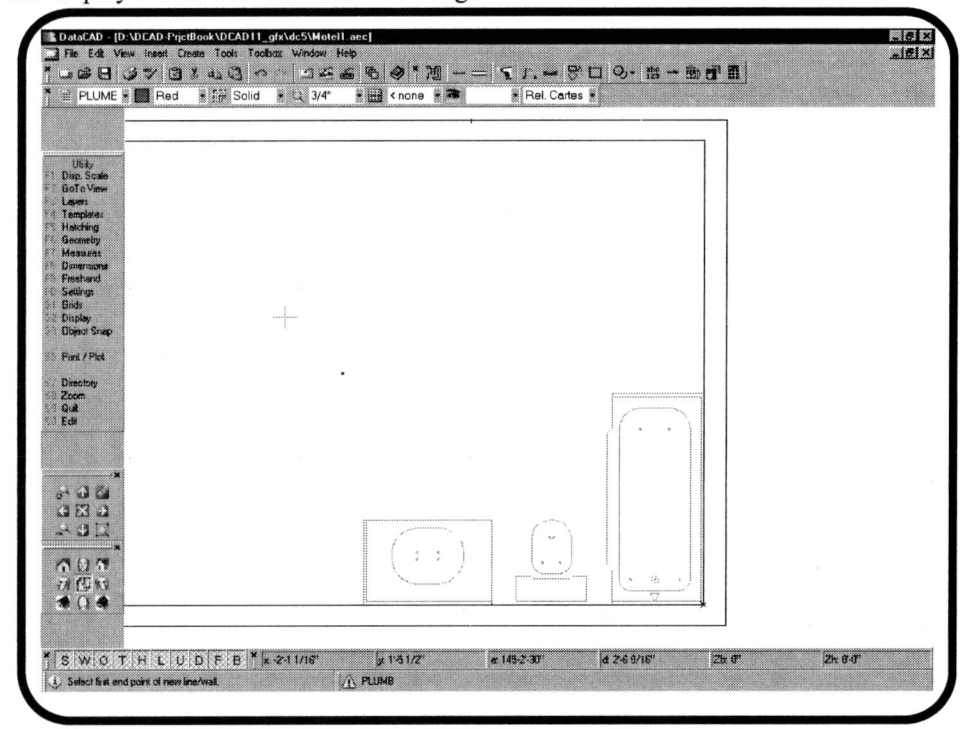

Figure 4-4, The bath group.

Adding Interior Walls

Next, some walls will be built around the bathroom. The walls, of course, live on the **PLAN** layer *not* the **PLUMB** layer. Since the bathroom walls are interior walls, they will be 4 inches thick instead of 6 inches like the outside walls.

Chapter 4, Page 89

The DataCAD 11 Project Book

INFO STOP

In many cases, architects would use a wall thickness of 4 1/2 inches. This project uses an even 4 inches just to keep things simple for beginners. As you gain experience and start on your own plans, you should select whatever wall thickness is appropriate. Don't try to get too fancy too soon, or you may become overwhelmed with details.

38.__ Press the **Tab** key until the **PLAN** layer is active.

> •• **NOTE:** The computer remembers the snap grid size separately for each drawing layer. This means that you could set the snap very large for a plot plan layer and very small for a detail layer.

39.__ Press the **s** key (*not* the S key) to jump into the snap grid setting menu.

40.__ Set the snap size for both X and Y to 2 inches (**0.2**).

41.__ Press the **a** key [not the **A** key] to jump into the **Architect** menu.

42.__ Select the **Width** option from the **Architect** menu and set the width to 4 inches (**0.4**).

43.__ Make sure that **Walls, 2 Line Walls,** and **Outside** are turned **on** in the **Architect** menu.

The walls will be located by entering the dimensions from the keyboard. This is easier and more precise than trying to carefully control the cursor with the mouse. The coordinate 0,0 location will be reset to the corner of the bathroom to make this job easier. The bath group symbol used was designed for a 8 foot by 5 foot bathroom.

> •• **NOTE:** If you want to verify the size of the plumbing group, reset 0,0 to the corner of the bath group. Then watch the coordinate tracking when the drawing cursor is at the opposite corner of the group.

44.__ Press the unshifted ~ key, and click the *object snap button* with the drawing cursor located in the *inside lower right* corner of the bathroom.

45.__ Press the *right* mouse button until the prompt *Select first end point of new line/wall*. appears. Then press the **space bar** to enter coordinate entry mode.

46.__ Enter values of **X=0.0** and **Y=5.0**.

Adding Interior Walls

•• **NOTE:** Move the cursor and verify that the anchor point is lined up with the top of the bath group. If it is not, press the *right* mouse button and repeat the process of setting a new reference 0,0 point.

INFO STOP Distances measured to the left and/or down must have a minus sign in front of them.

47.__ When the prompt *Select next end point of line/wall.* appears, press the **space bar** and enter (watch the minus sign) values of **X=-8.0** and **Y=0.0**.

48.__ When the prompt *Select a point to define the inside of the wall.* appears, locate the drawing cursor *above* the line just drawn, and then click the *left* mouse button *once*.

49.__ Move the drawing cursor to a point on the *inside* of the motel room wall directly *below* the end of the line just drawn to add the left side wall of the bathroom. *Click* the *left* mouse button *once* at this point.

50.__ Click the *right* mouse button *once* to stop creating lines.

Chapter 4, Page 91

The DataCAD 11 Project Book

The display should now look similar to Figure 4-5.

Figure 4-5, Adding bathroom walls.

Notice that the intersections of the new bathroom walls and the outside walls need to be cleaned up. The software gives us an easy way to do this.

51.___ Select the **Cleanup** option from the **Edit** menu.

52.___ Select the **T Intersect** option from the **Cleanup** menu.

53.___ Make sure that the **Layer Search** option is **off**.

Adding Interior Walls

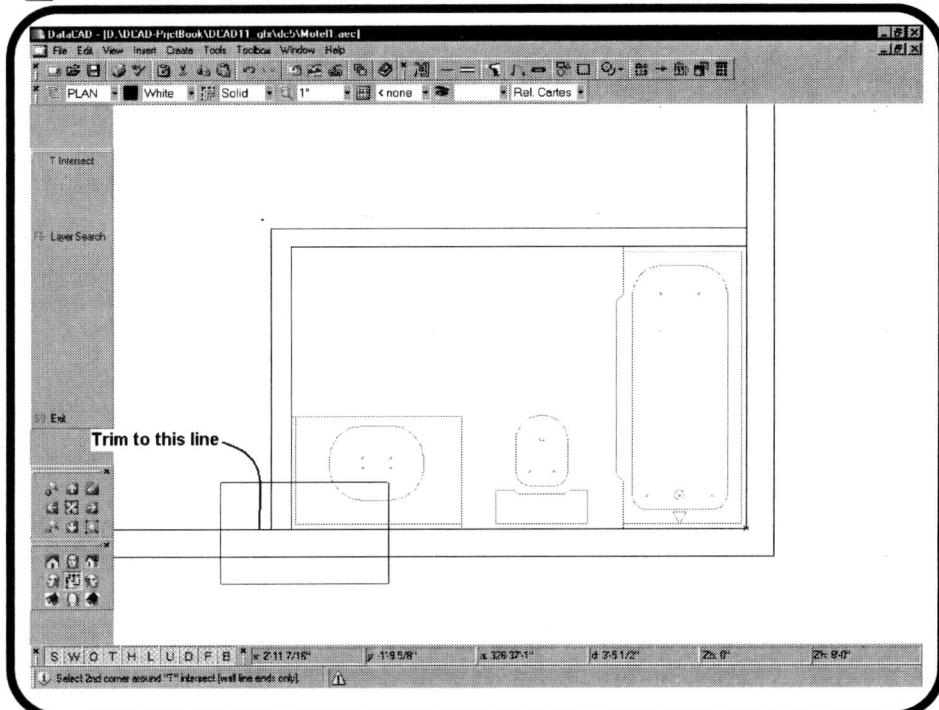

54. ___ Make a box around the wall intersection as shown in Figure 4-6.

Figure 4-6, Cleaning up wall intersections.

55. ___ At the prompt *Point to wall to trim to*, move the drawing cursor to the ***inside*** horizontal wall (the second horizontal line up from the bottom) and click the ***left*** mouse button ***once*** (see Figure 4-6).

56. ___ Repeat this process for the other intersection. Clean this intersection to the inside vertical wall.

INFO STOP
The cleanup function would get confused if it found too many lines in the area box. Even though there are some of the lines from the bath group in this box (as shown in Figure 4-6), the system does not see them because they are on a different drawing layer. This would not be the case if the layer search option was turned on.

Chapter 4, Page 93

The DataCAD 11 Project Book

••• **WARNING**••• Have you saved your drawing lately? (CTRL + S)

Next, a closet space will be added to the motel room. Once again coordinate entry mode will make the job easier. This time, an option will be used that will automatically clean up the wall intersections as the walls are drawn.

57.__ Select the **Zoom extents** button from the **Navigation Pad** [or press the / key and select the **Extents** option from the **WindowIn** menu] to get the whole motel room on the display; then return to the **Edit** menu.

58.__ Press the unshifted ~ key, and use the *object snap button* to reset the coordinate 0,0 to the *upper right inside* corner of the motel room.

59.__ Press the **a** [*not* the A] key to jump to the **Architect** menu. Toggle the **Clean** option **on**. Then press the *right* mouse button until the *Select first end point of new line/wall*. prompt appears.

60.__ Start the closet wall by pressing the **space bar** and entering values (watch the minus sign) of **X=0.0** and **Y=-1.10**. (Note that -1.10 is not the same as -1.1.)

If you move the mouse, you will see that a line has been anchored. There are several ways to go about aligning the end of the closet wall with the bathroom wall. The **big cursor** will help line things up. The **big cursor** extends from side to side and from top to bottom of the display. This is a big help in aligning things.

61.__ Press the + key to turn the **big cursor** on. (You can use either the + key in the numeric keypad or the + key on the top row of the keyboard.)

Adding Interior Walls

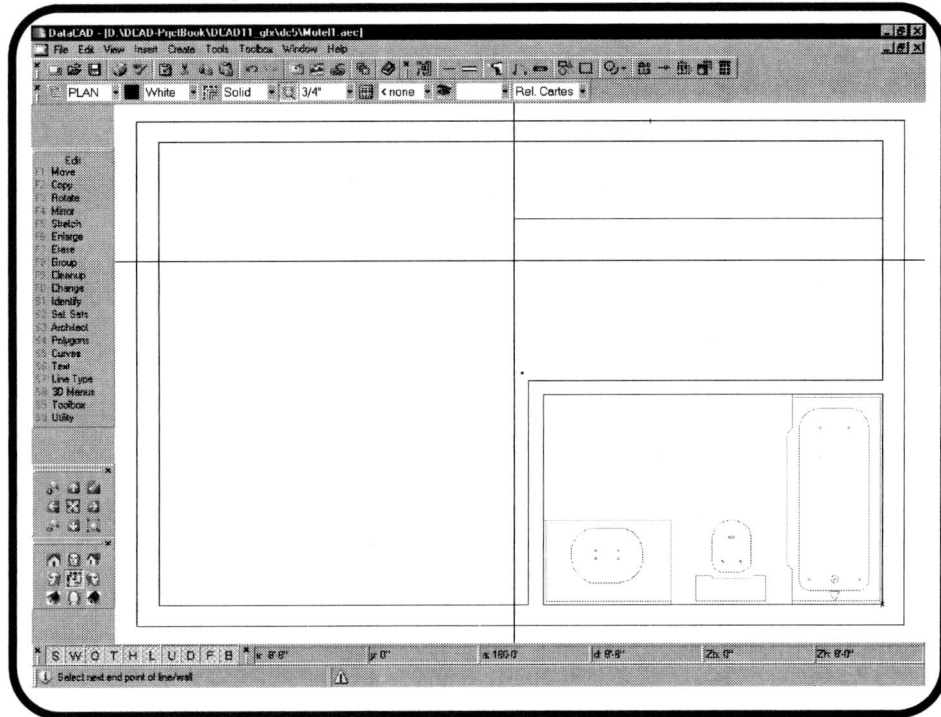

•• **NOTE:** The big cursor is shown in Figure 4-7.

Figure 4-7, Using the big cursor.

62.___ Use the big cursor to align the *left* closet wall with the *outside left* bathroom wall and click the *left* mouse button *once*.

63.___ At the prompt *Select a point to define the inside of the wall*, locate the drawing cursor *above* the line just drawn and click the *left* mouse button *once*.

64.___ Move the cursor line *straight up* to the inside of the motel room wall and click the *left* mouse button *once*.

65.___ *Click* the *right* mouse button to stop creating lines.

66.___ Press the [+] key to turn off the big cursor.

Chapter 4, Page 95

The DataCAD 11 Project Book

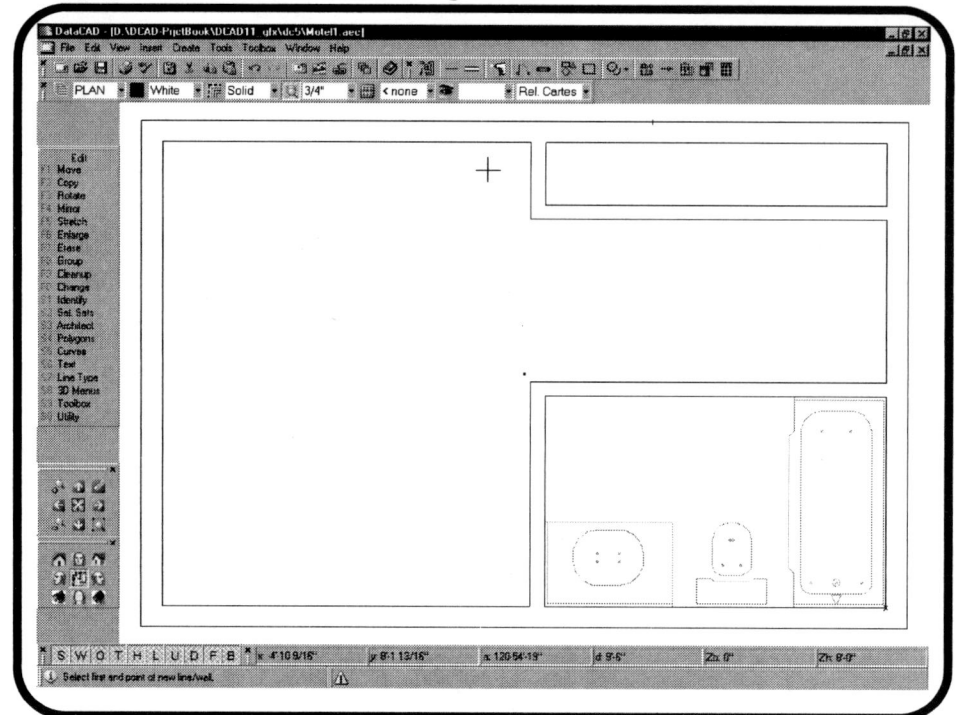

The display should now look similar to Figure 4-8.

Figure 4-8, Ferd's Motel Room, Phase 1.

Notice that the **Clean** option has automatically cleaned up the intersections.

>   •• **NOTE:** If the **Clean** option is **on**, and you end a wall close to the wrong side of the intersecting wall, the intersection will be cleaned to the wrong side of wall. If you make such a mistake, just press [CTRL] + [Z] to undo the mistake.

>   ••• **WARNING**••• Have you saved your drawing lately?

67.__ Press [CTRL] + [S] save your drawing now and take a break.

In the next phase of the motel room design, some doors and furniture will be added to the motel room.

Review and Important Points

• Press the **s** [*not* the S] key to set the snap size.

Chapter 4, Page 96

Adding Interior Walls

- Snap size is stored separately on each drawing layer.
- Use the **Tab** key to switch the active drawing layer.
- Use the **Cleanup** menu to fix intersecting walls.
- The **Clean** option in the **Architect** menu will automatically clean intersecting walls. If you use this option, be careful to click on the correct side of the intersecting wall, or you will create an editing nightmare.
- Press the unshifted ~ key to reset the 0,0 coordinate location, and then use coordinate entry mode to key-in accurate locations for design elements.
- The ⊞ key is used to toggle the **big cursor** on and off.

The DataCAD 11 Project Book

Practice Drawings

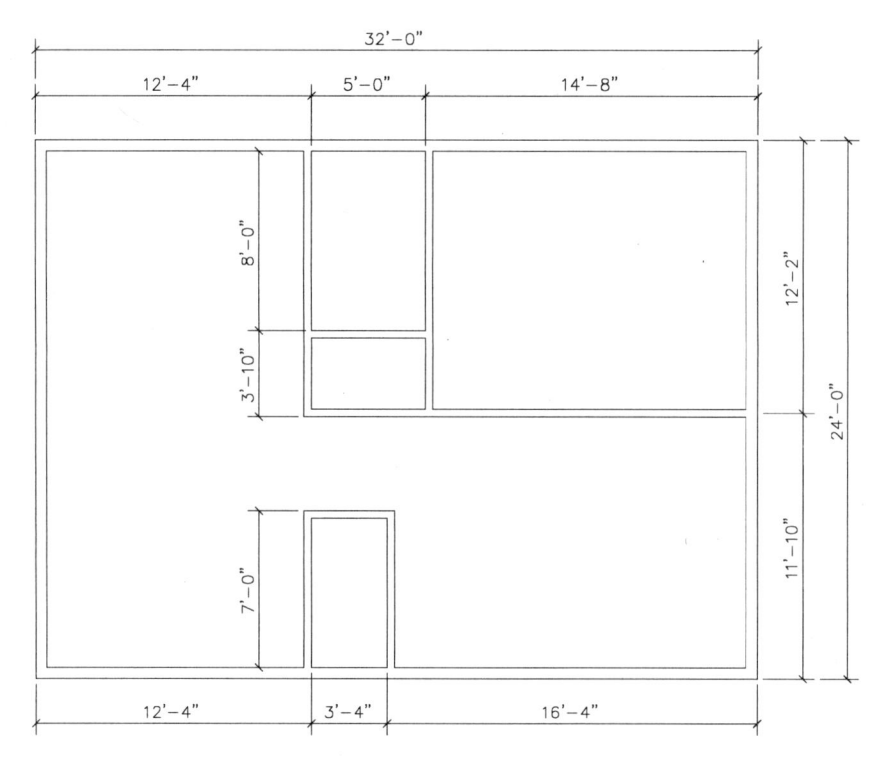

Project Hints
Assignment 4-A, Apartment

- Create the walls for a floor plan for an apartment using the dimensions shown above.
- Make the outside walls 6 inches and the inside walls 4 inches thick.
- The doors and window openings will be added in Assignment 5-A.
- Note that the bathroom has *interior* dimensions of 5'x 8' to fit the standard plumbing group symbol that will be added in Assignment 5-A.

The DataCAD 11 Project Book

Notes

The DataCAD 11 Project Book

CHAPTER 5

Ferd's Motel Room, Phase 2

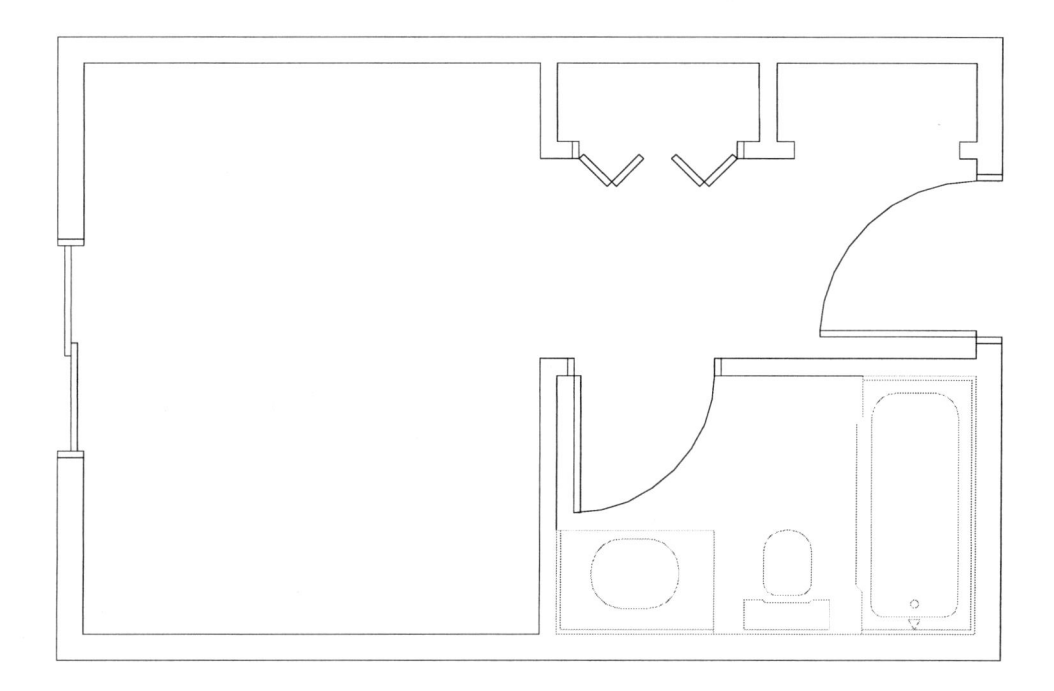

Project Goals

When you have completed this project, you will be able to:

1. Select and use different object snap options.
2. Locate features using a combination of object snap, coordinate entry, and mouse pick.
3. Use the **Cut Wall** function to cut an opening in a wall.
4. Select and place different door types.
5. Use the **Save As** function to create a copy of a drawing with a new file name.

The DataCAD 11 Project Book

Prerequisite: Chapter 4

Introduction

In Phase I of Ferd's Motel project, you created the basic walls, added the bathroom fixtures, and constructed the interior walls for the motel room. In Phase 2 of the motel room design, you will add several different door types to the room.

1.__ Start the **DataCAD** software and **Open** the **Motel1** drawing file.

2.__ Arrange the display so that it is similar to Figure 5-1.

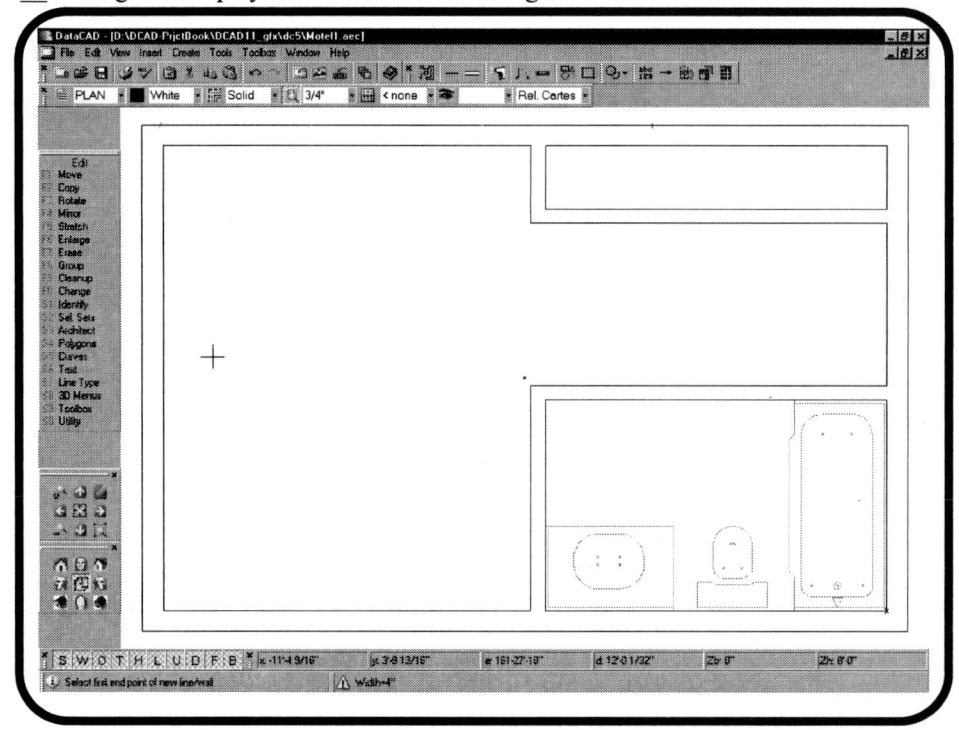

Figure 5-1, Ferd's Motel Room.

3.__ Make the **PLAN** layer active. (If PLAN does not appear in the menu reminder, press the **Tab** key until it appears.)

Adding a Standard Door

Adding a Standard Door

The first door to be added is a standard 2'8" bathroom entrance door. Since this door is to be placed 4 inches from the upper left inside corner of the bathroom, it is a good idea to reset the coordinate 0,0 location to help in positioning the door.

4. __ Press the unshifted ~ key and click the *object snap button* once when the drawing cursor is near the *upper left inside* corner of the *bathroom*.

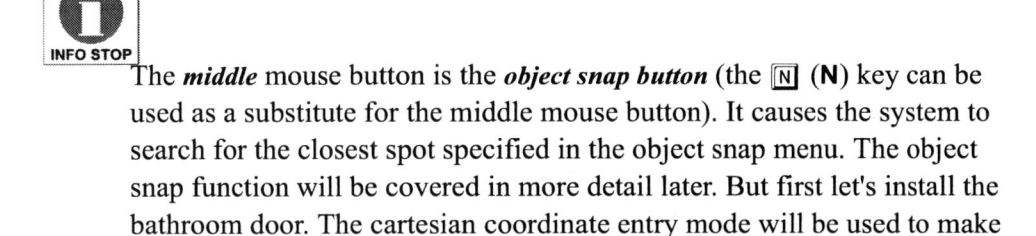

INFO STOP

The *middle* mouse button is the *object snap button* (the [N] (**N**) key can be used as a substitute for the middle mouse button). It causes the system to search for the closest spot specified in the object snap menu. The object snap function will be covered in more detail later. But first let's install the bathroom door. The cartesian coordinate entry mode will be used to make sure the door is in the correct place and that it is the proper size.

5. __ Jump to the **Architect** menu by pressing the **a** [not the A] key.
6. __ Select the **Door Swing** option from the **Architect** menu.
7. __ Select **Door Style** and be sure that the **Single** option is toggled **on**.
8. __ Press the *right* mouse button *once* to return to the **Door Swing** menu.
9. __ Be sure that the **Sides** option is turned **on** (doors will be defined by jambs).
10. __ At the prompt *Select hinge side of door* press the **space bar** and enter an **X** value of **0.4** (which means 0 feet four inches), and a **Y** value of **0.0**.
11. __ At the prompt *Select strike side of door* press the **space bar** and enter an **X** value of **2.8** and a **Y** value of **0.0**.
12. __ At the prompt *Select direction of door swing* move the drawing cursor somewhere *inside* of the bathroom and click the *left* mouse button *once*.
13. __ At the prompt *Select any point on the outside of the wall* move the drawing cursor *outside* of the bathroom (just above the door opening) and click the *left* mouse button *once*.

The DataCAD 11 Project Book

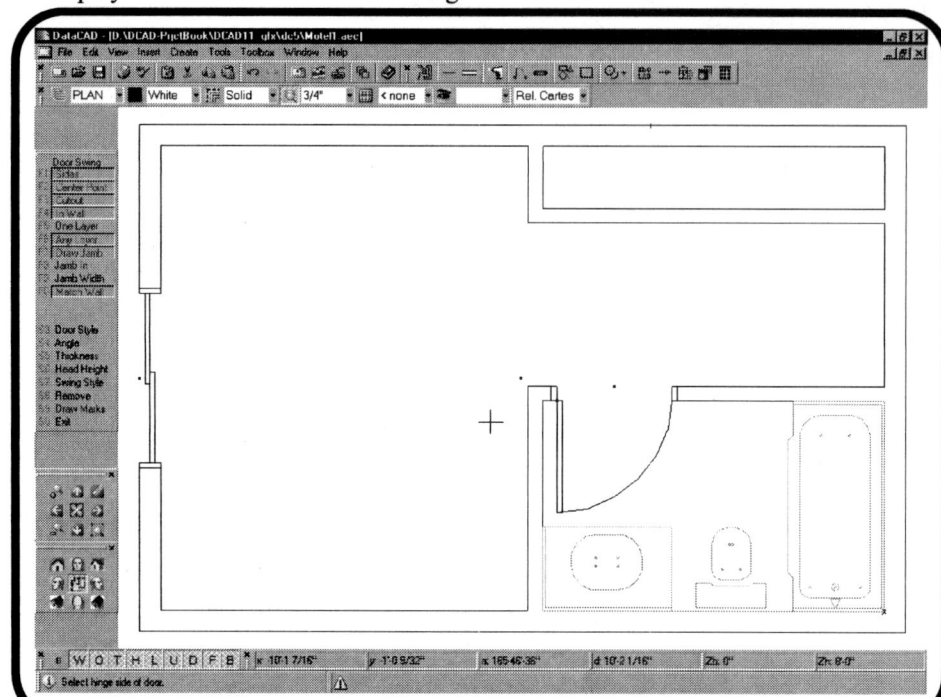

The display should now look similar to Figure 5-2.

Figure 5-2, Installing the bathroom door.

INFO STOP

What do you do if you have to erase a misplaced door? Sooner or later you will make a mistake and have to fix a door or window that is the wrong size or is in the wrong place. Both the **Door Swing** and **Windows** menus have a **Remove** option. To remove a door or window, select **Remove** and capture the door or window in an area box. If you place a door with the hinge exactly at the intersection of two walls, you will be in trouble. The easiest way out of such a mess is to start over. It is possible (but not easy) to erase lines and fix the wall with the **Cleanup** options. Remember, the walls are 3D and have lots of lines to be fixed. Just press [CTRL] + [Z] as many times as it takes to undo the mistake.

Centering a Door

••• **WARNING**••• Have you saved your drawing lately?

Centering a Door

Next, the entrance door to the motel room will be installed in the hall between the bathroom and closet. This door should be nicely centered in the hall. This is where the object snap function is very handy.

14.__ Press the **X** key (not the x key) to jump to the **Object Snap** menu.

The object snap menu is shown in Figure 5-3. The default object snap option is the **End Point** option. When this is **on** and the *object snap button* is pressed, the computer jumps to the end of the nearest line (within the snap miss distance). The option that will help locate the door in the center of a wall is the **Mid Point** option (the **Center** option is used to find the center of arcs or circles). The door to be installed is 3 feet wide with the hinge near the closet wall.

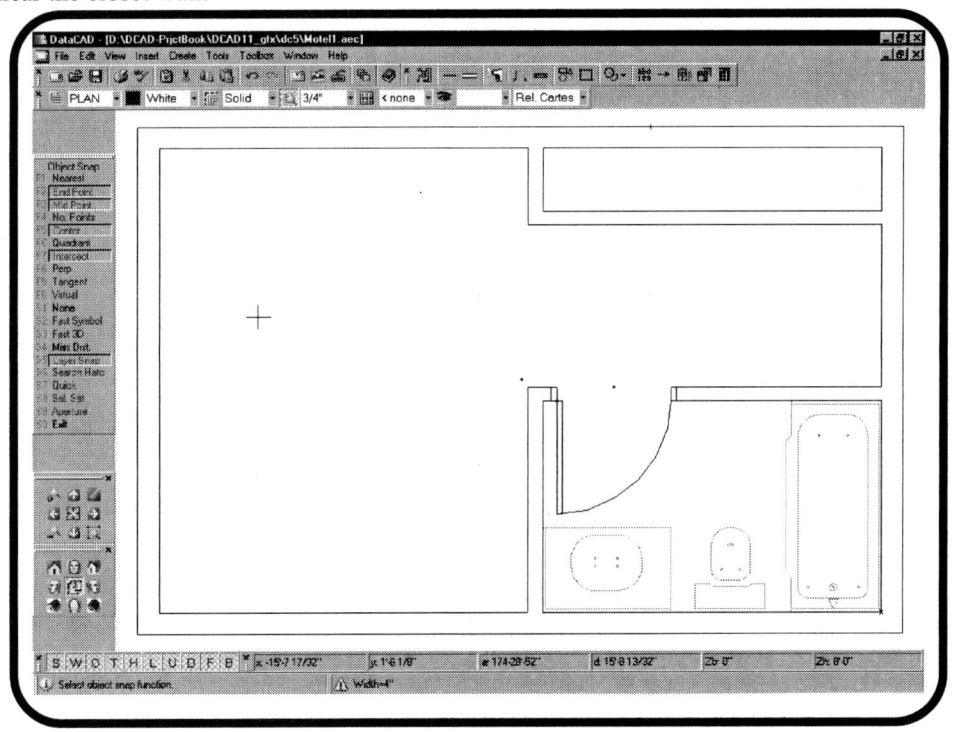

Figure 5-3, The object snap menu.

15.__ Make sure the **Mid Point** option in the **Object Snap** menu is **on**.

The DataCAD 11 Project Book

16.___ Press the **a** key [*not* the A key] to jump to the **Architect** menu.

17.___ Select the **Door Swing** option from the **Architect** menu.

18.___ Turn **off** the **Sides** option in the **Door Swing** menu (doors defined by center and strike side).

> •• **NOTE:** The **n** [or the **N**] key can be used in place of the *middle* mouse button.

19.___ At the prompt *Select center of door* move the drawing cursor near the *center* of the *inside* hall wall and click the *object snap button* once.

20.___ At the prompt *Select strike side of door* press the **space bar** and enter an **X** value of **0.0** and a **Y** value of **1.6**. (Y is 1.6 because the strike side is above the center of the door, and 1'6" is half of the door width.)

21.___ At the prompt *Select direction of door swing* move the drawing cursor *inside* the hall and click the *left* mouse button *once*.

22.___ At the prompt *Select any point on the outside of the wall* move the drawing cursor *outside* of the room and click the *left* mouse button *once*.

Chapter 5, Page 106

Centering and Cleaning a Wall

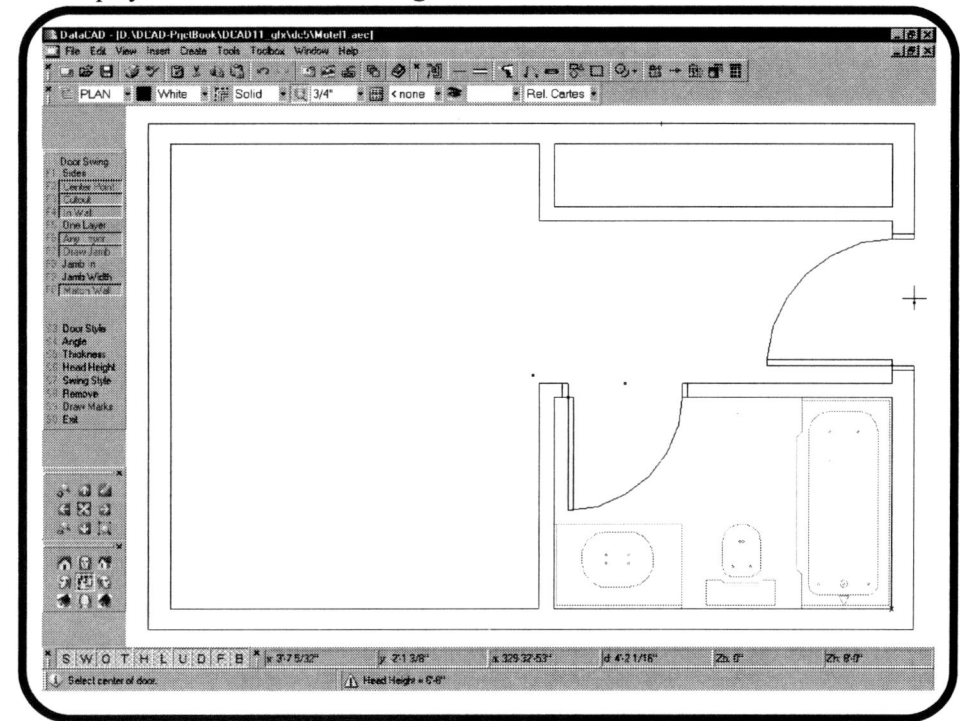

Your display should look similar to Figure 5-4.

Figure 5-4, Centering the entrance door.

Using a combination of object snap and coordinate entry mode makes the job of adding a door to the center of a wall quite easy. Many beginners try to use the mouse for everything, instead of taking advantage of object snap and coordinate entry mode options. As a result, they spend a lot of time fooling around with the mouse trying to get things in the proper place.

••• **WARNING**••• Have you saved your drawing lately?

Centering and Cleaning a Wall

The object snap function can also be used for drawing lines or creating walls. Next, the closet will be divided into two parts to illustrate how object snap can help anchor lines to a midpoint.

23.__ Make sure the drawing layer is set to **PLAN**.

The DataCAD 11 Project Book

•• NOTE: To insure that the interior closet wall is centered, the **Center Wall** option will be used.

24.___ Check the **Architect** menu and set the **2 Line Walls, Center Wall**, and **Clean** options **on**.

25.___ Press the *right* mouse button until you see the prompt *Select first end point of new line/wall*.

26.___ Click the *object snap button* near the *center* of an **inside** closet wall. Then click near the opposite inside wall. Click the *right* mouse button to stop adding walls.

Your display should look similar to Figure 5-5.

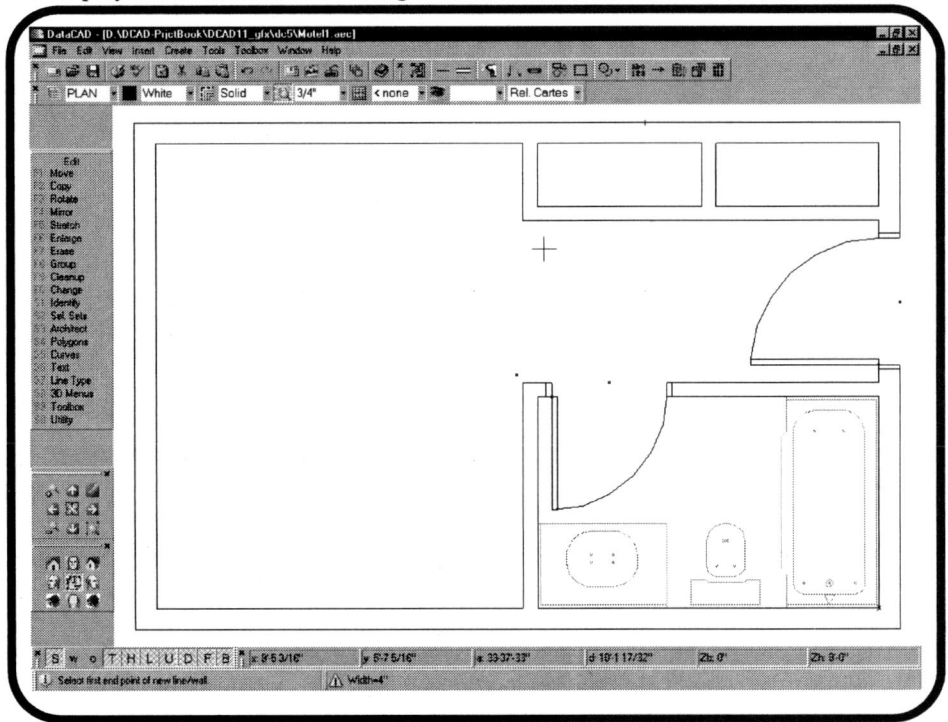

Figure 5-5, Adding a closet wall.

Adding a Cut Wall Opening

Now that the closet is divided into two sections, two different door options will be used. In one closet the **Cut Wall** option will be used to make an opening. In the other closet a bi-fold door will be installed.

Adding a Bi-fold Door

27.___ **Zoom to area** on the closet area as shown in Figure 5-6.

28.___ Select the **Cut Wall** option from the **Architect** menu.

29.___ Press the unshifted ~ key and use the *object snap button* to set 0,0 at the lower left inside corner of the *right* side closet.

30.___ At the prompt *Select first point on wall* press the space bar and enter **0.4** for **X** and **0.0** for **Y**.

31.___ Press the **space bar** and enter **3.2** for **X** and **0.0** for **Y**.

Your display should look similar to Figure 5-6 when you have completed this step.

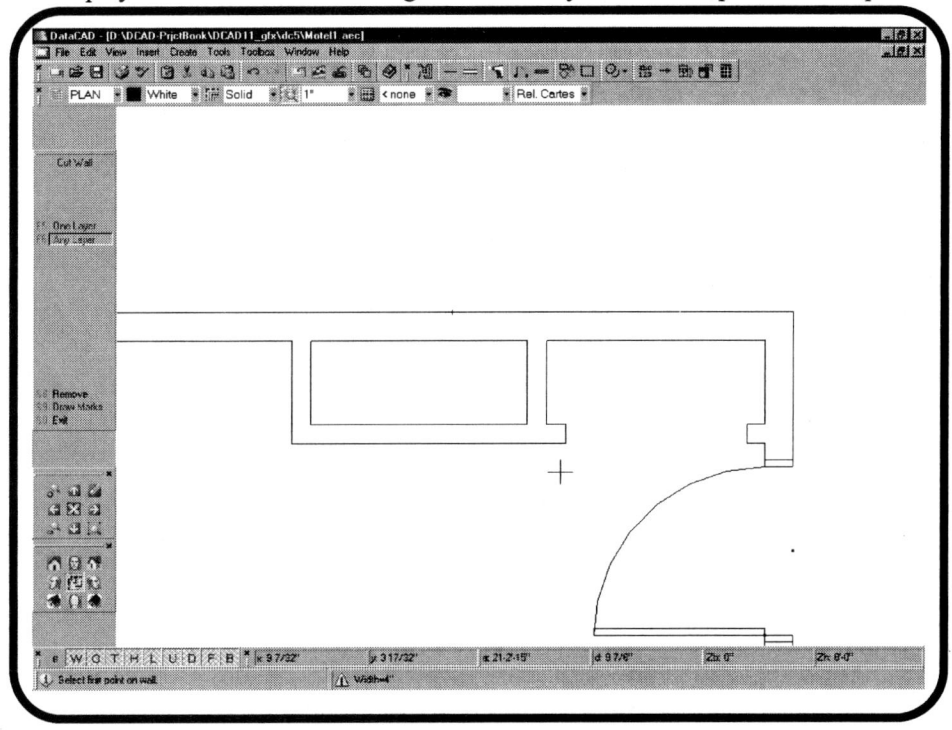

Figure 5-6, Adding a cut wall opening.

Adding a Bi-fold Door

32.___ Select the **Door Swing** option from the **Architect** menu and make sure that the **Sides** option is **off**. (Doors will be defined by center and strike side.)

33.___ Select **Door Style**, and select the **Bi-Fold** option.

The DataCAD 11 Project Book

34.___ Select **Angle**, and enter **45.0**.

35.___ Click the *object snap button* near the *center* of the lower *inside left* closet wall.

36.___ At the prompt *Select other side of door* press the **space bar** and enter an **X** value of **1.6** and a **Y** value of **0.0**.

37.___ Click the *left* mouse button **outside** of the closet for **both** the direction of door swing and outside of wall.

The display should now look similar to Figure 5-7.

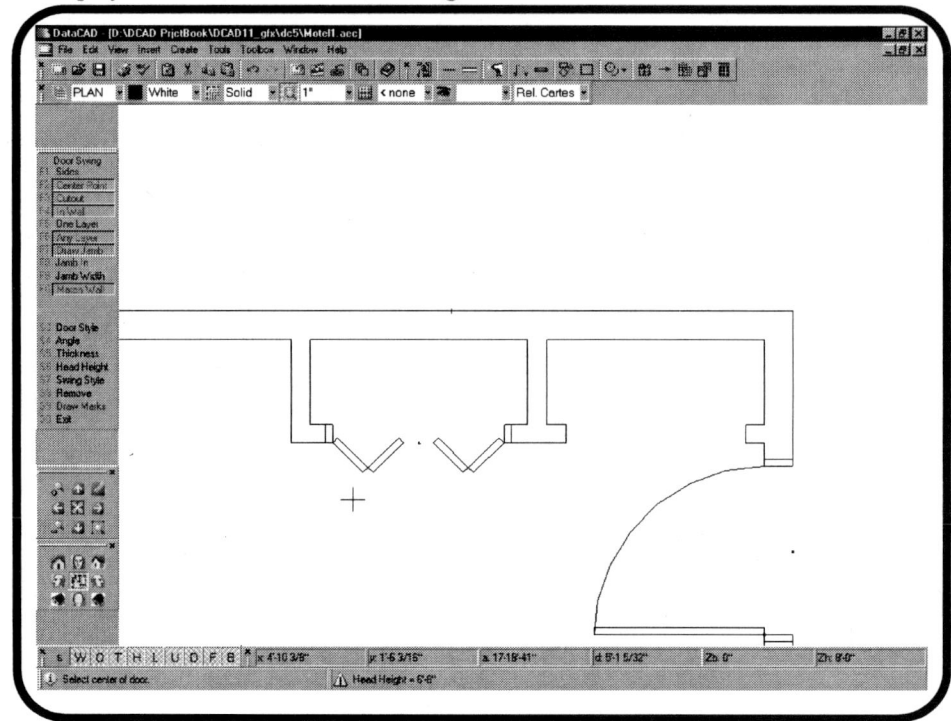

Figure 5-7, Adding a bi-fold door.

As you can see, using DataCAD to draw a bi-fold door centered in a closet wall is cake compared to doing it with pencil and paper. One more door will be added to the room. This door will be a sliding door. A four foot wide sliding door will be centered on the outside wall of the motel room.

Adding a Sliding Door

38.___ Use the **Navigation Pad** buttons to make the motel room fill the display.

Adding a Sliding Door

39.__ Select the **Sliding** door option from the **Door Style** menu and set the **Angle** to **0**.

40.__ Click the *object snap button* near the *center* of the *left* wall of the motel room.

41.__ At the prompt *Select the other side of the door.* press the **space bar** and enter an **X** value of **0.0**, and a **Y** value of **2.0**.

42.__ At the prompt *Select any point on the outside of the door.* click the *left* mouse button somewhere *outside* (and to the left of) the motel room.

The display should now look similar to Figure 5-8.

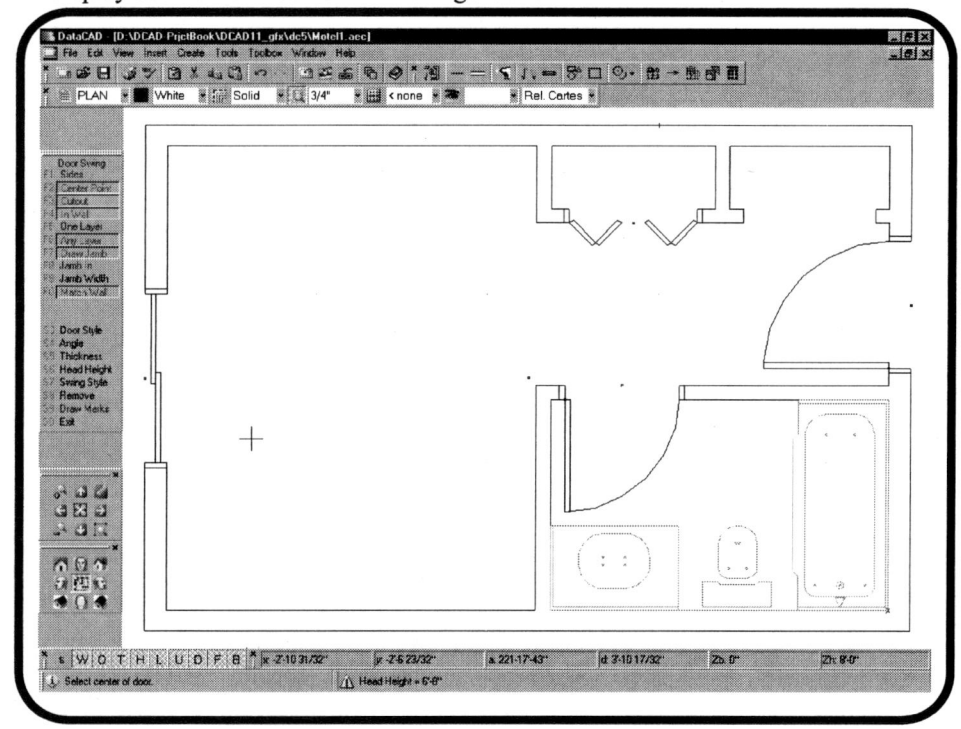

Figure 5-8, Motel room with doors.

INFO STOP

Remember that you can use the **Door Swing, Remove** option to remove a door and heal a wall. Also, the **Angle** option controls the angle for single and bi-fold doors, and also controls the amount a sliding door is open.

Chapter 5, Page 111

The DataCAD 11 Project Book

43.__ Press [CTRL] +[S] (or select **File, Save**) and save your **Motel1** drawing.

In the next chapter some furniture will be added to the motel room. In a later chapter, this same motel room will be used to create an entire motel complex.

Because the furniture will not be included in all of the rooms (if you've seen one room you've seen them all), it would be nice to have a copy of the unfurnished room to use.

Creating a New Drawing File

The **File, Save As** function provides a way to create a copy of the drawing that has a different name. A copy of the unfurnished motel room will be made for use in a later project.

> ••• **WARNING**••• If you intend to use your current drawing in the future, make sure you first **Save** the drawing before going onto the next step. If you don't then all the changes you made since your last Save will be lost.

44.__ Select the **File, Save As** option from the **Menu Bar**.

45.__ Use the *Save file as...* dialog box to save the file with the new name **Motel8**.

46.__ Select **File, Exit** to end the session.

In the next project, some furniture will be added to Ferd's motel room.

Review and Important Points

- To automatically locate the center of a line, toggle the **Mid Point** option on in the object snap menu; then click on the line with the *object snap button*.
- The **n** [or the **N**] key can be used in place of the *middle* mouse button.
- The **X** [not the x] key jumps to the object snap menu.
- When centering doors or windows, the **Sides** option must be toggled **off**.
- Use the **File, Save As** option to make a copy of a drawing with a new file name.

Creating a New Drawing File

Practice Drawings

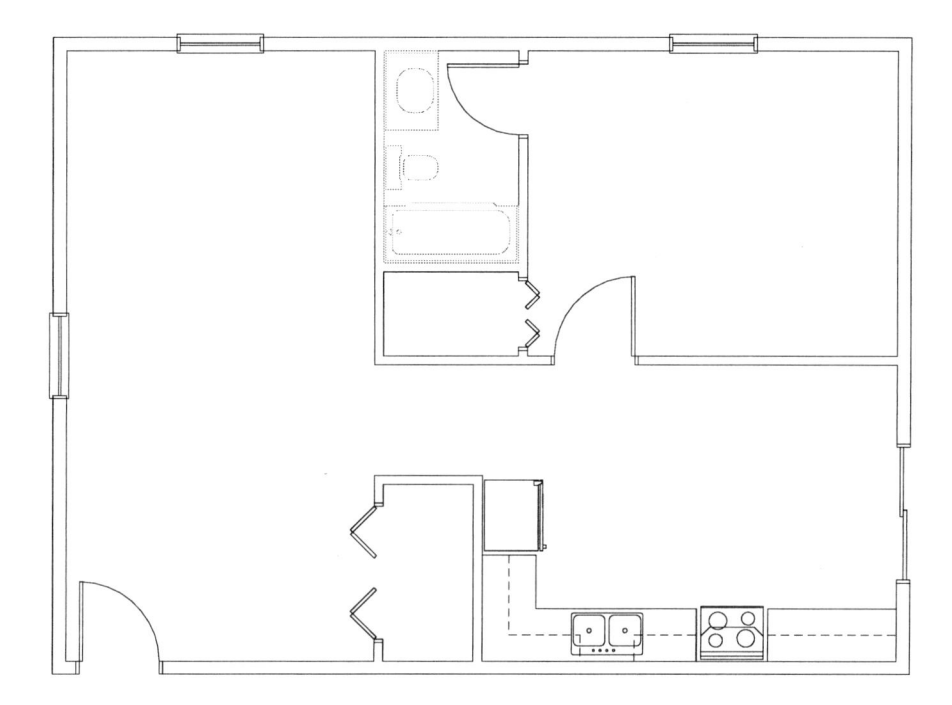

Project Hints
Assignment 5-A, Apartment design

- Start with the apartment floor plan created in Assignment 4-A.
- Add doors, windows, bathroom plumbing group, and kitchen to the apartment.
- Look for the kitchen symbols with the path **C:\DCADWIN\Symbols\Kitchen\Equipment**.
- The cabinet size was adjusted by turning the **Explode** option *on* before placing the symbol, then using the **Edit, Stretch** function to change the length of the counter top.

The DataCAD 11 Project Book

Notes

CHAPTER 6

Ferd's Motel Room, Phase 3

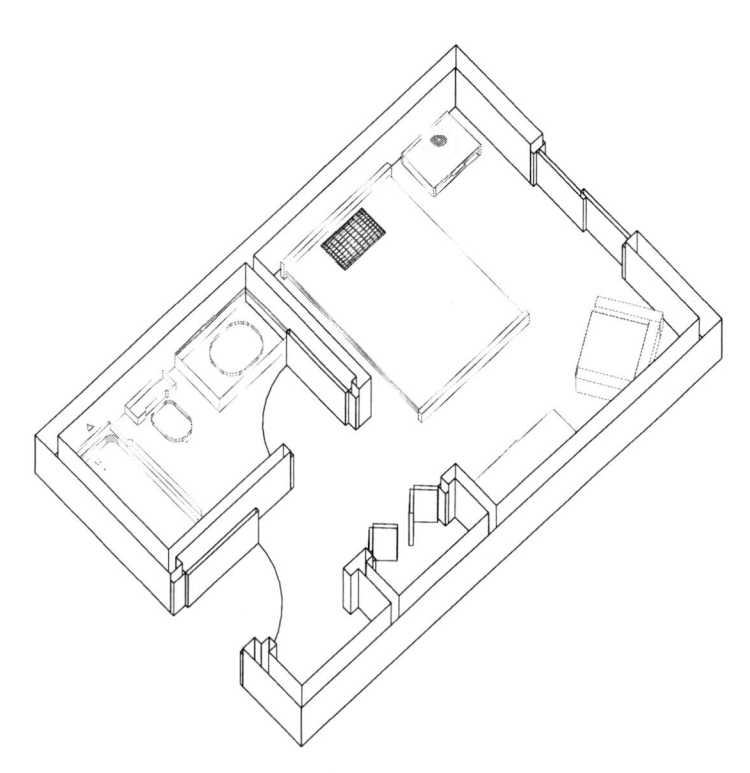

Project Goals

When you have completed this project, you will be able to:

1. Select different disk drive paths for templates.
2. Add symbols to a drawing using a fixed rotation angle.
3. Add symbols to a drawing using dynamic rotation.
4. Change the **Z-Base** (elevation) for positioning symbols.
5. Change the display scale.
6. Complete a pictorial drawing of a plan with hidden lines removed.

The DataCAD 11 Project Book

Prerequisite: Chapter 5

Introduction

In Chapter 5 you completed the motel room walls and doors. In this project you will learn different ways to add symbols to drawings. In this case, the symbols will be different pieces of furniture.

1.__ Start **DataCAD** and **Open** the drawing file named **Motel1**.

Adding Furniture

The DataCAD software stores the symbol files in a subdirectory (which is sort of like storing papers in an envelope, which is in a folder, which is in a drawer, which is in a file cabinet, which also has other drawers and folders and envelopes and papers). To view and select symbols you use the **Symbol Browser**.

2.__ Hold down the [CTRL] key and press the **T** [or the **t**] key to enter the **Symbol Browser** menu.

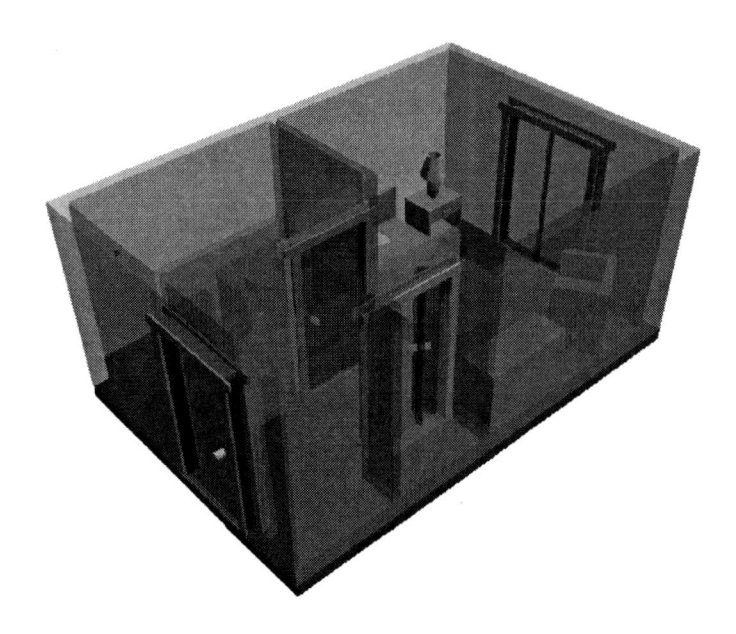

Adding Furniture

3. __ Select the symbol path **C:\DATACAD\Symbols\Bedroom\3D Beds** (see Figure 6-1).

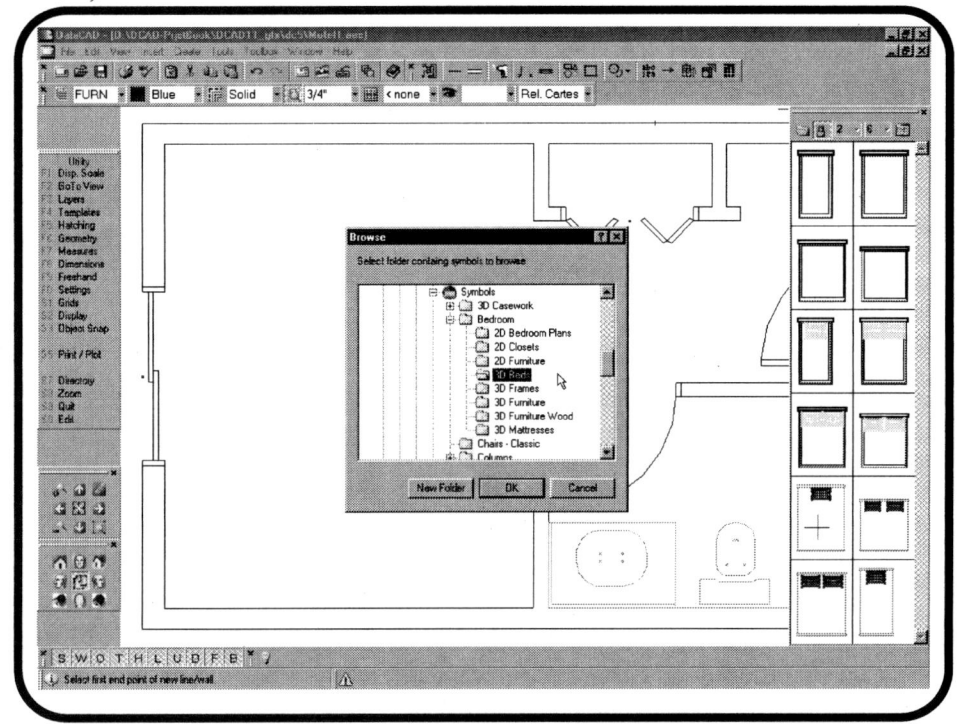

Figure 6-1, Setting the template path.

If everything has worked properly, the display should look similar to Figure 6-2. If you do not see the list of file names shown in the dialog box in Figure 6-1, ask your teacher for help.

The DataCAD 11 Project Book

The **3D Beds** template is shown in Figure 6-2. Remember, you will want to created a special drawing layer for furniture. Be sure that layer is active before adding any furniture to the drawing. A bed will be added using the fixed rotation angle option.

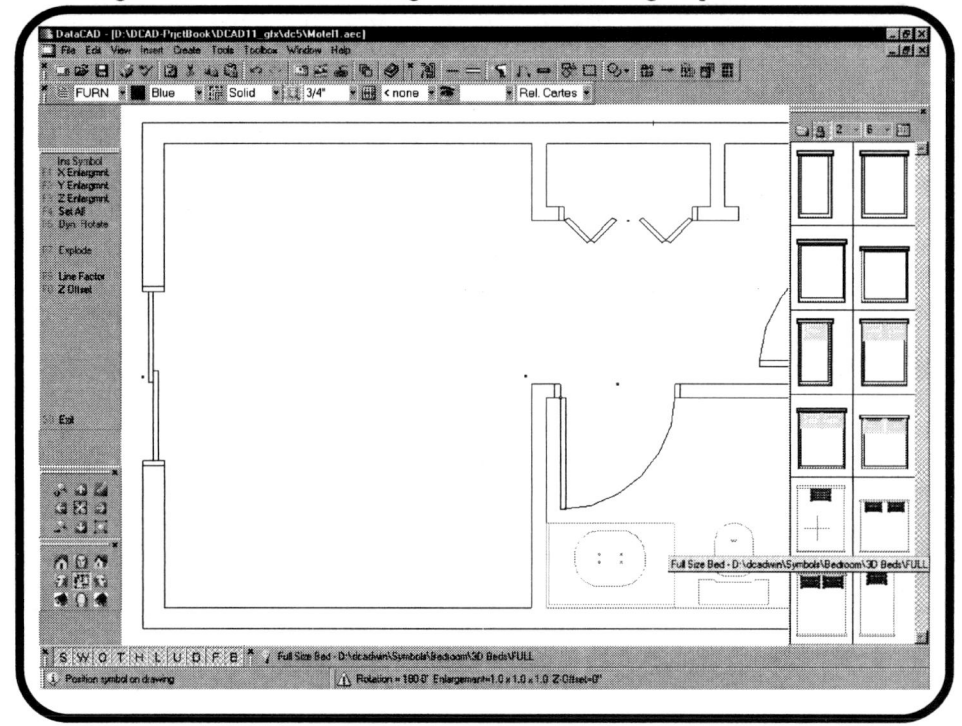

Figure 6-2, The BEDS symbols.

4. __ Press the **Tab** key until the **FURN** layer is active.

5. __ Select the **FULL** bed symbol by *clicking* on it with the *left* mouse button.

6. __ In the **Ins Symbol** menu that pops up on the left side of the screen, if the **Dyn. Rotate** option is NOT already selected, select it twice to enter the angle specification mode. If the **Dyn. Rotate** option IS already selected, select it once to enter the angle specification mode.

7. __ Enter a rotation angle of **180** degrees.

Adding Furniture

To locate the bed exactly 2 inches away from the outside wall and 1 foot from the bathroom wall, the coordinate 0,0 will be reset, and then coordinate entry mode will be used. Remember that the computer remembers the snap grid size separately for each drawing layer so snap will have to be set to 2 inches for the **FURN** layer. Also, the layer snap option must be **on** for the system to snap to an entity on the **PLAN** layer when the **FURN** layer is active.

8.___ Press the **s** key [*not* the S key] and set the snap size to 2 inches (**0.2**) for both **X** and **Y**.

9.___ Press the **X** [*not* the x] key and toggle **Layer Snap on** in the **Object Snap** menu.

10.___ Press the unshifted ~ key and then click with the *left* mouse button *once* in the lower right corner of the room shown in Figure 6-3.

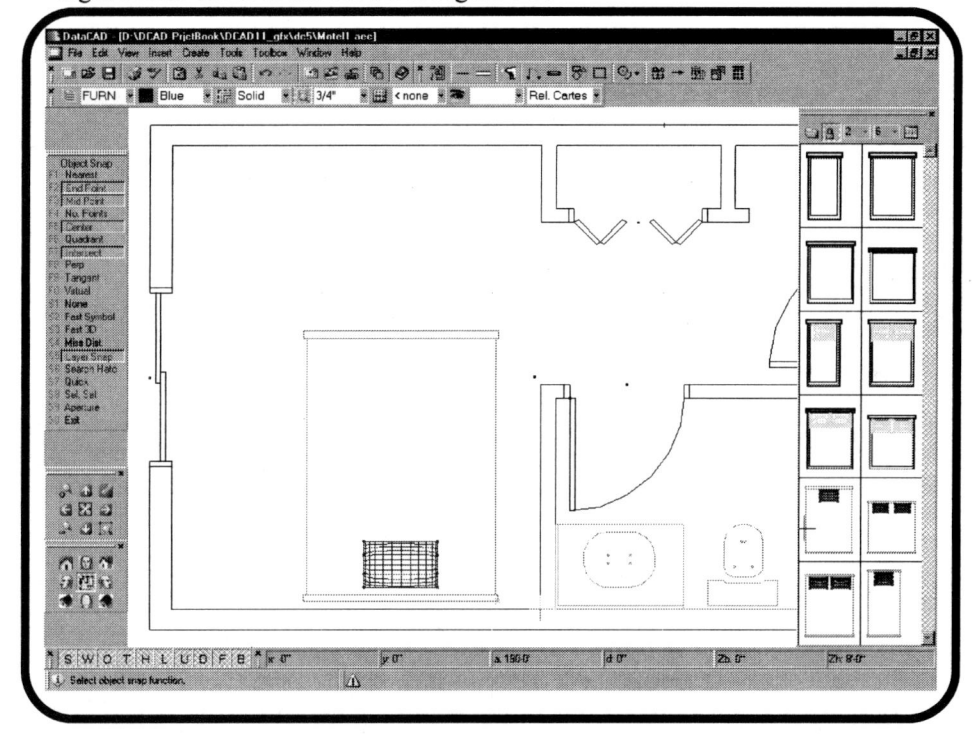

Figure 6-3, Placing a bed.

11.___ Press the **space bar** and enter values (watch the minus sign) of **-1.0** for **X** and **0.2** for **Y**.

12.___ Press the *right* mouse button *once* to drop the bed symbol.

Chapter 6, Page 119

The DataCAD 11 Project Book

The display should now look similar to Figure 6-3. Next, a night table will be placed beside the bed. The night table symbol is found in the **Symbols\Bedroom\3D Furniture** directory.

13.__ Click with the *left* mouse button on the **NITETABL** symbol.

14.__ Place the night table between the bed and the outside wall as shown in Figure 6-4.

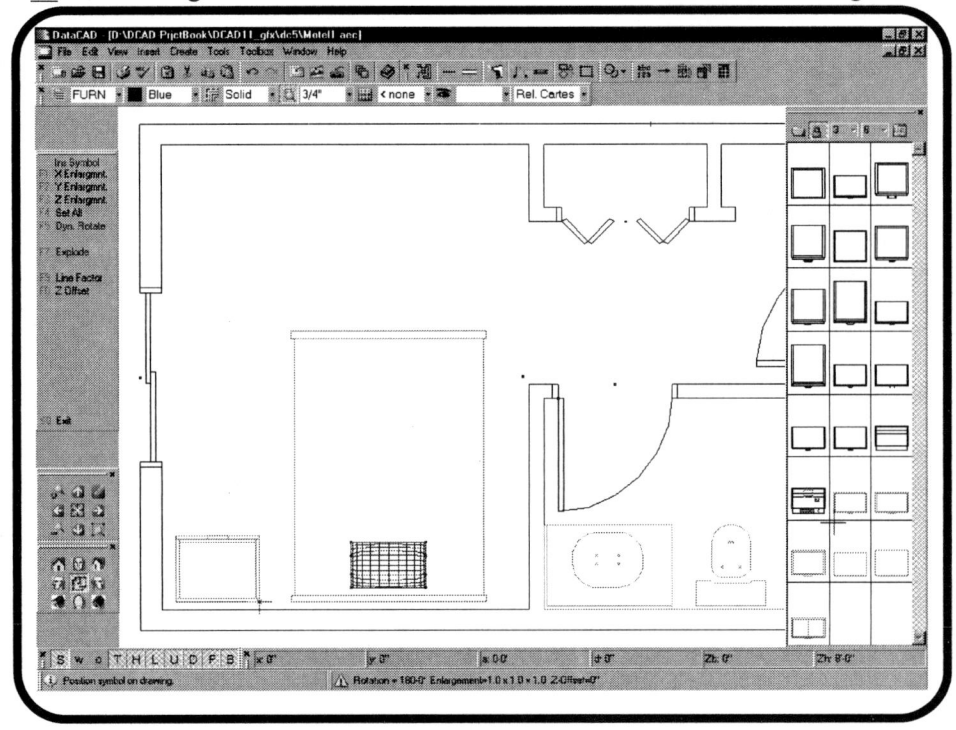

Figure 6-4, Adding a night table.

INFO STOP

If you put a symbol in the wrong spot, you can erase it with a quick dual key press. Hold down the [CTRL] key and press the [Z] key (the standard Windows undo command). You can also use the **Edit, Undo/Redo** function from the **MENU BAR**.

Chapter 6, Page 120

Changing the Symbol Scale

Next a *low dresser* will be added to the room. This will fit in the room the same way it was created, so the rotation angle must be changed back to 0 degrees. Adding the dresser gives you the chance to see another feature of the software. You can change the size of any symbol in either the X, Y, or Z direction before it is placed on the drawing. The dresser will be made a little longer in the X direction than the original.

15.___ Select the **LOWDRESR** symbol by *clicking* on it with the *left* mouse button.

16.___ In the **Ins Symbol** menu that pops up on the left side of the screen, if the **Dyn. Rotate** option is NOT already selected, select it twice to enter the angle specification mode. If the **Dyn. Rotate** option IS already selected, select it once to enter the angle specification mode.

17.___ Enter a rotation angle of **0** degrees.

18.___ Select **X Enlargmnt** and enter a value of **1.5**.

19.___ Place the **LOWDRESR** symbol next to the closet as shown in Figure 6-5.

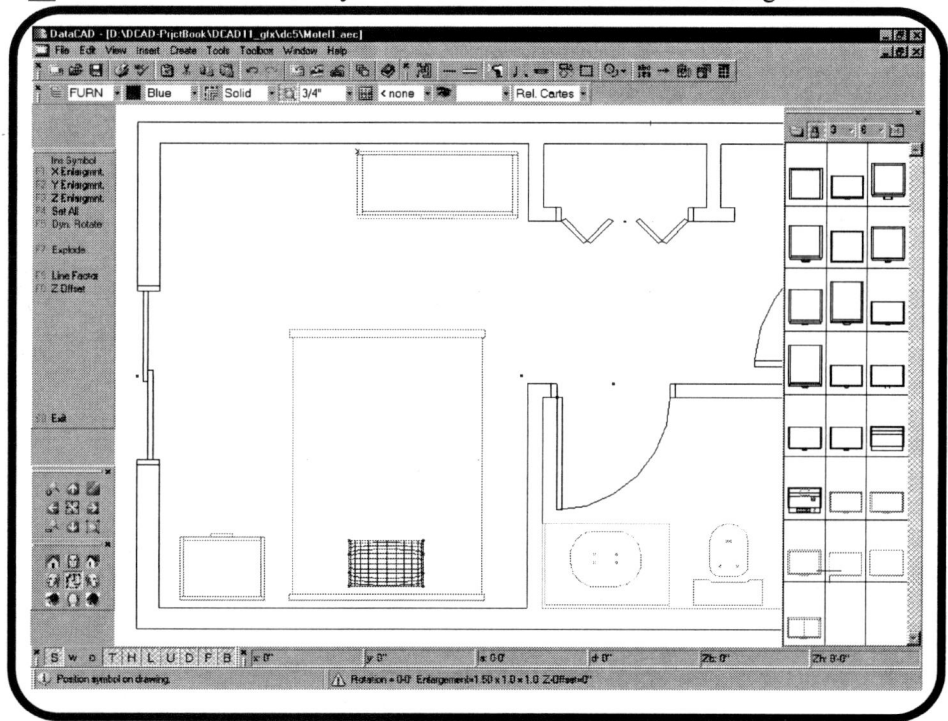

Figure 6-5, Adding a dresser.

Chapter 6, Page 121

The DataCAD 11 Project Book

Using Dynamic Rotation

Next a chair will be placed in the corner of the room. To get the chair to be at an angle, the dynamic rotation option will be used. When dynamic rotation is used, the first click of the mouse anchors the "handle" of the symbol. Moving the mouse then causes the symbol to rotate around the anchor point. The second click of the mouse places the symbol on the drawing.

20. __ In the **Symbols\Chairs - Classic** folder select the **COMFY** chair symbol by *clicking* on it with the *left* mouse button.

21. __ Select **X Enlargmnt** and enter a value of **1.0**.

22. __ In the **Ins Symbol** menu that pops up on the left side of the screen, if the **Dyn. Rotate** option is NOT already selected, select it. If the **Dyn. Rotate** option IS already selected, leave it as it is.

> **INFO STOP**
> If you pick a symbol with the *left* mouse button, the anchor point will be at the insertion point of the symbol (as selected by the author of that symbol). If you pick a symbol with the *object snap button*, the anchor point will be at the object snap point. In the next step, it may be easier to control placement by setting the anchor point at the upper left corner of the chair. If the **Ortho** mode is toggled **on**, you will be locked into 45 and 90 degree rotation positions.

23. __ If it is not already **off**, press the [O] **O** key to toggle **Ortho** mode **off**.

24. __ Click the *object snap button* on the *upper left* corner of the **Comfy** chair symbol.

Changing the Z-Offset

25.__ Locate the corner of the **Comfy** chair as shown in Figure 6-6 and click the *left* mouse button *once*.

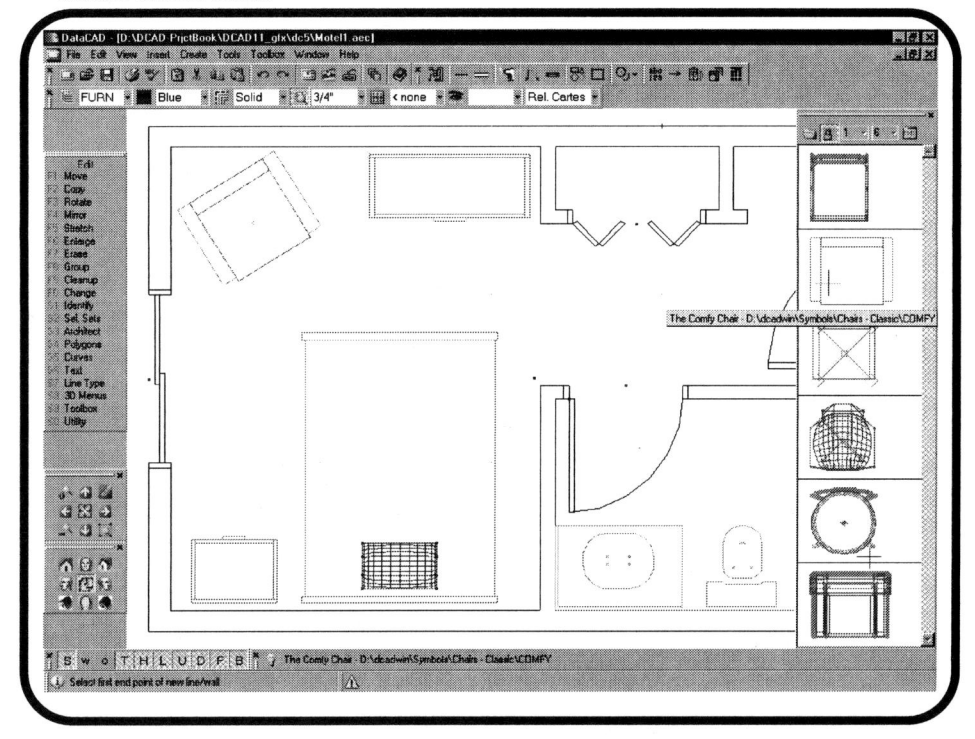

Figure 6-6, Adding a chair.

26.__ Move the mouse around until the outline of the **Comfy** chair is where you want it. Then click the *left* mouse button *once*.

> •• **NOTE:** Remember, you can always do a quick **undo** and place the symbol again if you don't like where it is after the first try.

Changing the Z-Offset

The last piece of furniture to be added is a table lamp. All of the symbols used have been created in 3D. (Remember the 3D views of the chairs in Ferd's Playhouse?) Everything added to the room so far has been placed on the floor, so there was no worry about the third dimension. The table lamp, however, will have to be placed on the top of the table, not on the floor.

Chapter 6, Page 123

The DataCAD 11 Project Book

DataCAD uses the **Z-Offset** to control the elevation of symbols. The problem is that you don't know how high the top of the table is. There is a function you can use to find elevation information. The **Identify** function provides information about entities.

27.___ Select the **Identify** button on the **Navigation Pad** (the one that looks like a pointing finger). Then *click* on the the top of the night table.

The selected night table is shown with dashed lines, and the **Identify** menu is displayed (see Figure 6-7). Note that the information line shows that **Z-Max** for the night table is 2'1". This means that if you want to place an object on top of the night table, the **Z-Offset** must first be set to 2'1".

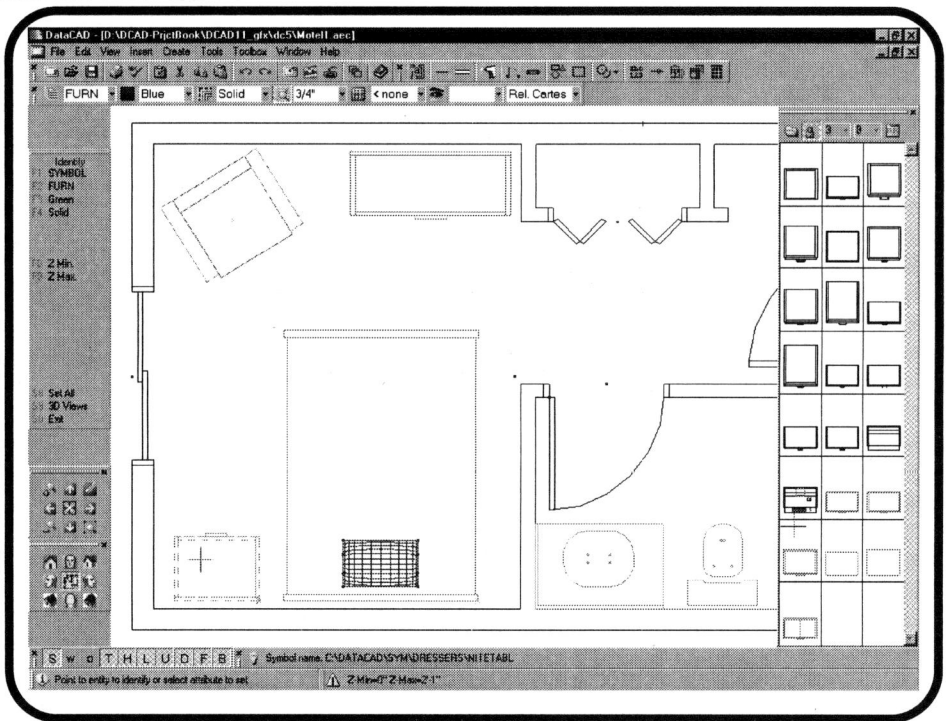

Figure 6-7, Identifying night table symbol height.

28.___ In the **Symbols\Lights\3D Lighting Fixtures** folder select the **TABLE** lamp symbol by *clicking* on it with the *left* mouse button.

Chapter 6, Page 124

Changing the Z-Offset

After the lamp symbol is selected, the proper elevation (known as the **Z-Offset**) must be set. There are several options for setting this option, but we'll focus on the one called **Z-User 1** (one of two available user-defined z-heights). Also, the table lamp symbol is a little too big for the table so the enlargement factor will be changed before it is placed in the room.

29.___ Select the **Z Offset** option from the **Ins. Symbol** menu. Then select the **Set Z User 1** option.

30.___ Enter a value of **2.1**.

31.___ At the top of the **Ins. Symbol** menu select the **Z-User 1** option.

INFO STOP

The **Set Z User 1** option that you first selected will remain at the selected value of 2.1, in the current drawing only, unless you change it. The **Z-User 1** option is one of four options available for where to locate the symbol in the z-direction. Selecting it will cause the symbol to be placed at the value set by **Set Z User 1**. **Z-Base** will put a symbol at the current drawing Z-base elevation, while the **Z-Height** option will put a symbol at the current drawing Z-height elevation. The fourth option, **Z-User 2**, allows you to select a second z-offset height, without having to change the Z-User 1 value.

32.___ *Right-click* or select **Exit** to back out of the **Ins. Symbol** menu.

33.___ Select the **Set All** option from the **Ins. Symbol** menu (this will set all the X, Y and Z enlargment values at once).

34.___ Enter a value of **0.65**.

35.___ Turn the **Dyn. Rotate** option **off**.

Chapter 6, Page 125

The DataCAD 11 Project Book

36.___ Click the *left* mouse button to place the lamp symbol as shown in Figure 6-8.

Figure 6-8, Adding a table lamp.

The last thing to do in this project is to create a pictorial view of part of the motel room.

37.___ Turn **off** the **Symbol Browser** menu by *clicking* on the **X** in the upper right corner of the **Symbol Browser**, or by pressing `CTRL` + **T**.

Changing the Z-Offset

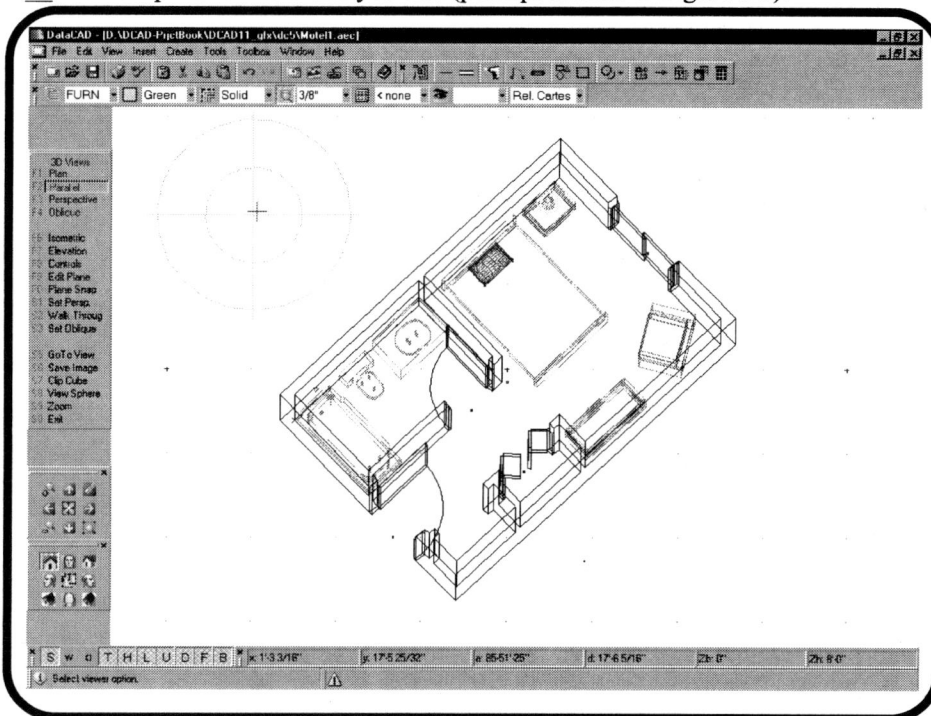

38. __ Create a pictorial view that you like (perhaps similar to Figure 6-9).

Figure 6-9, The furnished motel room.

> **INFO STOP**
> If you have the time, you might want to create a new drawing layer that contains a view with the hidden lines removed. (Review the steps in Chapter 3 that explain how to do this.) One caution before you do any hidden line removal: adding the furniture added a lot of lines to the drawing. The length of time the computer takes to remove hidden lines depends on how fast the computer processor chip is and how many lines it has to deal with. If you are using an older PC, removing the hidden lines might take too long for a normal class period. Check with your instructor before you try to do hidden line removal on the furnished motel room.

In the next project you will create Ferd's Motel 8 from the motel room.

Chapter 6, Page 127

The DataCAD 11 Project Book

••• **WARNING**••• Don't forget to select **File, Exit** before you end the session. If you turn off the computer without doing a proper **Exit**, you will lose most of the changes you made to the drawing this session.

Review and Important Points

- Use coordinate entry mode to place a symbol at a precise location.

- If you want to erase the last symbol placed, hold down the [CTRL] key and press the [Z] key (the standard Windows undo command).

- Use the four **Enlargmnt** options on the **Ins. Symbol** menu to change the X, Y, or Z (or all 3 with **Set All**) size of a symbol.

- To select the default anchor point on a symbol, select it by clicking the *left* mouse button.

- To change the anchor point for a symbol, click on the symbol in the **Symbol Browser** with the *object snap button* near the end of the entity where you want the anchor point.

- To rotate a symbol around its anchor point, toggle **Dyn. Rotate** on.

- To determine the maximum **Z-height** of a symbol, select the **Identify** button on the **Navigation Pad** and then *click* on the symbol. Read the **Z-Max** value on the information line.

- To change the height where the base of a symbol will be placed, use the **Z Offset** option in the **Ins. Symbol** menu.

- Remember that it is possible to add a large number of lines to a drawing very quickly by placing symbols. A large number of symbols on a drawing can slow display refresh time and create large drawing files, though it shouldn't make much of a difference unless you have an old, slow computer.

Changing the Z-Offset

Practice Drawings

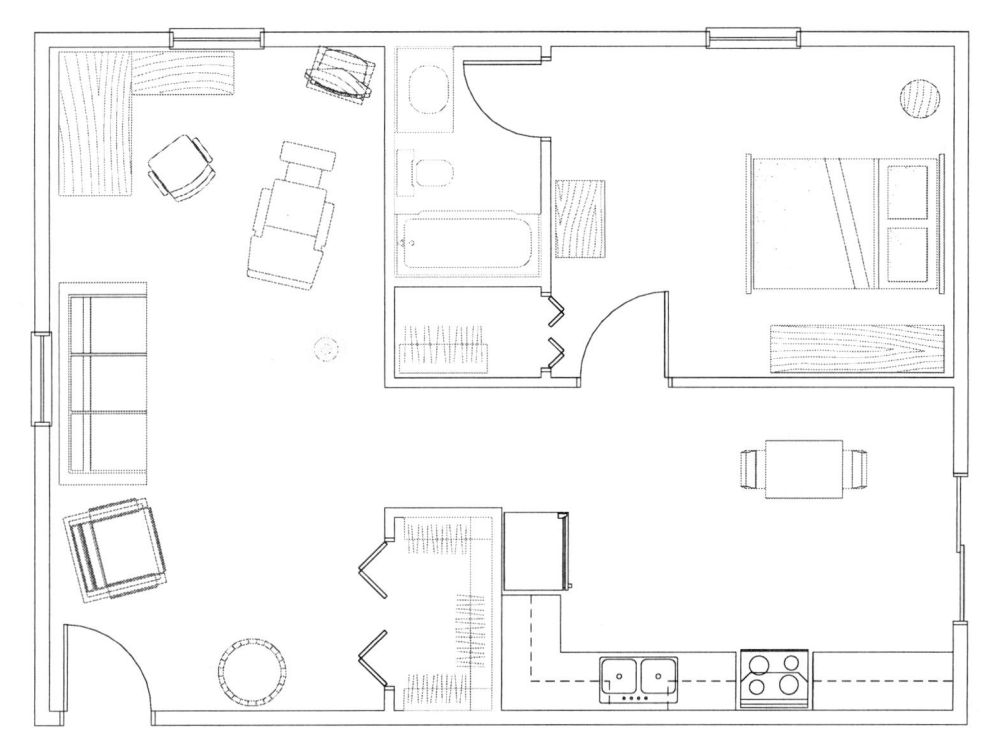

Project Hints

Assignment 6-A, Apartment furnishings

- Furnish the apartment created in Assignments 4-A and 5-A.
- Not all of the symbol files supplied with DataCAD are in 3D.
- Switch to the Isometric view after placing each symbol to see how it looks.

The DataCAD 11 Project Book

Practice Drawings

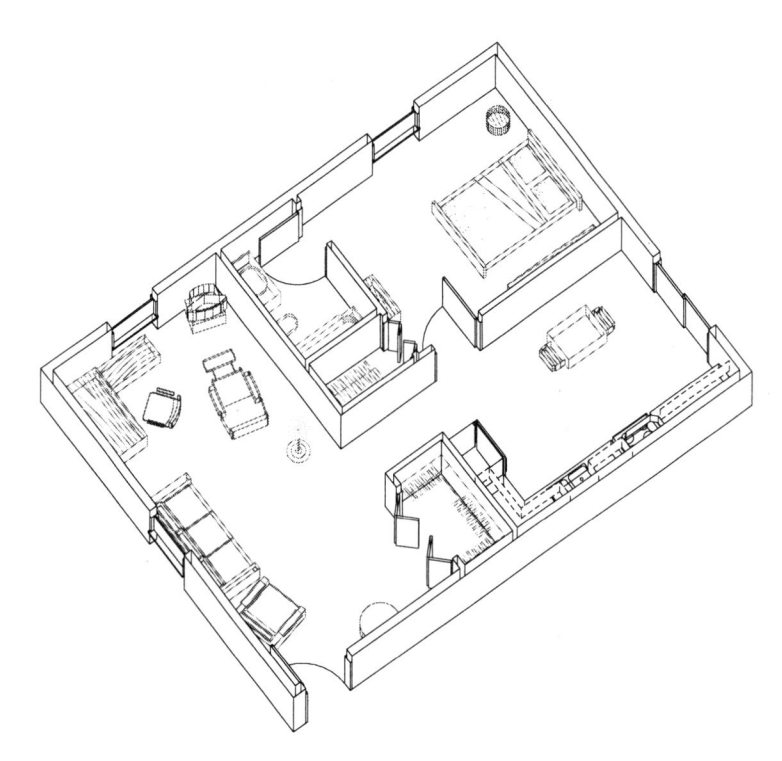

Project Hints
Assignment 6-B, Pictorial rendering

- Create a hidden line pictorial rendering of your apartment layout.

Notes

The DataCAD 11 Project Book

CHAPTER 7

Ferd's Motel 8, Mirror & Group

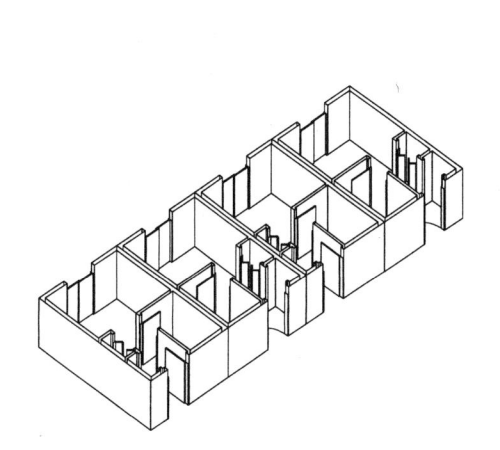

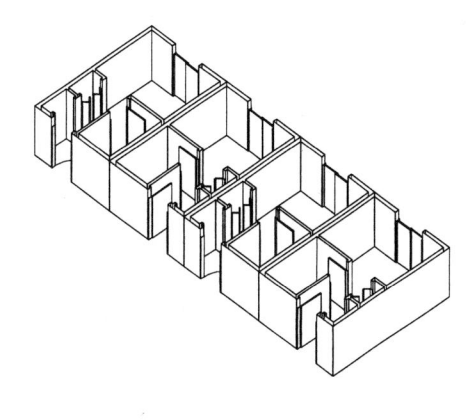

Project Goals

When you have completed this project, you will be able to:
1. Use the DataCAD **Mirror** function to copy objects.
2. Use the **Rotate** function to rotate objects.
3. Specify the rotation angle for changing the orientation of objects.
4. Create a design by making and manipulating copies of an existing plan.

The DataCAD 11 Project Book

Prerequisite: Chapter 5

Introduction

At the end of Chapter 5 you used the **File, SaveAs** option from the **Menu Bar** to save a copy of the unfurnished motel room as **Motel8**. In this chapter, you will turn that single motel room into an 8 room motel without drawing any new lines. This will illustrate one of the powerful concepts of using CAD. That is, that you should never have to draw anything twice. After you have created something once, you can get the computer to do the busy work of drawing it again.

1.__ Start the **DataCAD** software and **Open** the drawing named **Motel8**.

To keep the size of the drawing file small, and to keep the redraw speed from slowing down, the **PLUMB** layer will be turned **off** before more rooms are added to the motel.

2.__ In the **STATUS AREA**, click on the **Layer Manager** icon. Use the **Layer Manager** to turn **off** all of the **layers** except the **PLAN** layer.

The Mirror Function

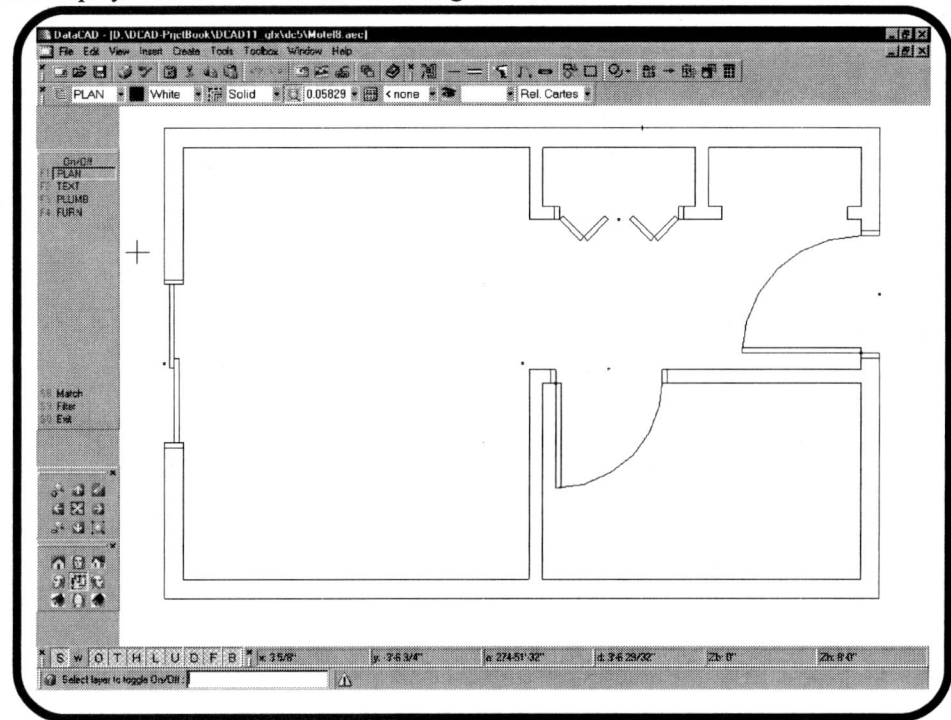

The display should now look similar to Figure 7-1.

Figure 7-1, The unfurnished motel room.

The Mirror Function

Next the **Mirror** function will be used to make a copy of the motel room. To use the **Mirror** function, you first establish a line of reflection and then tell the computer what to include in the mirror. DataCAD has an option for just reflecting things, and another option to both reflect and copy. The **And Copy** option will be used to add a new room to the motel.

3.__ Select the **Mirror** option from the **Edit** menu [or press Alt+M].

The DataCAD 11 Project Book

4.__ Follow the prompts and make the reflection line be the *lower horizontal outside* wall of the motel room (see Figure 7-2). Make sure **Ortho** is turned on so you get a nice straight line

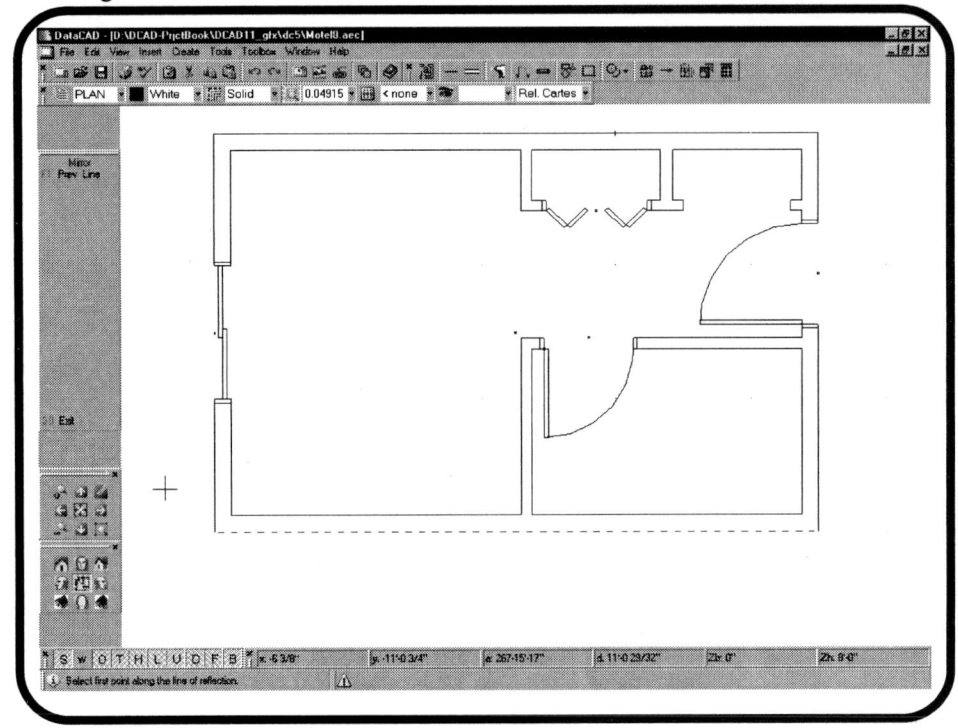

Figure 7-2, Specifying the reflection line.

5.__ Make sure that the **And Copy** option in the **Mirror** menu is turned **on**.

Chapter 7, Page 136

The Mirror Function

6. __ Select the **Area** option from the **Mirror** menu and capture the motel room in an area box as shown in Figure 7-3.

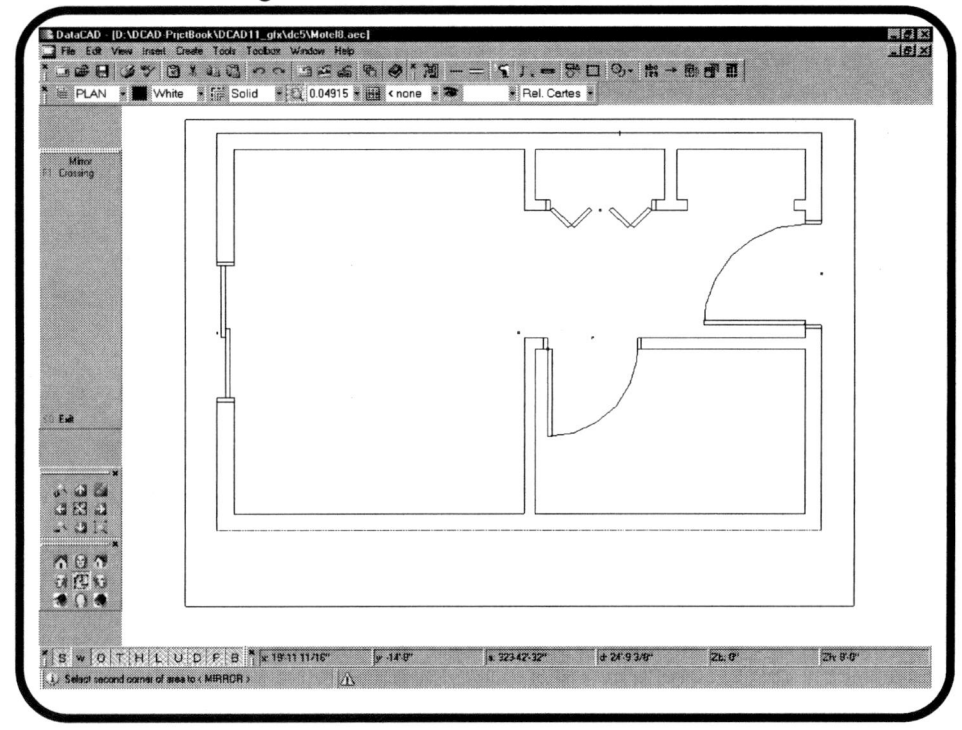

Figure 7-3, Capturing an area to mirror.

The DataCAD 11 Project Book

7.__ Arrange the display similar to Figure 7-4 (use the **Zoom** menu or the **Navigation Pad** buttons).

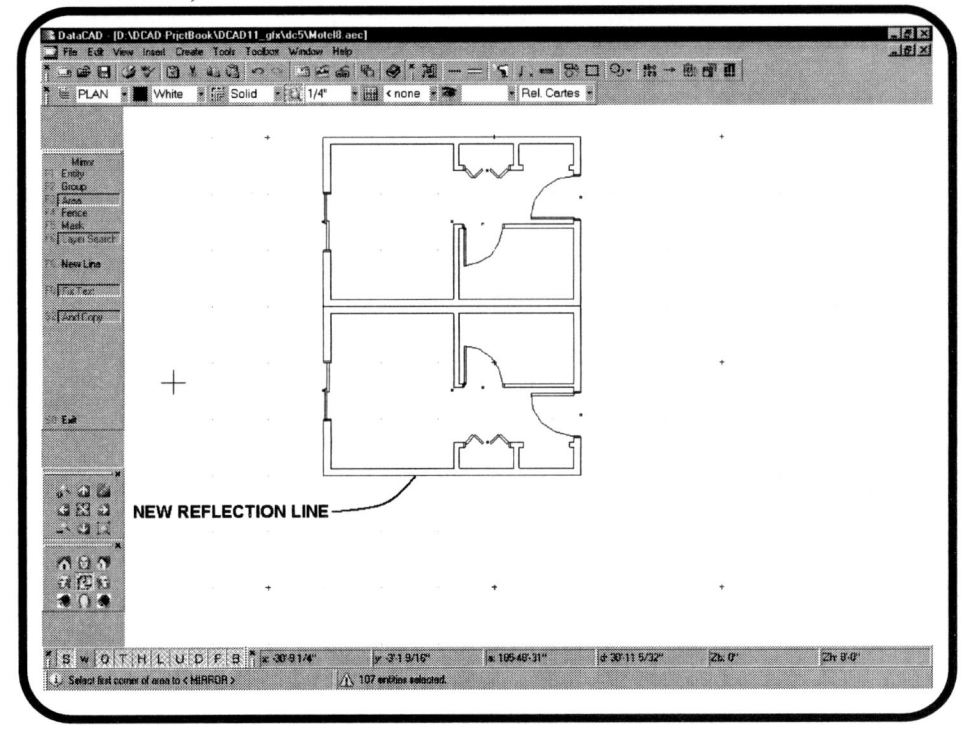

Figure 7-4, The two room motel.

The display should now look similar to Figure 7-4. There are now two motel rooms with a thick wall between them to contain the plumbing and also to act as a fire and noise break. By using the mirror function again, a four room motel can be created.

> •• **NOTE:** The scale of the drawing is making it harder to pick exact points with the drawing cursor. To pick the line of reflection in the next step, you should use the object snap function to lock on to the end points of a line.

8.__ Select the **New Line** option from the **Mirror** menu.

9.__ Use the *object snap button* to pick the two ends of the lower horizontal wall of the motel as the line of reflection.

10.__ Capture the two motel rooms in an area box.

Chapter 7, Page 138

The Mirror Function

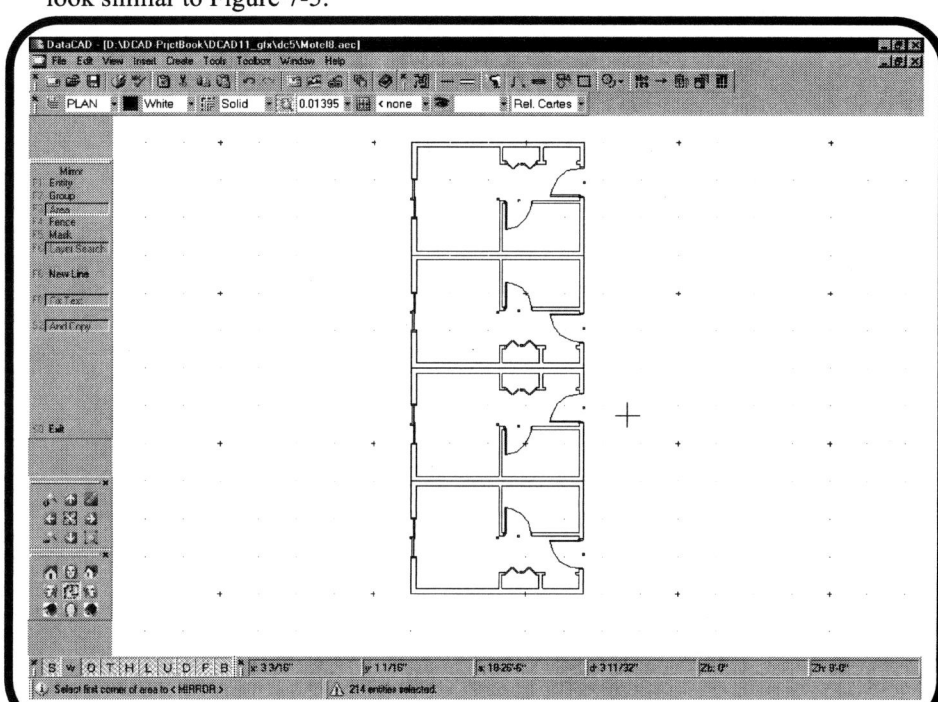

11.__ Use the **Zoom** menu options (or the **Navigation Pad** buttons) to make the display look similar to Figure 7-5.

Figure 7-5, The four room motel.

The **Mirror** function will be used one more time to go from four rooms to eight rooms. This time the line of reflection for the mirror will be a vertical line. As the scale of the drawing gets smaller, it gets harder to pick exact spots with the mouse. This is where the object snap function is very handy.

Object snap will be used to place the 0,0 coordinate at the upper right corner of the motel. Then coordinate entry mode can be used to enter the ends of the new reflection line. By placing the reflection line two feet from the motel rooms, a four foot wide hall between the rooms will be created.

12.__ Press the unshifted ~ key and then *click* the *object snap button* near the *upper right* corner of the motel.

13.__ Select the **New Line** option from the **Mirror** menu.

Chapter 7, Page 139

The DataCAD 11 Project Book

14.__ Check the information line to make sure the coordinate entry mode is set to **Relative Cartesian** (if not, press the **Insert** key to change it).

> •• **NOTE:** From author number two: I am not a fan of the Relative Cartesian input method, though I know others who are. I prefer to use the Relative Polar method, where you enter a distance and then an angle. Relative Cartesian is OK for drawing lines that are perpendicular to one another, or for drawing roof pitches. Relative Polar is best for everything else. To each their own, they say, so feel free to experiment with both methods to decide which you like best. See Appendix D for a more thorough discussion of your input options.

15.__ Press the **space bar** and enter an **X** value of **8** and a **Y** value of **0**.

16.__ Move the mouse straight down (as shown in Figure 7-6) and click the *left* mouse button *once* to complete the line of reflection.

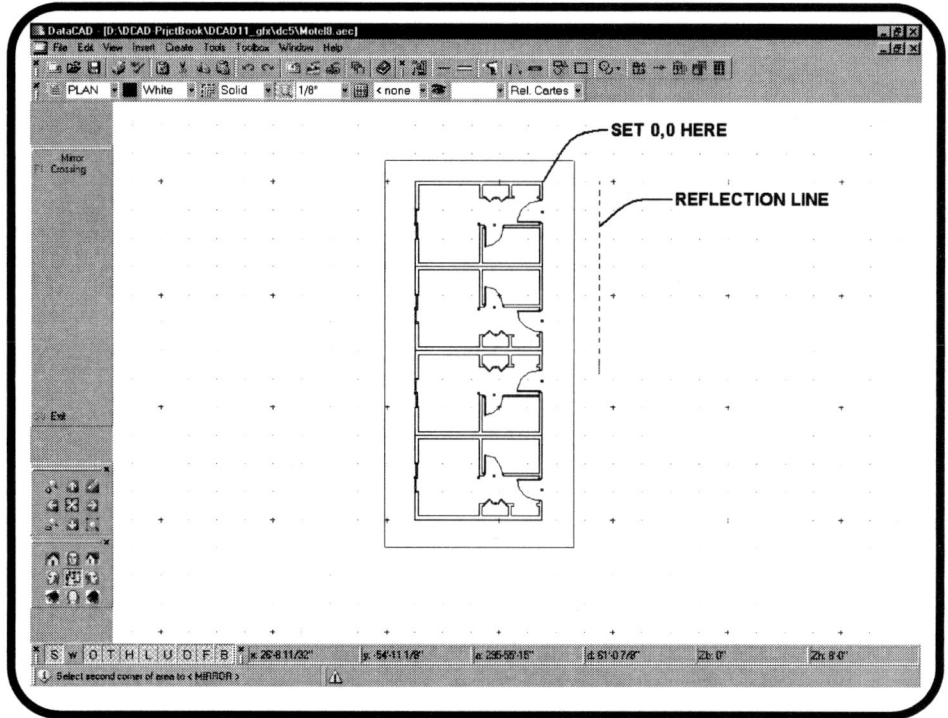

Figure 7-6, Creating a new reflection line.

17.__ Capture the *four* motel rooms in an area box.

The Rotation Function

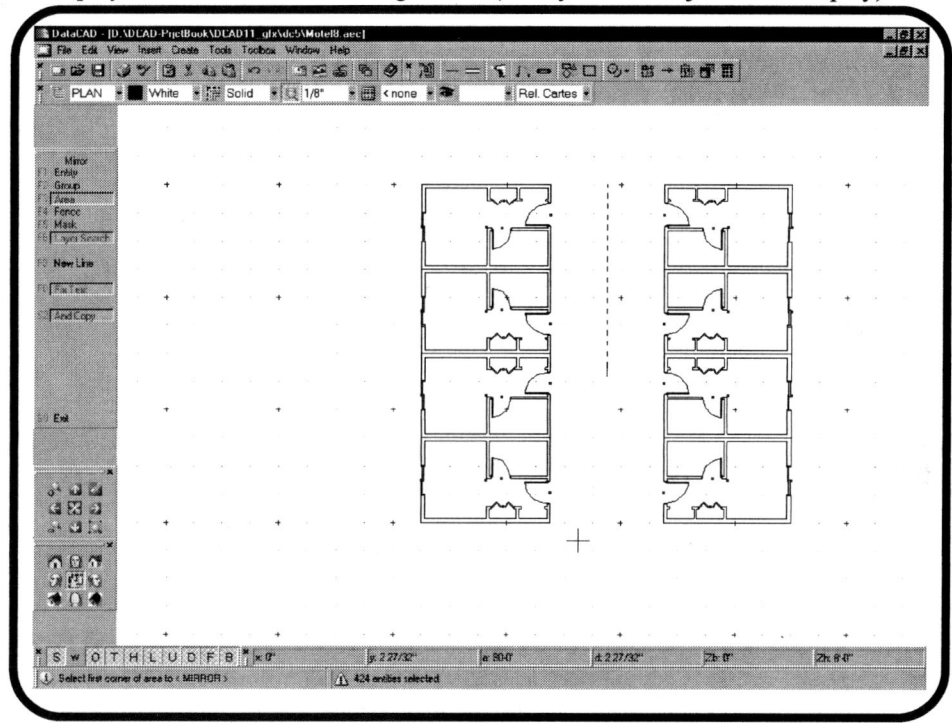

The display should look similar to Figure 7-7 (after you have adjusted the display).

Figure 7-7, The eight room motel.

Without having to create any new lines, you have gone from one room to a complete motel. The number of rooms could be doubled until the computer ran out of memory, but this is enough to make the point. (The rooms could be copied in the Z direction to make a multi-story motel, but that is a later discussion.

••• **WARNING**••• Have you saved your drawing lately?

The Rotation Function

To create a more interesting motel design (and to demonstrate another DataCAD function), four of the rooms will be rotated. To rotate objects in DataCAD, a center of the rotation must be established, an angle specified, and then the entities to rotate are selected.

18.___ From the **Edit** menu, select the **Rotate** option [or press the R or the r key].

The DataCAD 11 Project Book

19. __ Set the center of rotation at the *upper right* corner of the *left* bank of motel rooms (see Figure 7-8) by clicking with the *object snap button*.

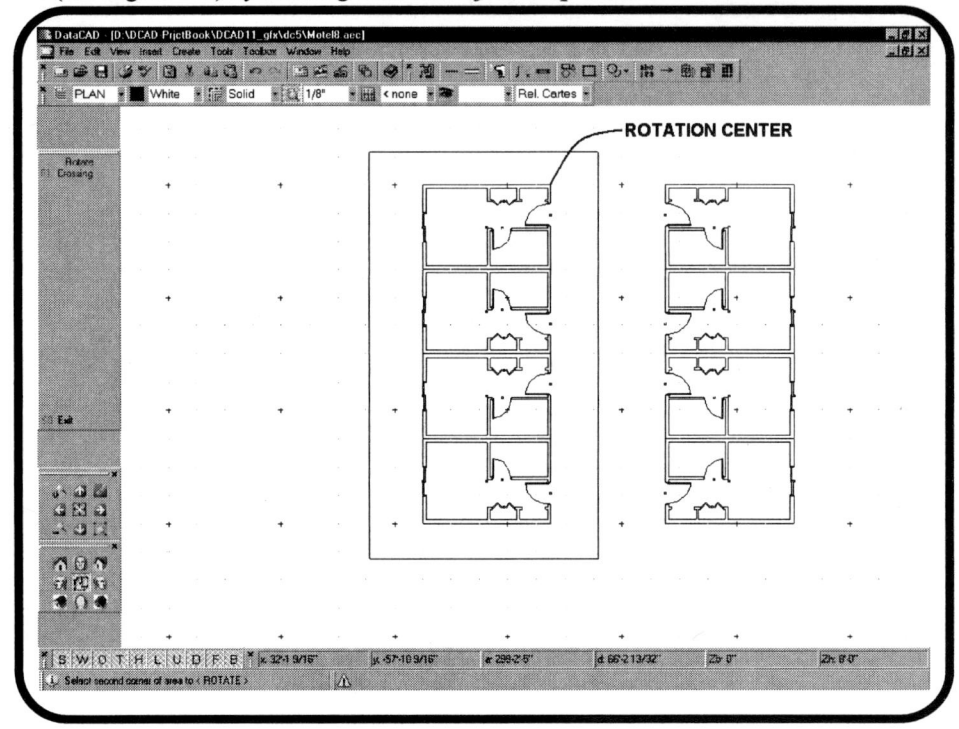

Figure 7-8, Setting a rotation center.

20. __ Select the **New Angle** option from the **Rotate** menu, and (watch the minus sign) enter a value of **-90** degrees.

21. __ Select the **Area** option from the **Rotate** menu, and capture the *left* four motel rooms in an area box.

The Rotation Function

The display should now look similar to Figure 7-9. This would be a good time to save the changes to the drawing.

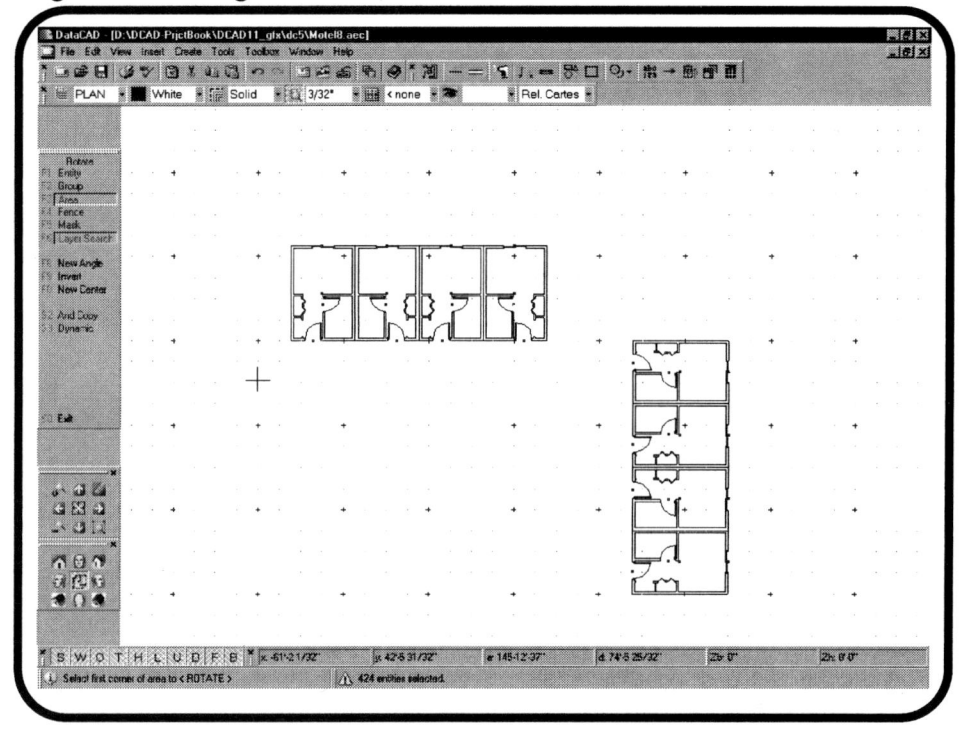

Figure 7-9, The new motel 8.

22.___ Press [CTRL] + [S] to save your drawing.

As you can see, DataCAD provides a lot of power to multiply and rearrange the parts of architectural designs. You can have some fun by creating different 3D Views of your motel design. Notice that the system takes a lot more time to redraw the display for each new view. This is because of the large number of lines now in the design. Imagine how many lines there would be if the plumbing fixtures and furniture would have been left in the room before it was copied.

Chapter 7, Page 143

The DataCAD 11 Project Book

23.__ Experiment with different 3D views of the motel. (The parallel projection view shown in Figure 7-10 was created by clicking on the place shown in the viewing sphere, and then using the **Hide** function.)

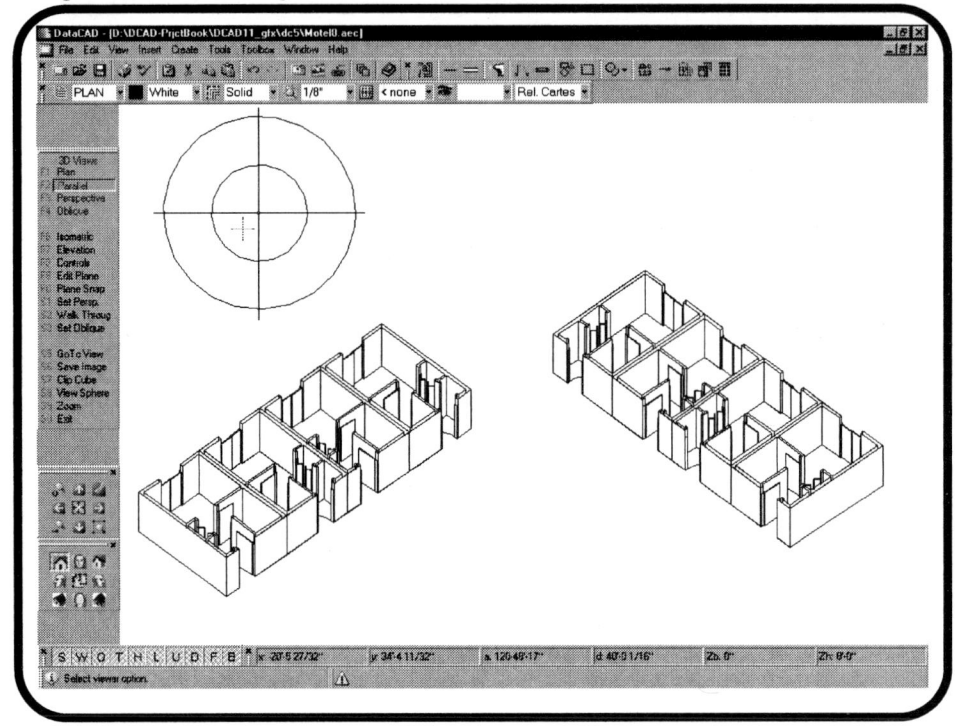

Figure 7-10, A pictorial view of the motel.

INFO STOP

Removing hidden lines in this design requires a large number of calculations. The calculations will be much faster if the **HLR Partial** option is toggled **on** in the **Hide** menu before starting the hide process. Some computers will take too much time for this operation when there are this many lines in the drawing. The good news is that the price for a computer with enough power to handle hiding lines in this design is always lower than it was a few months ago.

The Rotation Function

Review and Important Points

- The **Mirror** function can be used either with or without the **And Copy** option. The **And Copy** option leaves the original entities in place, and copies them around a line of reflection.

- The **Rotate** option is used to spin selected entities around an axis of revolution.

- The DataCAD angle convention causes positive rotation angles to be counter clockwise.

- The **Rotate** option allows for rotation by either a fixed angle or by dynamic rotation using the mouse.

The DataCAD 11 Project Book

Practice Drawings

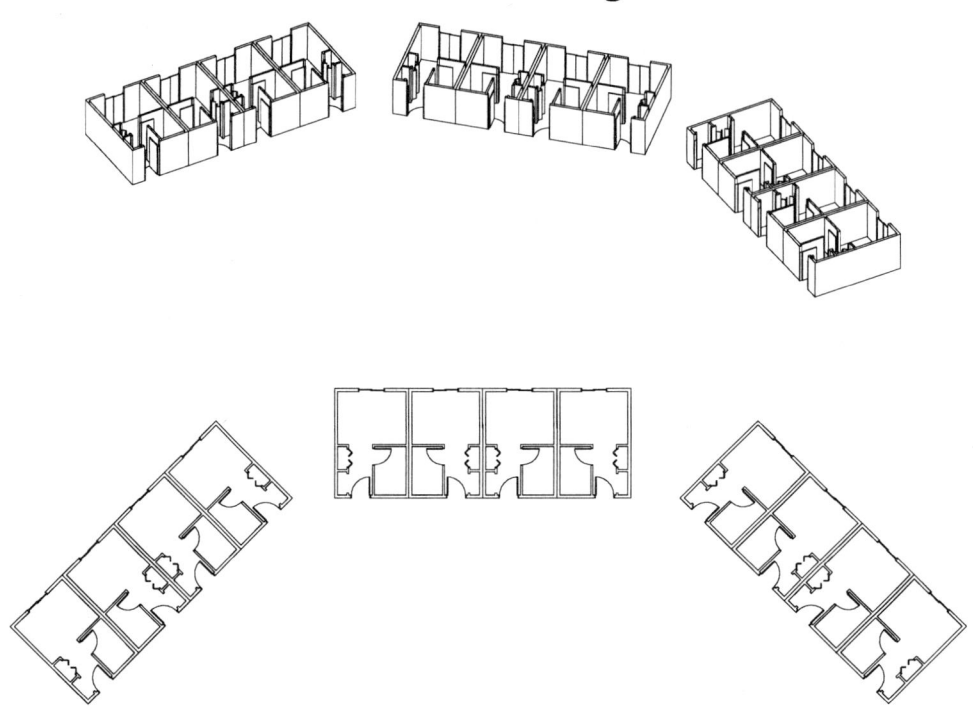

Project Hints
Assignment 7-A, Motel design

- Start with the motel room created in Project 4 and use the **Mirror, Copy, Rotate**, and **Move** commands to create your own motel design similar to that shown above. (Be sure to use **File, Save As** to create a new file before making any changes.)

- Be careful. It is easy to create large files very quickly using these techniques. Also, they can take a great deal of time to render.

- Make sure that you do not get too ambitious for the system and storage space you are using.

CHAPTER 8

Editing and Dimensioning

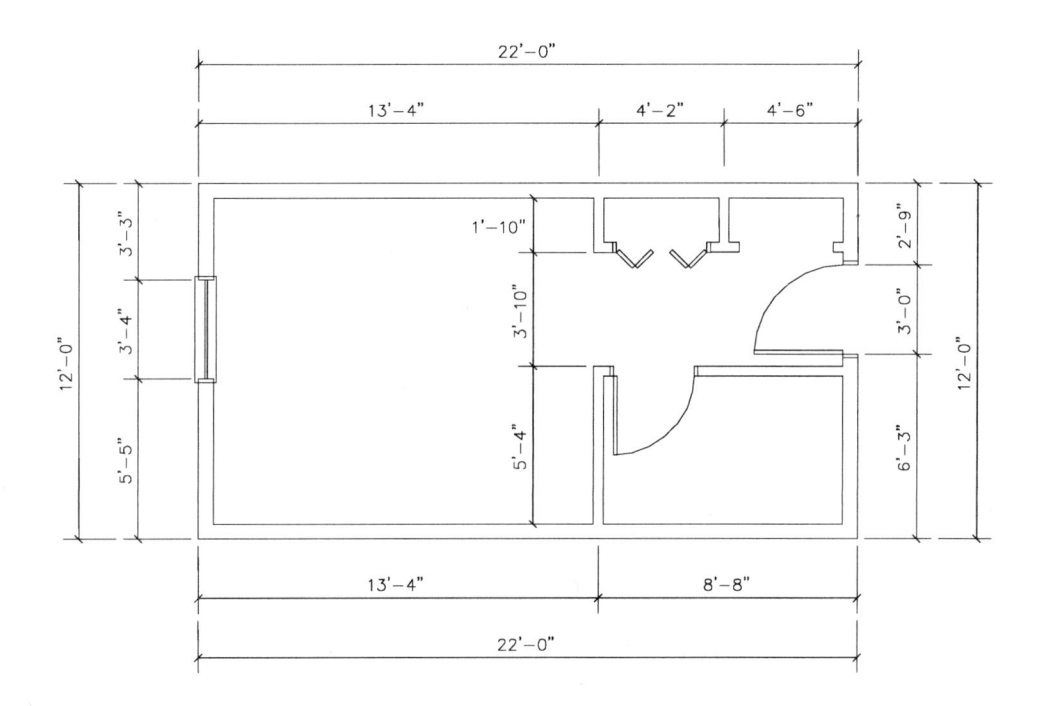

Project Goals

When you have completed this project, you will be able to:

1. Remove doors and fix walls with the **Remove** function.
2. Use the **Stretch** function to change the size or location of entities.
3. Adjust the text and dimension style options.
4. Select an arrowhead style for dimensions.
5. Dimension drawings using the **String Line** and **Overall** options.

The DataCAD 11 Project Book

Prerequisite: Chapter 6

Introduction

Oh, No! Just when Ferd's architectural firm had completed the motel room design, the client came in with some changes. Your job is to edit the motel room drawing to incorporate the changes, and then add the dimensions so that the contractors will know what size to build it.

DataCAD provides some very powerful functions which help make the job of changing drawings and adding dimensions much easier than if you had to do the same thing with pencil and paper. In this project you will make a new version of the motel room. Then you will add enough dimensions to see how the computer automates the dimensioning process.

1. __ Start **DataCAD** and load the drawing named **Motel1**.
2. __ Select **File, Save As** and save the current drawing as **Motel2**.
3. __ If you have a pictorial view of the motel room showing on the display, from the **MENU BAR** select **View, Plan**.
4. __ In the **STATUS AREA**, click on the **Layer Manager** icon. Use the **Layer Manager** to turn **off** all of the **layers** <u>except</u> the **PLAN** layer.
5. __ Select the **New** option from the **Layer Manager** and create **1** new layer.
6. __ Double click on the new layer name and change the name of the new layer to **DIM**.

Erasing Layers

> **INFO STOP**
>
> The **Erase** option in the **Layer Manager** does not delete the layers, it only erases all of the lines on the layer. Since the motel room will be changed quite a bit, the furniture and plumbing from the old design should be removed.
>
> •• **NOTE:** In the **Layer Manager** you can highlight several different layers at the same time by holding the [CTRL] key down while clicking on each layer you want to select. This makes it possible to erase the contents from more than one layer at a time.

Removing a Door

7.__ In the **Layer Manager** dialog box, highlight the **PLUMB** layer. Then use the **Erase** option from the **Layer Manager** to erase all entities on the layer. Repeat this process to erase entities on the **TEXT**, and **FURN** layers.

The display should now look similar to Figure 8-1.

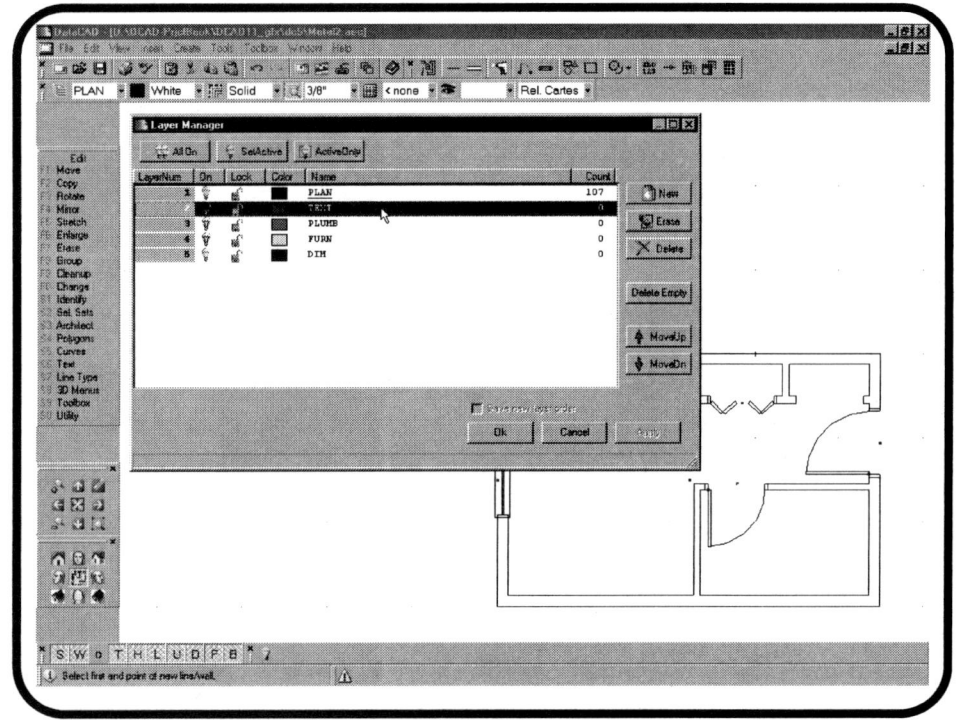

Figure 8-1, Erasing by layers.

8.__ Press [CTRL] + [S] to update the **Motel2** drawing file.

Removing a Door

The first change the customer wants is to replace the sliding glass door with a window. This means that the door must be removed, the hole in the wall fixed, and then the new window must be added. If you were drawing with pencil and paper, this would take a while. But it is easy to do with DataCAD.

9.__ Press the **a** [*not* the A] key to jump into the **Architect** menu.

10.__ Select the **Door Swing** option from the **Architect** menu.

The DataCAD 11 Project Book

11.___ Select the **Remove** option from the **Door Swing** menu. Capture the sliding door on the *left* motel room wall in an area box.

- •• **NOTE:** If the door was not removed, check to make sure that the **PLAN** layer is active.

Well, that got rid of the door in a hurry, and it also fixed the hole in the wall. Next, a window will be added to this wall. The client wants the window to be 3'4" wide.

12.___ Press the **X** key [not the **x** key] to jump to the **Object Snap** menu. Make sure that the **Mid Point** option is turned **on**.

13.___ Press the **a** [*not* the A] key to jump to the **Architect** menu.

14.___ Select the **Windows** option and make sure that the **Sides** option is turned **off**.

15.___ Click on the *left outside* motel room wall with the *object snap button*.

16.___ Press the **space bar** and enter an **X** value of **0.0** and a **Y** value of **1.8**.

17.___ Click to the *left* of the motel room *once* with the *left* mouse button.

The Stretch Function

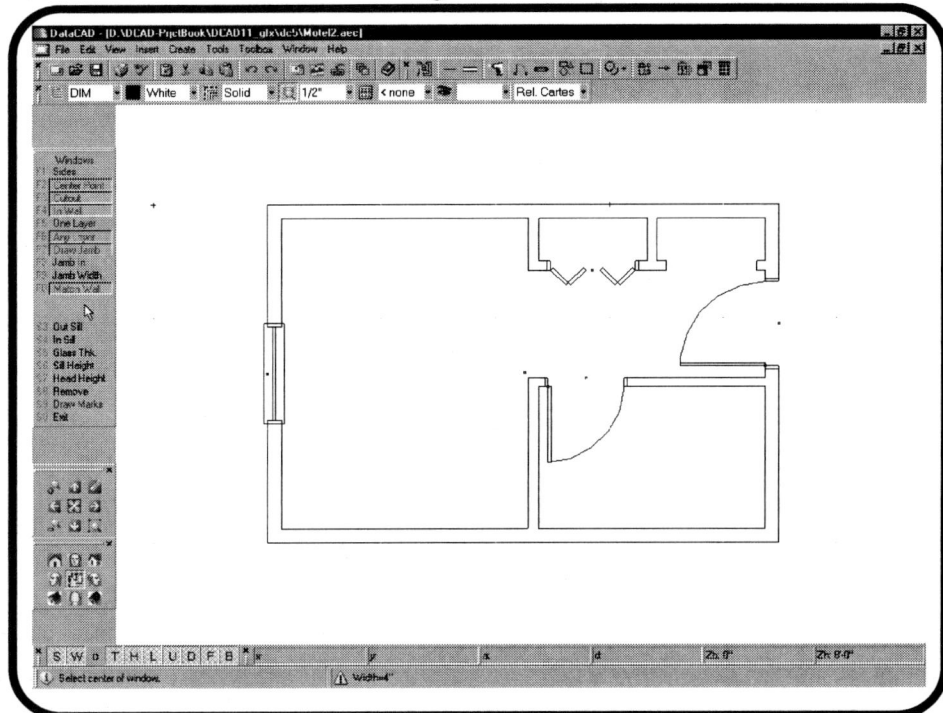

The modified motel room is shown in Figure 8-2.

Figure 8-2, The modified motel room.

The Stretch Function

The original room was 18 feet long. The client wants this to be changed to 24 feet. With pencil and paper, this would be a bother. With DataCAD it is not much trouble at all.

18.___ Select the **Stretch** option from the **Edit** menu.

19.___ At the prompt *Select first point of the distance to stretch*. click the *left* mouse button *once* on the left *outside* wall.

20.___ At the prompt *Select second point of the distance to stretch*. press the **space bar** and (watch the minus sign) enter an **X** value of **-4.0** and a **Y** value of **0.0**.

The DataCAD 11 Project Book

21.__ Select the **Area** option from the **Stretch** menu and capture the *left* end of the motel room in an area box as shown in Figure 8-3.

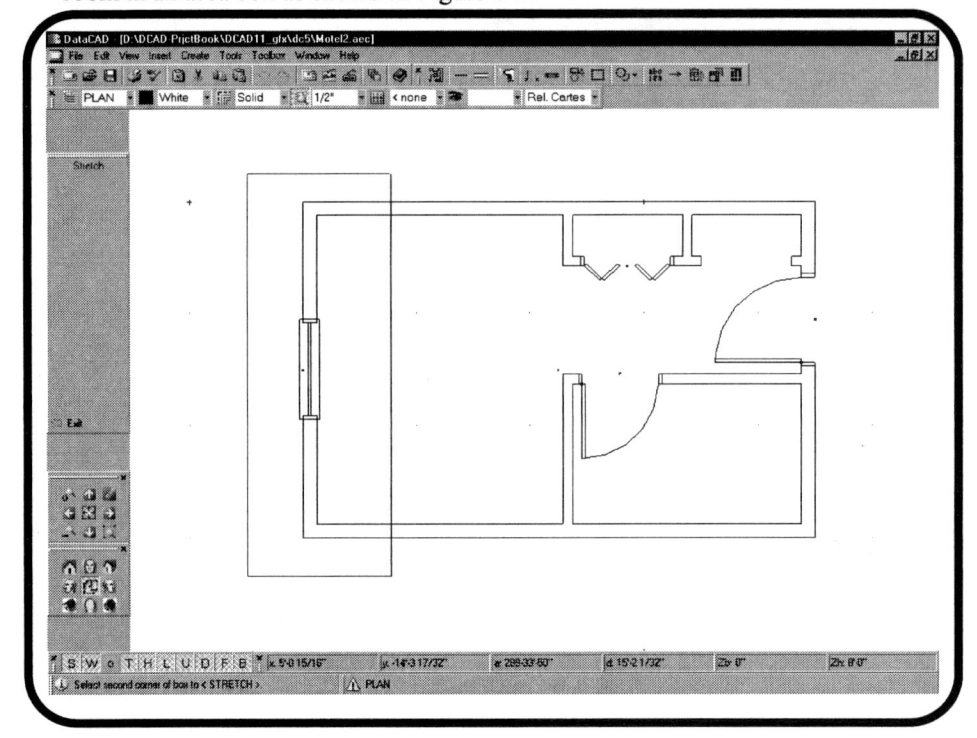

Figure 8-3, Stretching the motel room.

22.__ Adjust the display with the **Zoom** menu or **Navigation Pad** buttons.

Notice how the computer has not only moved things over, it has also fixed up the walls that were stretched. Now some dimensions will be added to the drawing. It is not unusual with pencil and paper drawing for the dimensions to take more time than the design. The computer can take care of a lot of the dimensioning busy work and speed things up.

Adding Dimensions

There are a few things that have to be set up before starting to add dimensions. The **DIM** layer must be activated, the text size specified, and a text style selected.

23.__ Press the **Tab** key until the **DIM** layer is active. Press the **k** [or the **K**] key to select a color for dimensions.

Adding Dimensions

24. __ Jump to the **Dimensions** menu by pressing the **d** [or the **D**] key, or by picking it from the **Utility** menu.
25. __ Select the **Linear** option from the **Dimensions** menu. Make sure that the **Vertical** and **Associative** options are turned **on**.
26. __ Select the **Text Style** option from the **Linear** menu.
27. __ Select the **Text Size** option from the **Text Style** menu and enter a size of 6 inches **(0.6)**.
28. __ Make sure that the **Out Horz.**, **Above**, and **Auto** options are turned **on** in the **Text Style** menu.
29. __ Press the *right* mouse button *once* to return to the **Linear** menu.
30. __ Select the **Dim. Style** option from the **Linear** menu and make sure that **Fixed Incr.** (fixed increment) option is turned **on** in the **Dim. Style** menu.
31. __ Press the *right* mouse button *once* to return to the **Linear** menu.
32. __ Select the **Arrow Style** option from the **Linear** menu and select the **Tick Marks** option.
33. __ Select the **Weight** option from the **Arrow Style** menu and set the weight to **1**.
34. __ Press the *right* mouse button *once* to return to the **Linear** menu.

Next, dimensions will be added to the motel drawing starting with the vertical wall on the left side of the motel.

35. __ Be sure that the **DIM** layer is active, and that **Vertical** in the **Linear** menu is **on**.
36. __ Click the *object snap* button *once* at the *outside lower left* corner of the motel.
37. __ Move the drawing cursor until it is on the *outside lower corner* of the window *opening* (not the window sill) and click the *object snap* button *once*.
38. __ Move the drawing cursor until the dimension line is to the *left* of the motel wall and click the *left* mouse button *once*.

Chapter 8, Page 153

The DataCAD 11 Project Book

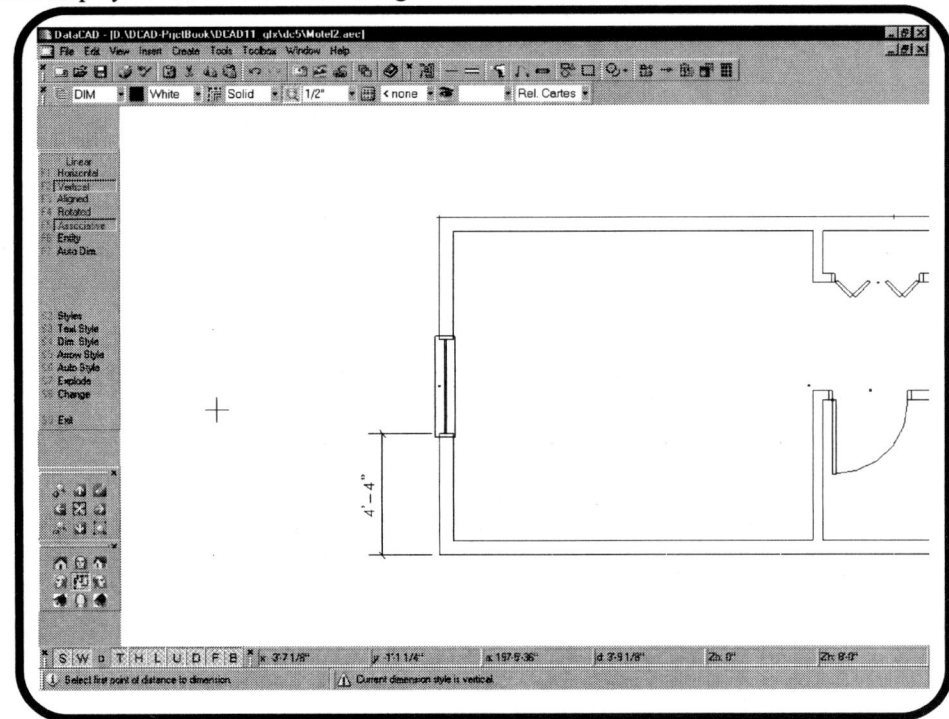

The display should look similar to Figure 8-4.

Figure 8-4, Starting a dimension string.

Notice that the computer has drawn the extension lines, the dimension lines, the tick marks, and has added the dimension text, all automatically. Also notice that as soon as you place a dimension on the drawing, a new option (**String Line**) is added to the menu. The string line option allows you to add more dimensions in a line with the first one placed.

39.___ Select the **String Line** option from the menu.

40.___ Move the drawing cursor to the *outside top corner* of the window opening and click the *object snap* button *once*.

41.___ Move the drawing cursor to the *outside upper left corner* of the motel room and press the *object snap* button *once*.

42.___ Click the *right* mouse button *once*.

The **Overall** option has now been added to the menu. This option will automatically add an overall dimension to the string of dimensions just completed.

Chapter 8, Page 154

Adding Dimensions

43.___ Select the **Overall** option from the menu.

The display should now look similar to Figure 8-5. Just like magic, all of that work has been done for you. Now move over to the right vertical wall and repeat the process.

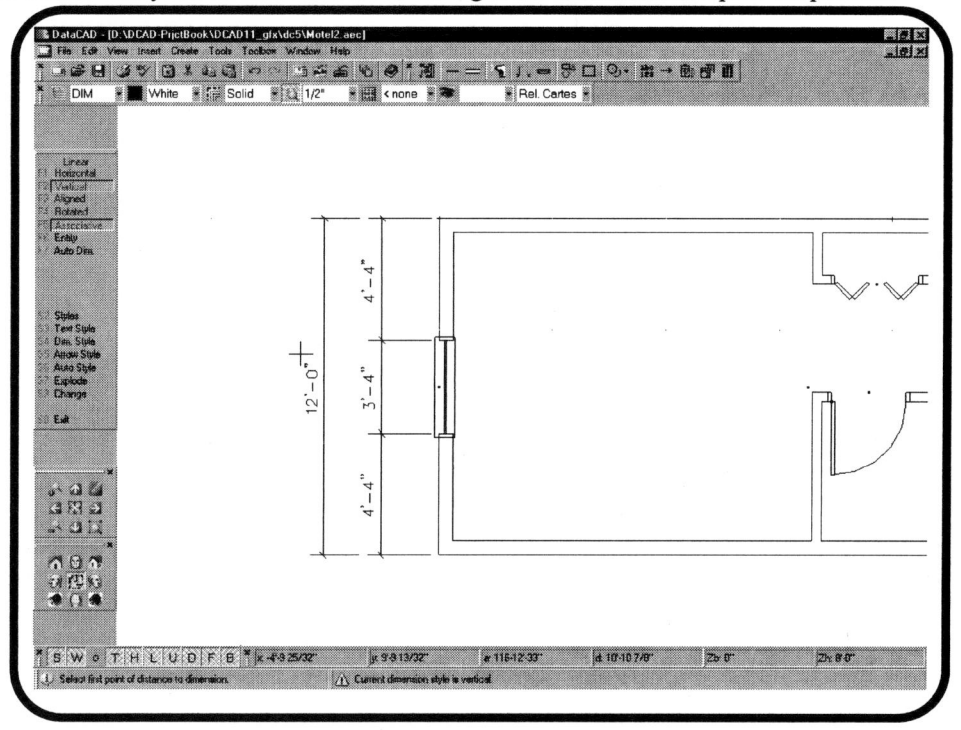

Figure 8-5, Adding dimensions.

44.___ Adjust the display so that you can work on the right side of the motel room.

Chapter 8, Page 155

The DataCAD 11 Project Book

45.__ Repeat the above process to dimension the outside of the right side wall as shown in Figure 8-6.

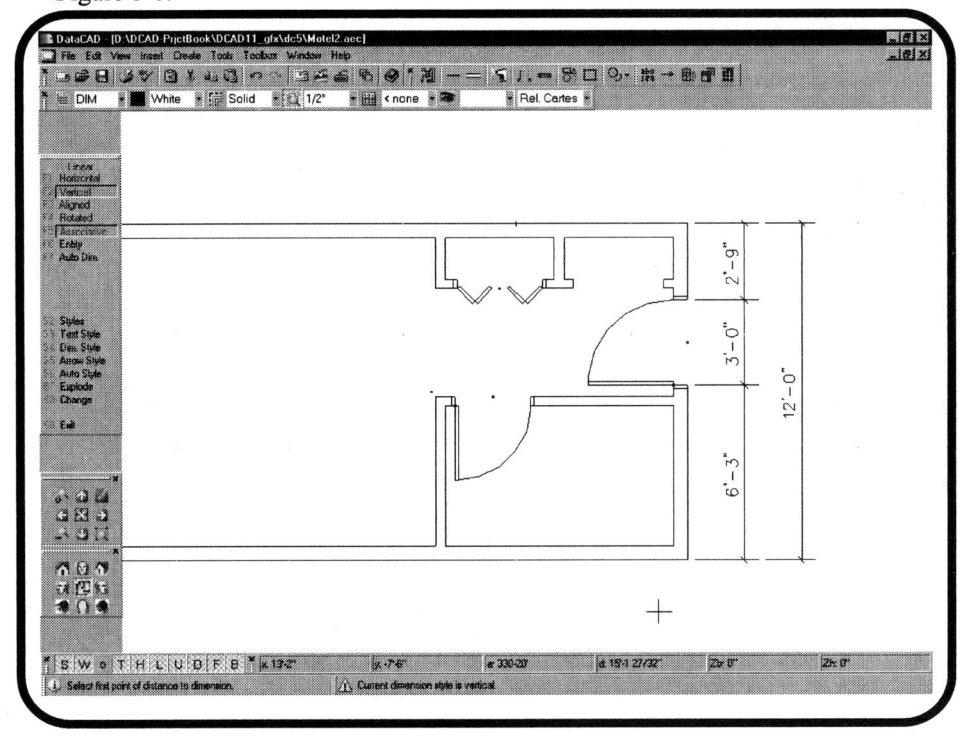

Figure 8-6, More dimensions.

INFO STOP
If the system has a question whether or not the text will fit properly, it will show a box and allow you to place the text in a different position. Or, if the space is too small, you can rotate the text 90 degrees to fit, using the **Rotate** option in the **Linear** menu. This provides you with a great deal of flexibility in placing dimension text.

Adding Dimensions

46.__ Use the **Vertical** and **String Line** options to add dimensions which show the size of the bathroom, hall, and closet width (see Figure 8-7).

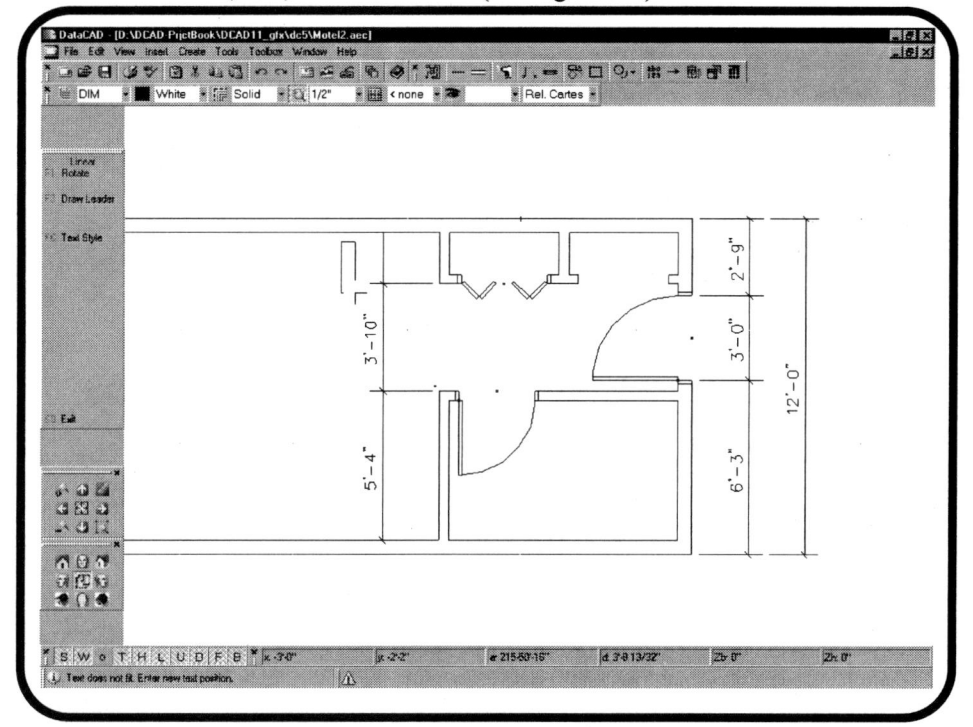

Figure 8-7, Adding dimensions.

INFO STOP

When you get to the closet width, the computer discovers that it does not have enough room for the text. It then waits to see what you want to do about it. If you move the mouse, you will see a text box that you can use to position the text somewhere else. You can also use the **Rotate** option on the menu to turn the text so that it will fit. The **Rotate** option will be used in this example.

Chapter 8, Page 157

The DataCAD 11 Project Book

47.___ Select the **Rotate** option from the menu. Enter an angle of **90** degrees and locate the text box as shown in Figure 8-8.

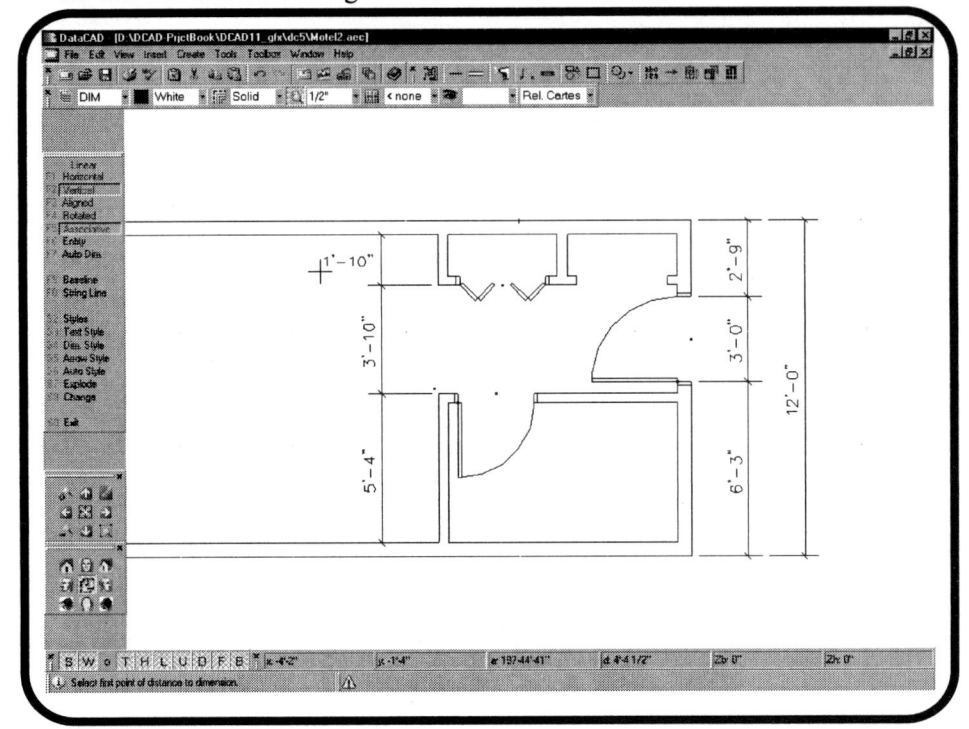

Figure 8-8, Rotated text.

Next the horizontal dimensions will be added. These are completed in the same manner as the vertical dimensions.

Adding Dimensions

48.__ Select the **Horizontal** option from the dimension menu and add the horizontal dimensions as shown in Figure 8-9.

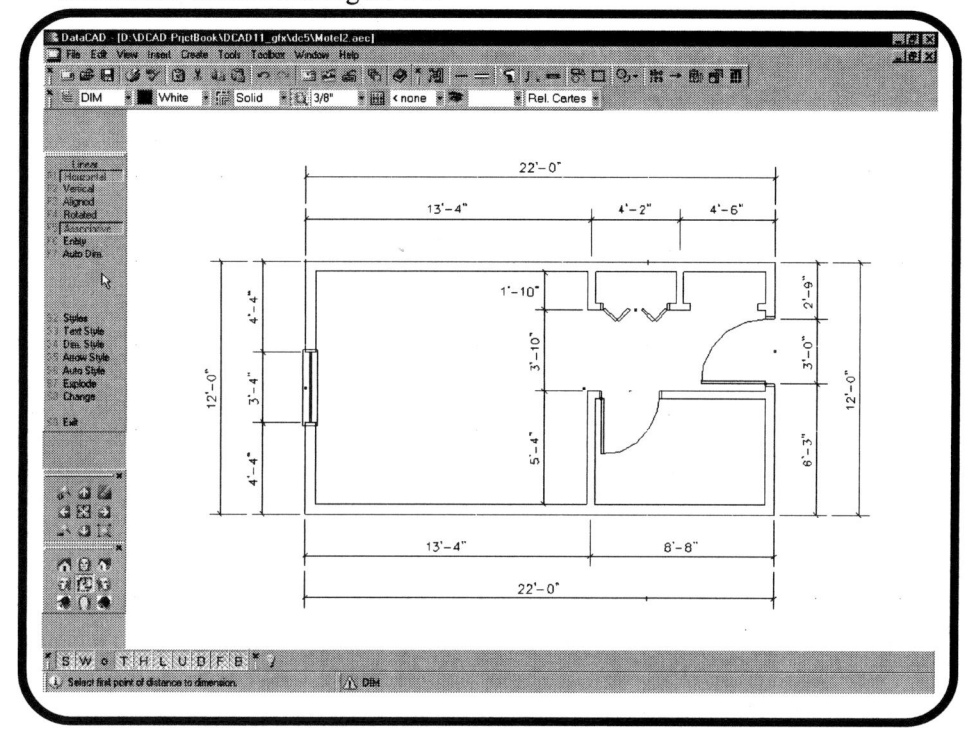

Figure 8-9, The dimensioned motel room.

INFO STOP You can adjust the display with the **Navigation Pad** buttons, or the arrow keys on your keyboard, even in the middle of placing a dimension. Sometimes it is necessary to zoom-in while you are dimensioning to locate an exact spot.

Oh No! The client just showed up and wants to move the window 1'1" toward the top of the display. If you were drawing with pencil and paper, you might be very angry because you would have to do a lot of erasing and redrawing. With DataCAD, this kind of change is no problem.

The DataCAD 11 Project Book

Design Changes

The **Stretch** function (that was used earlier to move the wall) can also be used to move the window. Because DataCAD uses something called "associative dimensioning", the dimensions will automatically be fixed.

49.___ Select the **Stretch** option from the **Edit** menu.

50.___ At the prompt *Select first point of the distance to stretch*. click the *left* mouse button *once* near the center of the window.

51.___ At the prompt *Select second point of the distance to stretch*. press the **space bar** and enter an **X** value of **0.0** and a **Y** value of **1.1**.

52.___ Make sure that the **Layer Search** option is toggled **on** in the **Stretch** menu. Select the **Area** option from the **Stretch** menu and capture the window in an area box. (The result is shown in Figure 8-10.)

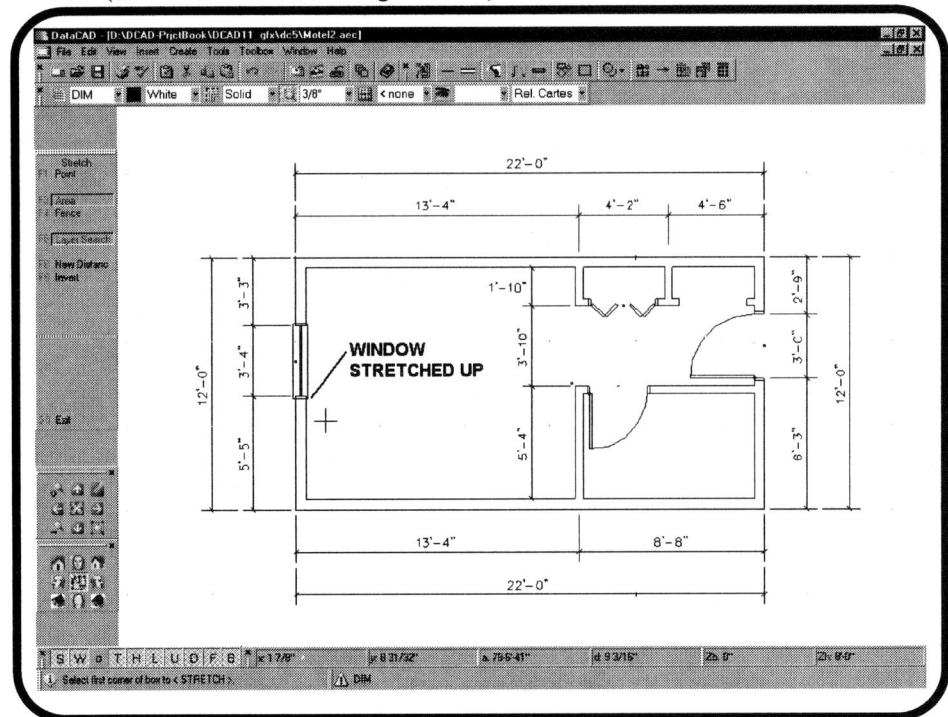

Figure 8-10, Using Stretch to make design changes.

53.___ Press the [ESC] (**Esc**) key to redraw the display.

Chapter 8, Page 160

Design Changes

••• **WARNING**••• Have you saved your drawing lately?

The functions demonstrated in this project give you an idea of how CAD can save time and make architectural designing much easier. Of course there are a lot more options than were explored in this project. If you have any extra time, you might want to experiment with some of the other menu options to see how they work.

•• **NOTE:** In the next two projects in this book, you will work with some of the DataCAD three-dimensional modeling functions for creating exciting architectural designs.

Review and Important Points

- The **Clear Layer** function in the **Layers** menu, or the **Erase** function in the **Layer Manager**, will delete all entities from a drawing layer without removing the layer or its settings.

- The **Delete Layer** function in the **Layers** menu, or the **Delete** option in the **Layer Manager**, *will* completely remove a drawing layer and all of its contents from a drawing.

- To remove a door (or window) and heal the opening in the wall, use the **Remove** option found in the **Door Swing** (or **Windows**) menu.

- The **Stretch** function will move entities (and/or dimensions) while preserving the integrity of the walls.

- Always use the object snap option when dimensioning to insure that the dimensions are locked to (and associated with) the entities being dimensioned.

- When placing a series of dimensions, use the **String Line** option after the first dimension is placed.

- Immediately after placing a string of dimensions, use the **Overall** option. This option will disappear as soon as a new dimension is placed.

- If a dimension will not fit between the extension lines, the system will change the cursor to a text size box and provide a text rotate option.

The DataCAD 11 Project Book

Practice Drawings

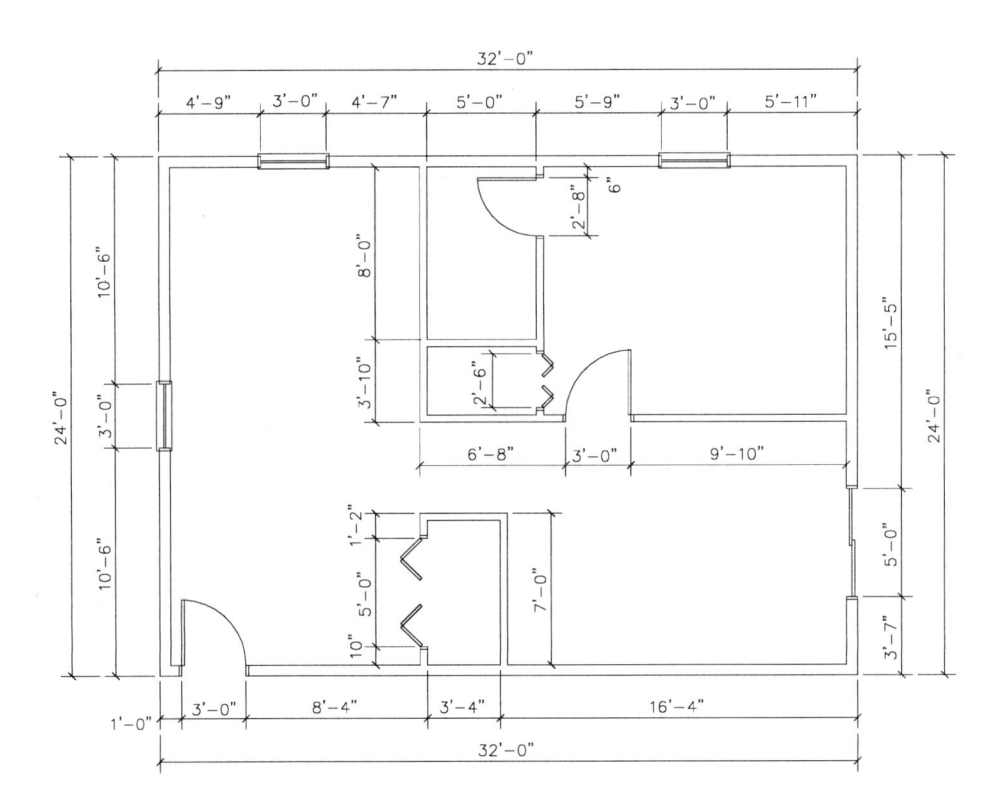

Project Hints
Assignment 8-A, Dimensioning

- Use the apartment floor plan created in Chapter 4.
- Add a new layer for dimensions and add dimensions similar to those shown above.

CHAPTER 9

3D Modeling, Ferd City

Project Goals
When you have completed this project, you will be able to:
1. Create a 3D "block city" model.
2. Use the **Copy** function to create a rectangular array of objects.
3. Use the **3D Entity** menu to create 3D objects.
4. Use the **Hide** function to remove hidden lines from a 3D model.
5. Save an image on a new drawing layer.

The DataCAD 11 Project Book

Prerequisite: DataCAD 11 interface experience

Introduction

In this project you will explore creating three dimensional objects using the 3D modeler that DataCAD provides. When architects plan large cities, they will frequently create a 3D model to see how the buildings will look together. It is possible to get a "feel" for how the city will appear by using simple blocks to represent the large buildings.

The first step in creating Ferd City is to layout some streets. This can be accomplished by creating an array of rectangles. Since one of the principles of CAD is that you should never draw anything twice, only one rectangle will be created and then the copy function will be used.

1.__ Start a new drawing with the file name **FerdCity**.
2.__ Select the **Save Delay** option from the **Settings** menu, and set the save delay time to **15** minutes.

As you have seen in earlier projects, the DataCAD software is always working in 3D. When you created the floor plans for the playhouse and the motel, walls were automatically drawn 8 feet high. There are times when 8 foot high walls or lines are not wanted. DataCAD provides a way to control the height of things.

> **INFO STOP**
> When you look at the **Plan** view shown on the display, note that the X direction is left and right on the display while the Y direction is up and down on the display. The third dimension (when you are looking at the **Plan** view) is in and out of the display. This is called the Z direction. By controlling things in the Z direction, the third dimension of the 3D world is controlled.

Changing the Z-base

Changing the Z-base

In the next exercise, rectangles will be used to represent city blocks. In order to look good in an illustration, these rectangles should be very thin. In other words, we want the Z dimension to be small. In the next step we will adjust the **Z-Base** (which controls where the lines will be drawn in the Z direction). In this example, 6 inch thick blocks will be used to represent the city blocks. The top of the city blocks will be at the bottom of the buildings. It seems reasonable to locate the bottoms of the buildings at an elevation of 0. This means that the bottom of the 6 inch thick city blocks must be at an elevation of minus 6 inches. (When you created the playhouse walls, the **Z-Base** was 0 and the **Z-Height** was set to 8 feet.)

3. __ Press the **z** key (or the **Z** key) to get to the Z-setting (or get there the long way by going to the **3D Menus** menu and picking **Settings**) and enter a value (watch the minus sign) of **-0.6** for the **Z-Base**. After pressing **Enter** you will be prompted to **Enter the new Z-height**. Enter **0.0** for the **Z-Height** and press **Enter**.

Because the scale for the city is large, the snap grid size must also be large.

4. __ Press the **s** key (*not* the S key) and set the snap size to **5.0** for both **X** and **Y**.

Creating 3D Blocks

Next, a 3D block will be created for the first city block. Ordinarily it would be drawn full scale, but for this model it will be 30 feet square instead of 300 feet. This will result in a better choice of plotting scales for small paper.

5. __ Press the **J** [*not* the j] key to jump to the **3D Entity** menu.
6. __ Press the **k** [or the **K**] key until the current color is **Green**.
7. __ Select **Blocks** from the **3D Entity** menu. Make sure that the **Rectangle, Auto Height**, and **Base/Height** options are toggled **on**.
8. __ Move the drawing cursor to a convenient point on the display and click the *left* mouse button *once*.
9. __ Press the **Insert** key until *Relative Cartesian* coordinate mode is selected, and then press the **space bar**. Enter a value of **30** for both **X** and **Y**.

The DataCAD 11 Project Book

This is the first city block (see Figure 9-1). You may have to adjust the display to see all of it.

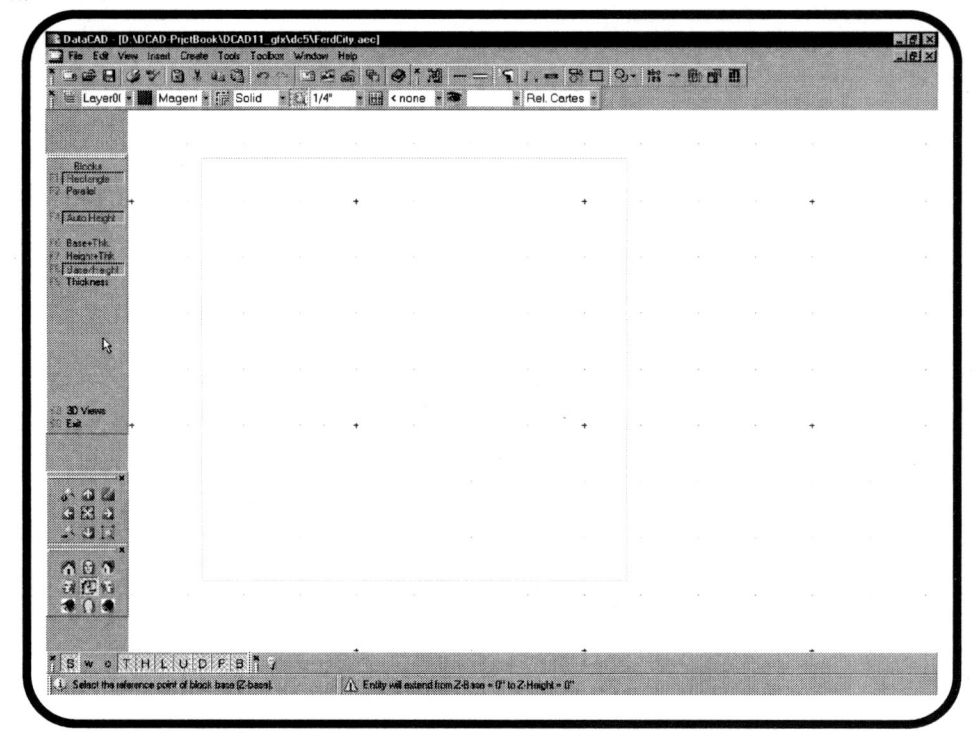

Figure 9-1, The first city block.

The Copy Function

Next, copies will be made using the rectangular array option.

10.___ Press the *right* mouse button *once*. Then select the **3D Menus** option [or press the **j** [*not* the J] key].

11.___ Select **Copy** from the **3D Menus** menu. Then select the **Rect. Array** option.

12.___ At the prompt *Select first point of distance to copy*. click the *left* mouse button *once* when the drawing cursor is positioned on the *lower left corner* of the first city block.

13.___ At the prompt *Select second point of distance to copy*, press the **space bar**. Then enter **40** for both **X** and **Y**. Enter **4** X repetitions, **3** Y repetitions, and **1** Z repetition.

14.___ Select the **Group** option and click on one of the lines in the first city block.

The Copy Function

After the display is adjusted, it should look similar to Figure 9-2. This array of rectangles will represent the city streets. A 3D view of the city will be used to make sure that the **Z-Height** was set to 0.

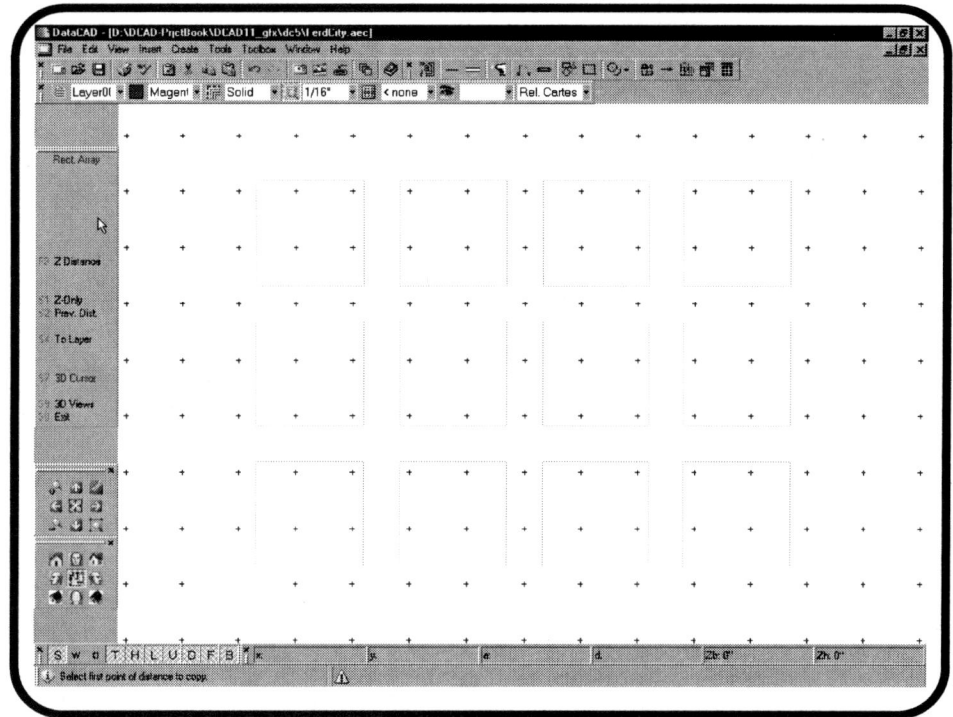

Figure 9-2, A rectangular array of blocks.

15.___ Press the **y** key [*not* the Y key] to jump to the **3D Views** menu.

16.___ Select the **Isometric** option from the **3D Views** menu.

The DataCAD 11 Project Book

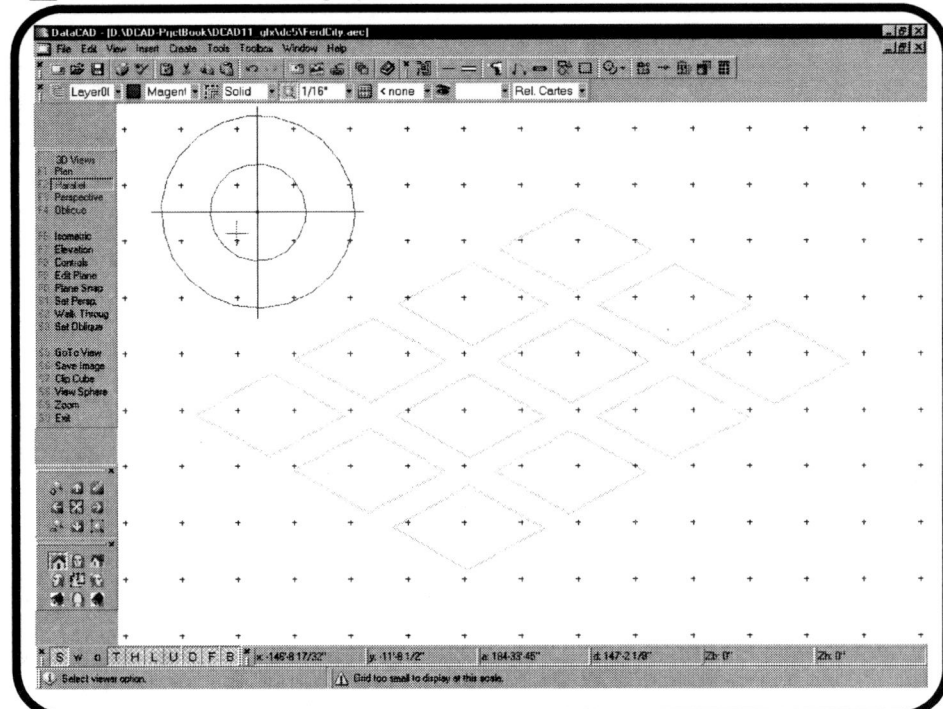

17.___ Check to see that the display looks similar to Figure 9-3.

Figure 9-3, The Ferd City blocks.

•• **NOTE:** If the display shows tall boxes instead of the rectangles shown in Figure 9-3, you did not set **Z-Base** and **Z-Height** properly. Select the **Plan** view, erase everything, and go back and start over.

Creating a Dome

Next, a dome and some block buildings will be added to the city. These will be created using options from the **3D Entity** menu.

18.___ Select the **Plan** button from the **3D Views** menu.

19.___ Press the **J** [*not* the j] key to jump to the **3D Entity** menu.

Creating a Dome

The **3D Entity** menu is shown in Figure 9-4. Take note of the different choices you have. Some day, when you have some extra time, you should experiment with the different options. For now, just concentrate on creating a dome.

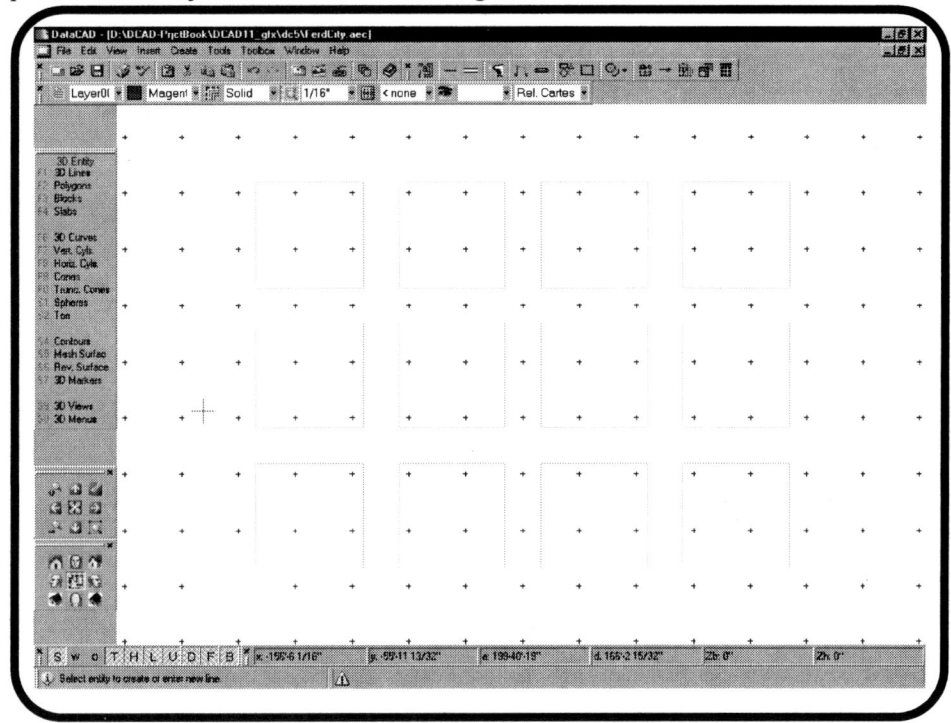

Figure 9-4, The 3DEntity menu.

20.___ Press the **z** [or the **Z**] key and set the **Z-Base** to **0**. (Don't worry about the **Z-Height**. The thickness option will be used to control the height of buildings.)

21.___ Select the **Spheres** option from the **3D Entity** menu. Check to see that the **2 Pt. Arc** and **Z-Base** options are toggled **on**.

22.___ Select the **Top Half** option from the **Spheres** menu.

A dome will be added to the lower left rectangular block in the city. Creating a dome is a three click process. The first click locates the center of the dome. The second click tells the computer the radius of the dome. The third click tells the computer how much of a dome you want. Since a full dome is desired, the second and third clicks will be in the same place.

The DataCAD 11 Project Book

23.___ Press the **k** [or the **K**] key to change drawing colors. Modify the display (window in on the lower left corner) so that it is similar to Figure 9-5.

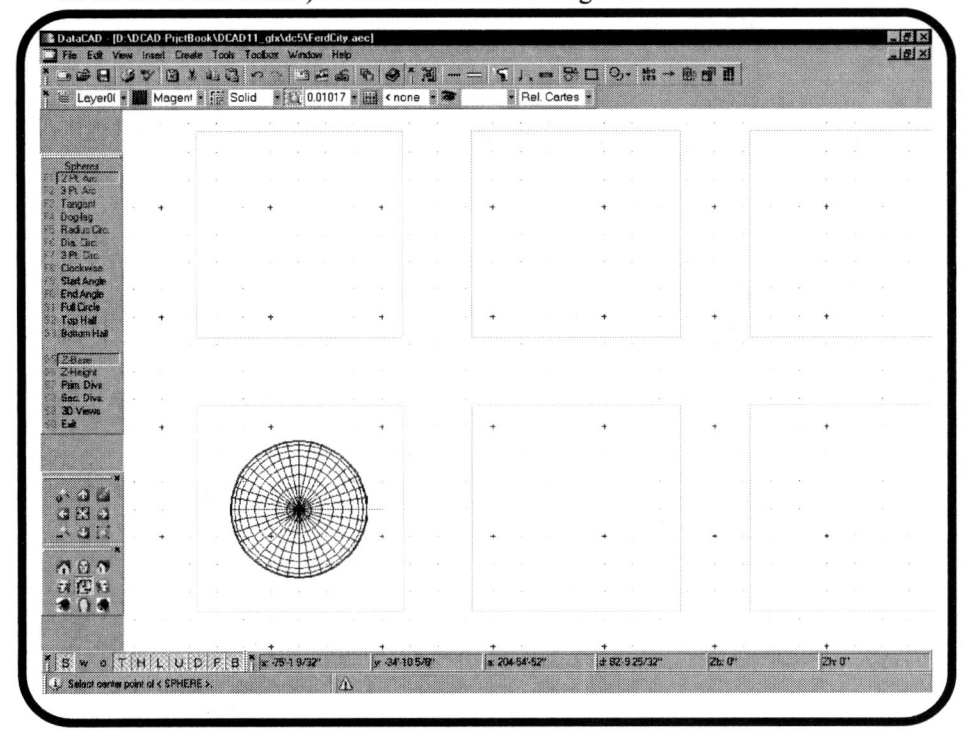

Figure 9-5, Creating the dome.

24.___ Click *once* with the *left* mouse button at the *center* of the *lower left* city block.

25.___ Move the drawing cursor to the right until it is near the inside edge of the rectangle and click the *left* button *twice*.

The display should look similar to Figure 9-5. What is not obvious in the **Plan** view is that the system created a 3D mesh representing the dome. Take a quick look at the isometric view to see this.

26.___ From the **MENU BAR**, select **View, Isometric**.

Chapter 9, Page 170

Adding Block Buildings

Figure 9-6 shows a close-up isometric view of the dome with the hidden lines (temporarily) removed. With only three clicks of the mouse button, the computer constructs the entire dome for you. The amount of drawing power just displayed by DataCAD is very impressive. (Especially if you try to duplicate this exercise by hand.)

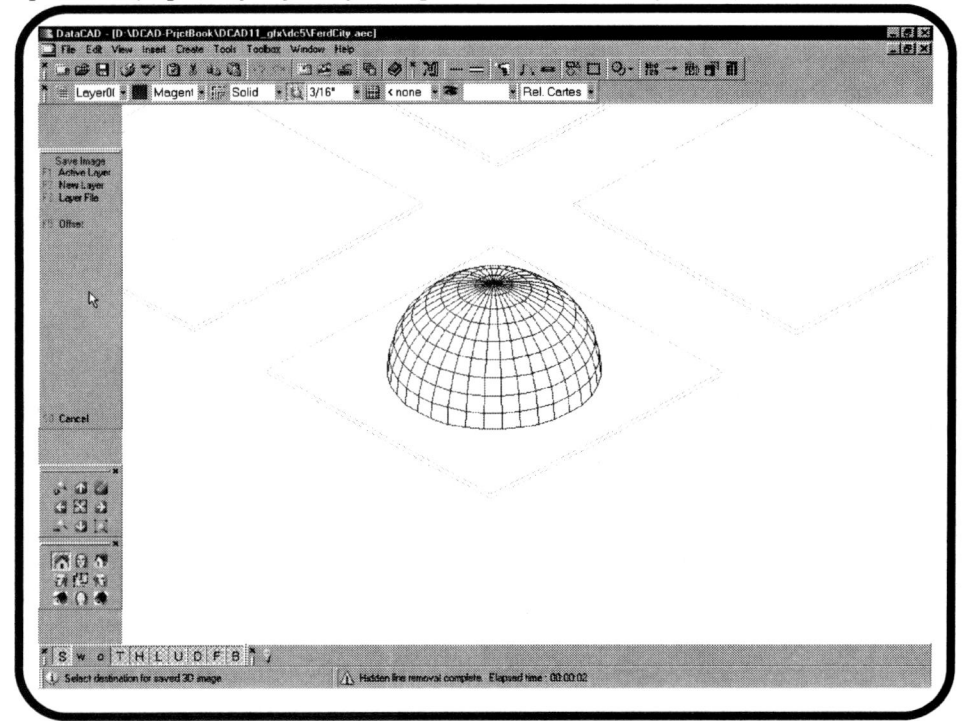

Figure 9-6, A pictorial view of the dome.

Adding Block Buildings

Now that a dome has been created to serve as a focal point for the city, let's add some skyscrapers. This will be done by creating some blocks that have different heights (in the Z direction).

27.___ From the **MENU BAR**, select **View, Plan**.

28.___ Select the **Blocks** option from the **3D Entity** menu.

29.___ Toggle the **Base+Thk.** option **on** in the **Blocks** menu.

30.___ Select the **Thickness** option from the **Blocks** menu and set the thickness to **8.0**.

The DataCAD 11 Project Book

•• **NOTE:** The block thickness will control the building height.

31.___ Press the **k** [or the **K**] key to change the drawing color.

32.___ Select a spot in one of the rectangles and create a block similar to one shown in Figure 9-7.

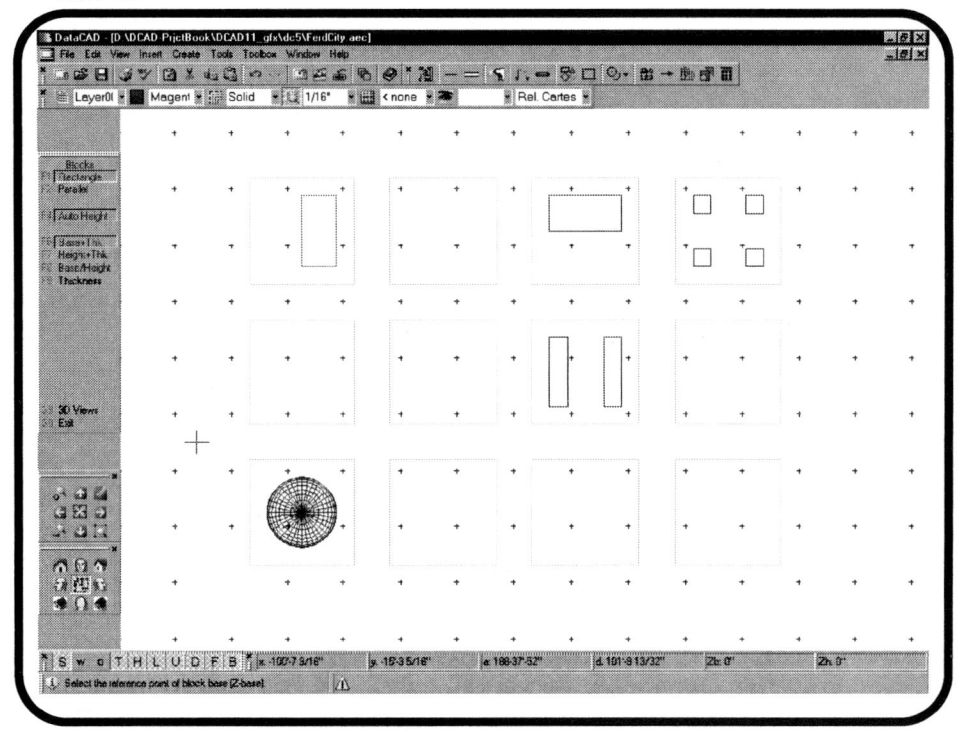

Figure 9-7, Adding block buildings.

33.___ Move to a different rectangle and create another.

34.___ Select **View, Isometric** to see how the city is developing. Select **View, Plan** before adding more buildings.

Adding Block Buildings

INFO STOP

You are creating buildings that have the same height. Now it is time to use your imagination and be creative. Sometimes there might be a row of buildings that are the same size and height. Sometimes there will be one large building in a block. Maybe you want to change the snap size to put buildings closer together. Make buildings of different colors.

35.__ Create four or five buildings of one height at different places in your city.

36.__ Change the thickness to **12** and create a few more buildings at different places.

37.__ Select **View, Isometric** to see how the city is developing. Select **View, Plan** before adding more buildings.

38.__ Change the thickness to **20** and add some more buildings.

Chapter 9, Page 173

The DataCAD 11 Project Book

When you are through adding buildings, create a perspective view of your city. You may want to save an image of the city that has the hidden lines removed. Figure 9-8 shows one version of Ferd City with the hidden lines removed.

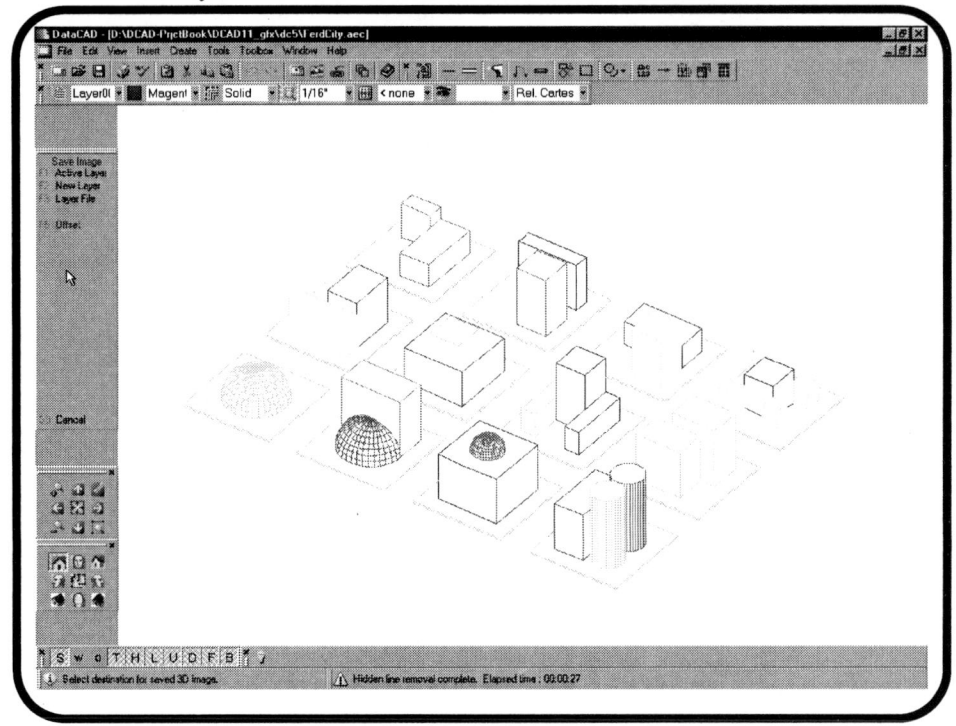

Figure 9-8, Ferd City.

- **NOTE:** You may want to take a simulated tour through the city by creating a perspective view and then selecting the **Walk Throug** option in the **3D Views** menu. Because of the scale of the city, you will want to change the walk and step distances.

Adding Block Buildings

39.__ Press the **j** [*not* the J] key to jump to the **3D Menus** menu. Select **Shader** (if available), select **Shade**, and examine a shaded image of your city. Try this with different views. Experiment with the different shader options.

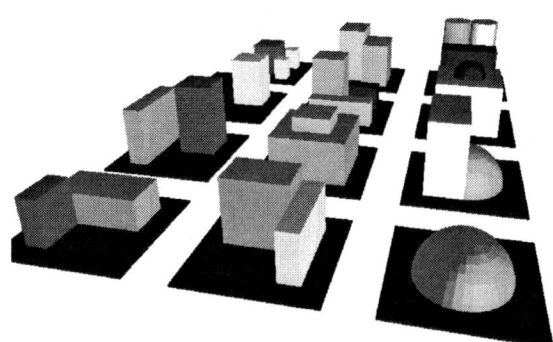

40.__ Create a 3D view of your city. Select **Hide** from the **3D Menus** menu. Toggle the **Save Image** and **HLR Partial** options **on**.

41.__ Select **Begin**. When the system has completed its calculations, select **New Layer** and enter a name for the new layer.

You can create several different layers containing different views of your city. By turning the different layers on and off, you can select from different views for plotting.

Now that you know the secret of creating 3D city plans, you can amaze your friends by producing a large variety of designs in a very short time.

Review and Important Points

- The value of **Z-Base** controls the vertical location of entities.
- To set the **Z-Base** and **Z-Height**, press the **z** [or the **Z**] key and follow the prompts.
- The vertical size of 3D **Blocks** can be specified by **Z-Base** plus **Thickness**, Z-Height plus **Thickness**, or **Z-Base** plus **Height**.
- The **Copy, Rect. Array** function allows the number of repetitions and distance to be specified in the X and Y distances.
- It is possible to copy entities in the Z direction. This can be useful in constructing multi-story buildings.
- DataCAD provides a variety of 3D entities and objects to be created.
- The **3D Menus** menu is very similar to the 2D **Edit** menu. You must be careful to select the proper menu for the editing task.

Chapter 9, Page 175

The DataCAD 11 Project Book

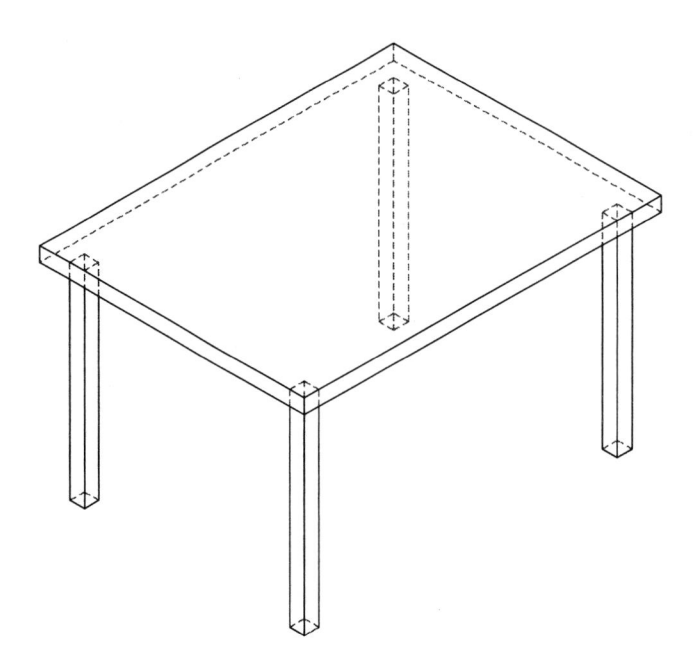

Practice Drawings

Project Hints
Assignment 9-A, 3D Construction

- Create a "3D "block table" as shown above.

- Use **3D Entity, Blocks, Base+Thk** to make a table leg 2 inches by 2 inches by 28 inches tall.

- Use **3D Menus, Copy, Rect. Array** and copy the leg 42 inches in the X direction and 30 inches in the Y direction, with 2 copies in each direction.

- Change the **Z-Base** to 28 inches and add a 2 inch thick 4 foot by 3 foot rectangular slab on top of the legs.

Chapter 9, Page 176

Adding Block Buildings

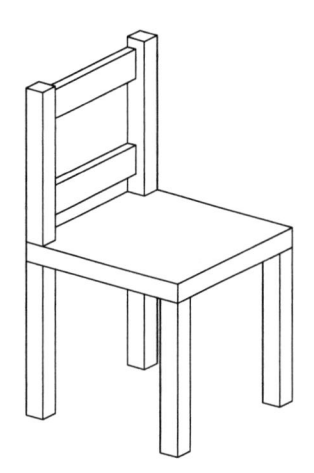

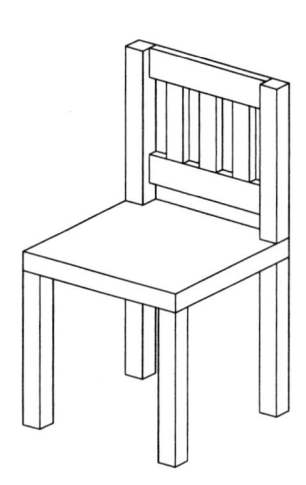

Practice Drawings

Project Hints
Assignment 9-B, 3D Modeling

- Use the entities in the **3D Entity** menu to create chairs similar to those shown above.
- The legs are 16 inches tall.
- Measure some chairs, or estimate sizes for backs and slats.
- Change the **Z-Base** and thickness to make blocks for the chair backs.
- Change the snap to 1/2 inch (0.0.1/2) to make it easier to center the 1 inch thick slats.

The DataCAD 11 Project Book

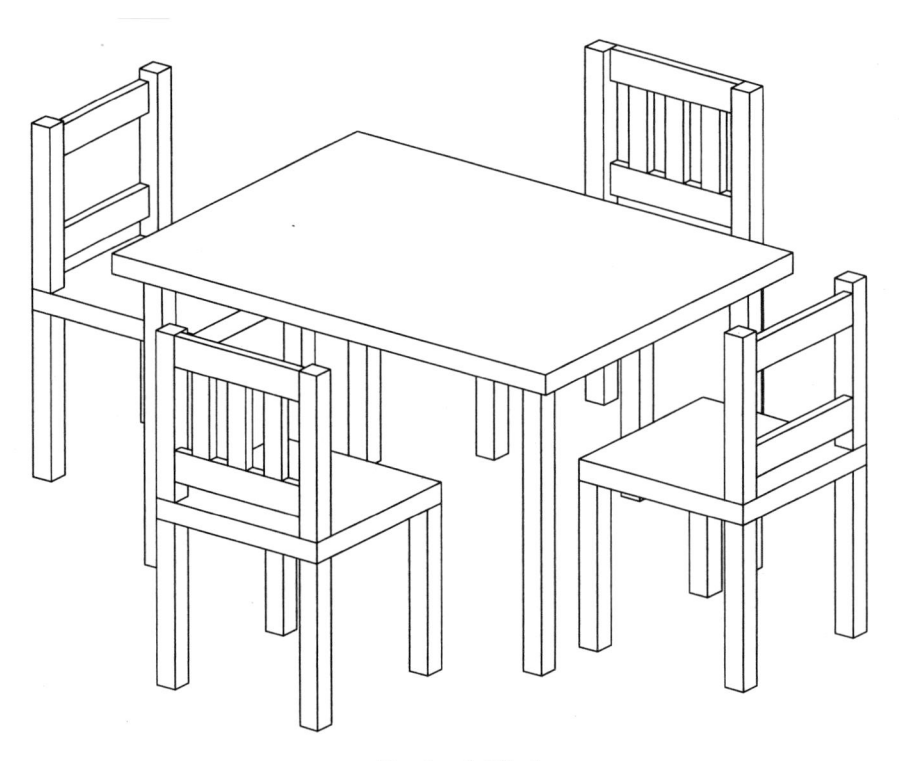

Practice Drawings

Project Hints

Assignment 9-C, 3D Modeling

- Create a pictorial view similar to that shown by copying, rotating and moving your chairs.
- Add a 48 inch square table.
- Be sure to save your drawing before making major modifications.

Chapter 9, Page 178

CHAPTER 10

3D Modeling, Using Macros

Project Goals
When you have completed this project, you will be able to:
1. Use DataCAD to create 3D views of house plans.
2. Use macros to add 3D doors and windows to designs.
3. Use the **ROOFIT** macro to add different styles of roofs to designs.

The DataCAD 11 Project Book

Prerequisite: Chapter 1

Introduction

In Chapter 3 you created pictorial views of Ferd's playhouse. Although these views were pretty exciting, there were two things that were not quite right. First, there were big spaces above and below the windows where the walls were cut. Second, there was no roof. In this project you will not only create 3D views of the playhouse that fix the holes in the walls, you will also explore the use of some macros that will make the 3D views look much better.

1.__ Start the **DataCAD** software and load the drawing file named **Play1**.

It is always a good idea to keep a copy of original designs (just in case you have to get back to an earlier version). A new copy of the Playhouse drawing will be made, and all of the design changes will be made on the copy.

2.__ Select **File, Save As** and enter the name **Play10**.

When you created the Playhouse in Chapter 1, you were not yet ready to deal with lots of different drawing layers. Now that you have some experience, you know how to create a different layer for each feature. Some new layers will be added for this new version of the Playhouse, and then the new layers will be used for various parts of the design.

3.__ Press the **l** key (or the **L** key) to jump to the **Layers** menu.

4.__ Select the **New Layer** option from the **Layers** menu and create **4** new layers.

5.__ Select the **Rename** option and name the new layers **FURN, DOORS, WINDOWS,** and **ROOF**.

6.__ Change the name of layer 1 to **PLAN**.

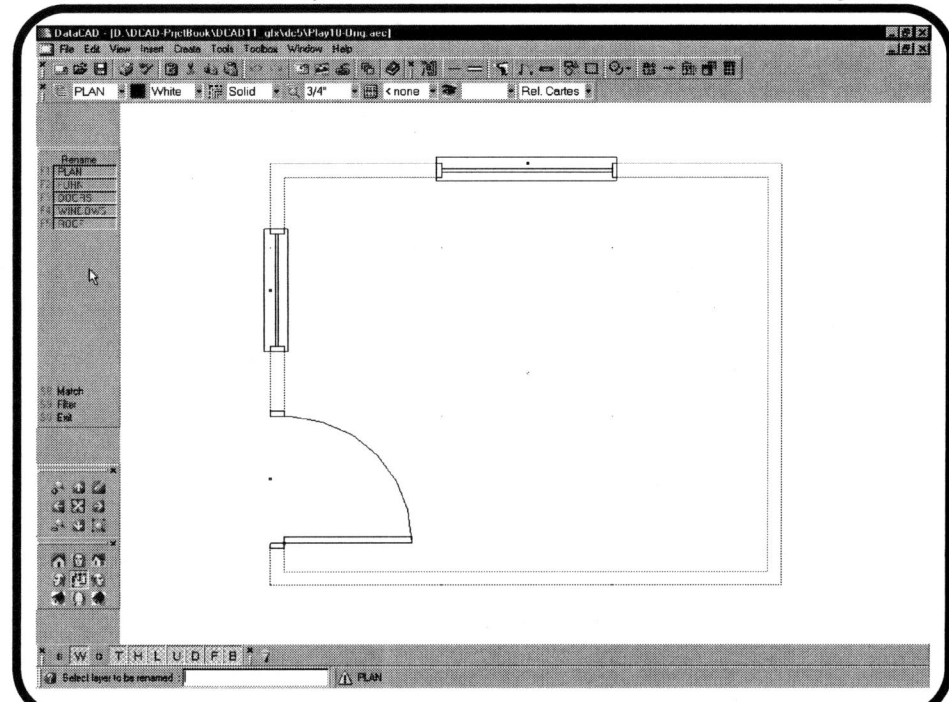

•• NOTE: Your list of layers should now look similar to that shown in Figure 10-1.

Figure 10-1, New drawing layers.

7.__ Press the **Tab** key to select a layer, and then press the **k** [or the **K**] key to change the color so that each layer has a different color.

When you first started using DataCAD and saw how quickly you could add doors and windows to a floor plan, you were probably very impressed. Especially if you have ever made the same kind of drawing with traditional tools. Even though the process of adding doors and windows to the Playhouse is quite impressive, DataCAD can provide much better looking 3D doors and windows. In the next part of this project, you will see how to use DataCAD macros to enhance the Playhouse design. First, the old door and windows will be removed.

8.__ Press the **Tab** key to make the **PLAN** layer active.

9.__ Press the **e** [or the **E**] key to jump to the **Erase** menu.

The DataCAD 11 Project Book

10. __ Make sure that the **Group** option is toggled **on** in the **Erase** menu. Then erase the door and two windows by clicking on the **Group** marker (see Figure 10-2) of each door and window.

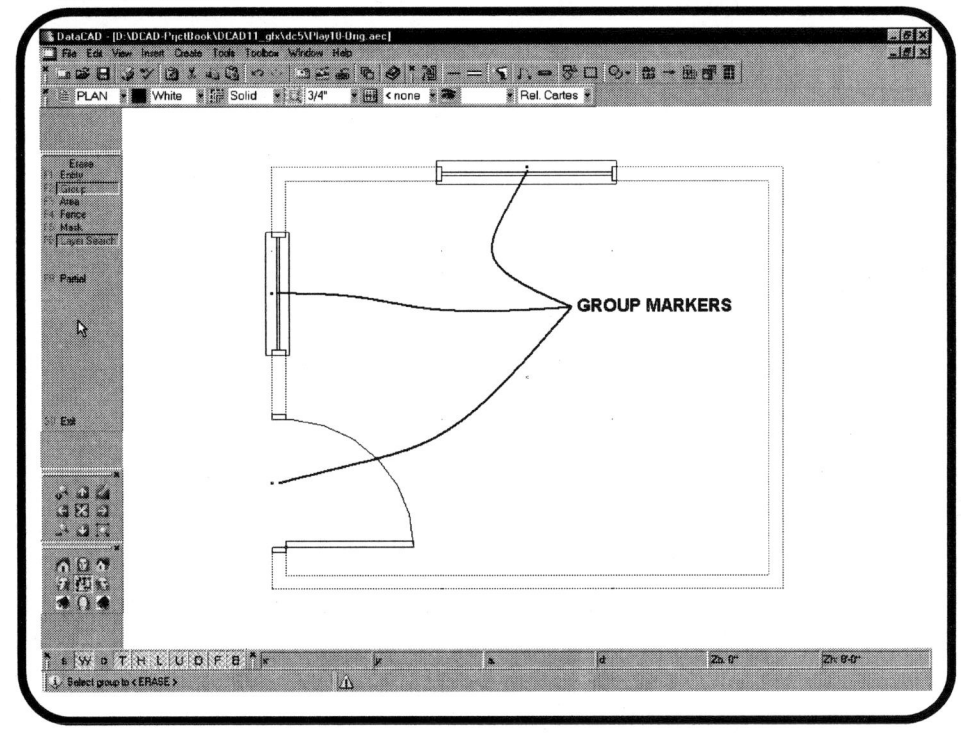

Figure 10-2, Door and window group markers.

11. __ When the door and windows are erased, press the **Esc** key to redraw the display.

Using the AEC_MODL Macro

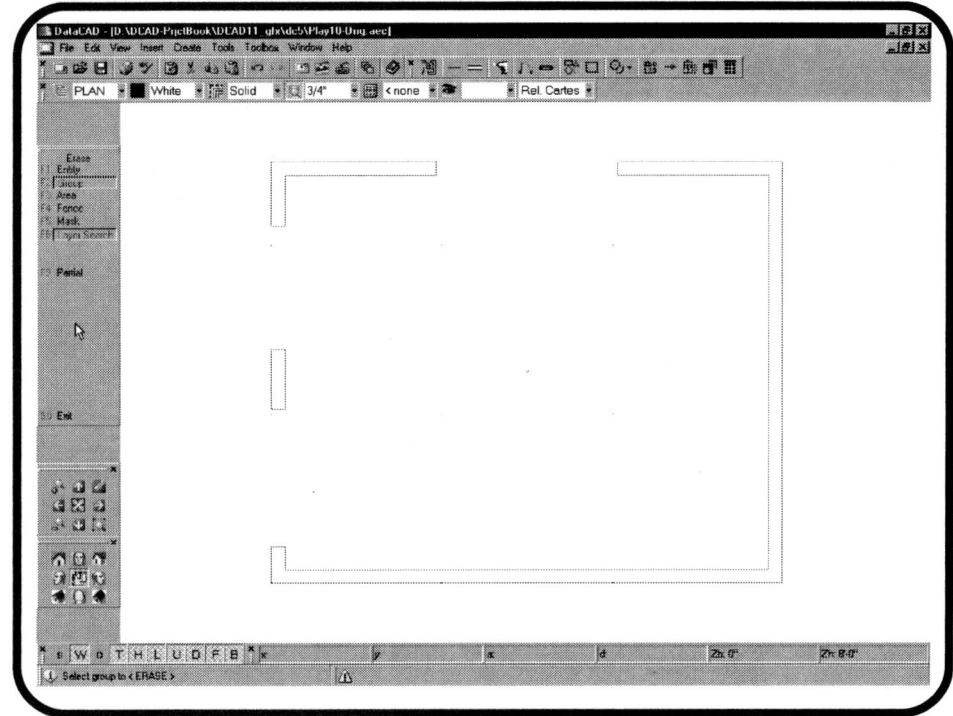

The display should look similar to Figure 10-3.

Figure 10-3, The Playhouse with openings.

Using the AEC_MODL Macro

Next, a macro will be used to install a new improved door. A macro is like a computer program someone has written to make DataCAD execute a series of drawing commands. A variety of macros are included with DataCAD that increase its power. One of the macros allows you to create different types of doors, including details such as the door knob and framing trim.

12.__ Press the **Tab** key to make the **DOORS** layer active.

13.__ Press the **s** (not the S) key and enter a snap size of 2 inches (**0.2**) for both X and Y.

14.__ Select **Toolbox** from the **Edit** menu. **Open** the **AEC_MODL** macro file, then select **Doors**.

Chapter 10, Page 183

The DataCAD 11 Project Book

Figure 10-4 shows the **DoorPlan** menu. There are a lot of options for controlling various door details. For this design, the default settings will be used. You do, however, have to select the type of door to be used.

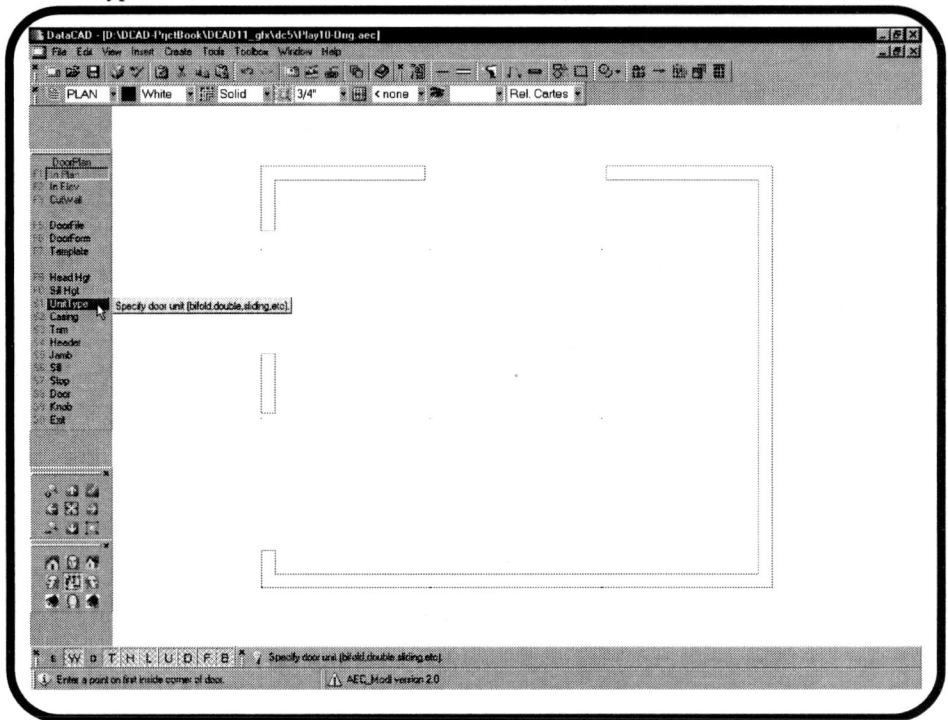

Figure 10-4, The DoorPlan menu.

15.___ Select **UnitType** from the **Doors** menu. Toggle the **Single** option **on**.

The doors macro allow you to specify what percentage the door will be open when it is drawn.

16.___ Select **% open** from the **DoorUnit** menu and enter **0**. Then press the *right* mouse button *once* to return to the **DoorPlan** menu.

17.___ Select **Door**, then select **HingRht**. Then press the *right* mouse button *once* to return to the **DoorPlan** menu.

When you installed doors and windows using the **Architect** menu options, you could click anywhere outside of the house to define the outside of the wall. The door and window macros require that you carefully locate both the inside and outside walls so that the details (like the jam and sill) will be the correct size.

Using the AEC_MODL Macro

18. ___ In the **DoorPlan** menu, toggle the **InPlan** option **on**.

19. ___ At the prompt *Enter a point on first inside corner of the door.* click with the *object snap button* at the lower inside corner of the door opening. Click the *object snap button* at the upper inside corner of the opening for the second point.

20. ___ At the prompt *Enter a point on outside surface of the wall.* click the *object snap button* on the outside corner of the door opening.

The macro will create a door with a lot of 3D detail for you. To appreciate how much the system has done, you need to look at a 3D view.

21. ___ From the **MENU BAR**, select **View, Isometric**.

Your display should look similar to Figure 10-5.

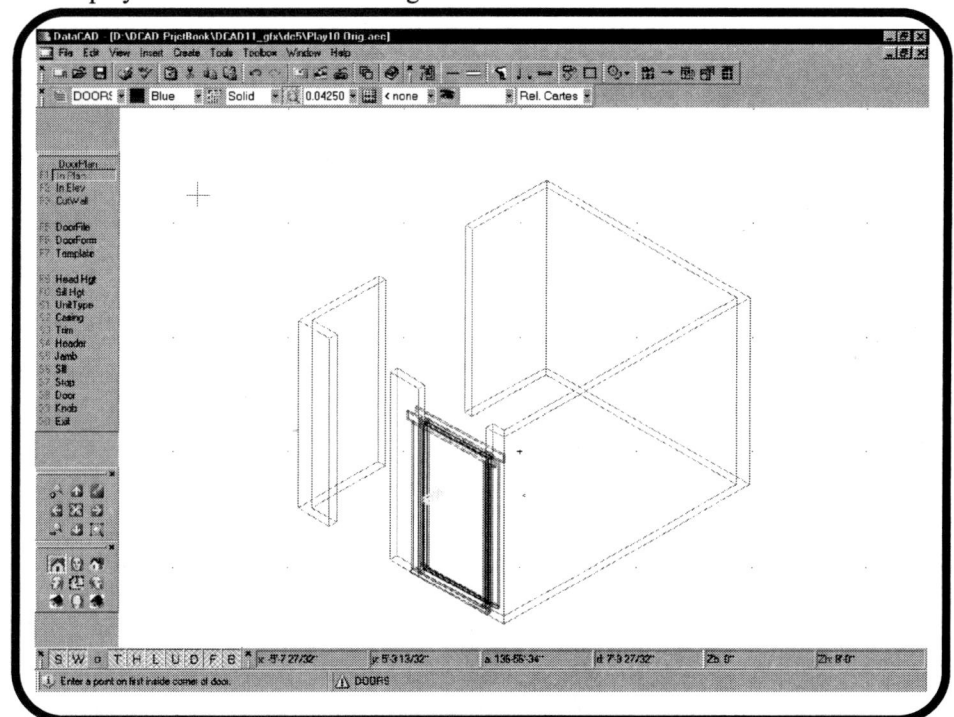

Figure 10-5, A 3D door.

This is so much fun, let's do a couple of windows.

22. ___ From the **MENU BAR**, select **View, Plan**.

Chapter 10, Page 185

The DataCAD 11 Project Book

23.__ Press the **Tab** key to make the **WINDOWS** layer active.

24.__ Press the **s** (not the S) key and enter a snap size of 2 inches (**0.2**) for both X and Y.

25.__ Press the **M** [not the m] key to jump to the **Toolbox** (just remember "M" for "macros"), and then select the **AEC_MODL** option.

26.__ Select **Windows** from the **AEC_Modl** menu.

27.__ In the **AEC_Modl** menu, toggle **In Plan on**.

28.__ Select **UnitType** from the **WindPlan** menu. Then toggle the **DblHung** option **on**. Press the *right* mouse button *once* to return to the **WindPlan** menu.

This time some of the options will be adjusted. Muntins are the horizontal and vertical dividers that are used to add trim to a window. (They actually simulate the small panes of glass that were used in the old-fashioned days.) The windows macro allows you to control how muntins are drawn.

29.__ Select **Muntins** from the **WindPlan** menu. Toggle **DoMuntn ON**. Then select **PaneHorz** and enter **3**, and **PaneVert** and enter **2**.

30.__ Select **Glass** from the **WindPlan** menu and toggle **DoGlass off**.

If the **DoGlass** option is toggled **on**, you will not be able to see through the windows when the hidden line removal function is used.

31.__ Follow the prompts and add windows to the openings in the playhouse walls. (Be careful to use the *object snap button* get the width correct.)

32.__ Select **View, Isometric** to see how your windows look in 3D.

INFO STOP

The door and window macros keep all of the entities in a door or window in a **Group**. If you want to experiment with the different types of windows, you can erase a window using the **Group** option.

Also note that the layer color does not control the colors of the different door and window components. The color of each component can be adjusted with a macro option. For example, if you wanted a different color sill, select the **Sill** option and then select the **Color** option from the **Sill** menu.

Chapter 10, Page 186

Fixing Wall Cutouts

When the door and windows were added to the original plan, the **Cutout** option was toggled **on**. This makes the plan view look good, but it is also the reason that the area above the door and windows is missing in the pictorial views. To fix this, walls will be added below the windows and above the windows and door. This is done by creating short wall pieces which have the proper **Z-Base** and **Z-Height**. The Z settings can be determined by checking the **Head Hgt** and **Sill Hgt** macro options.

33.__ From the **MENU BAR**, select **View, Plan**.

34.__ Select the **Head Hgt** option in the **WindPlan** menu and note the default setting on the prompt line (it will be 6.8 if it has not been changed).

35.__ Select the **Sill Hgt** option in the **WindPlan** menu and note the default setting on the prompt line (it will be 3.4 if it has not been changed).

36.__ Press the **a** [not the A] key to jump to the **Architect** menu.

37.__ Make sure that **Walls, 2 Line Walls,** and **Outside** are toggled **on**. Check to see that the **Width** is set to the Playhouse size of **4** inches.

38.__ Press the **Tab** key to make the **PLAN** layer active.

39.__ Press the **z** [or the Z] key and set the **Z-Base** to **0** and **Z-Height** to **3.4**.

40.__ Press the **l** [or the **L**] key to jump to the **Layers** menu and turn the **DOORS** and **WINDOWS** layers **off**.

41.__ Press the **Esc** key to redraw the display.

42.__ Switch to the **Plan** view (use the *object snap button* to lock on to the wall openings) and add short walls to fill the *window* openings. Check a 3D view to verify that the openings have been properly filled.

43.__ Press the **z** [or the Z] key and set the **Z-Base** to **6.8** and **Z-Height** to **8.0**.

The DataCAD 11 Project Book

44.__ Add short walls (use the *object snap button* to lock on to the wall openings) to fill the openings above the door and windows. Check a 3D view to verify that the openings have been properly filled (See Figure 10-6).

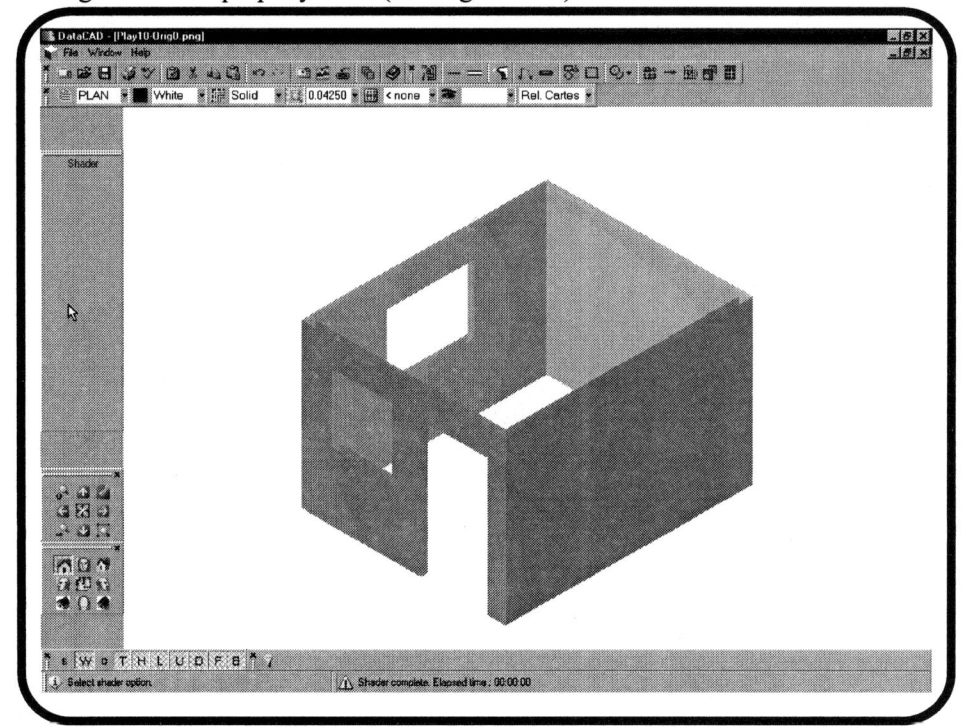

Figure 10-6, Adding filler walls.

45.__ Press the **z** [or the Z] key and set the **Z-Base** to **0** and **Z-Height** to **8.0**.

> **INFO STOP**
> It is a good habit to change the **Z-Base** and **Z-Height** back to normal as soon as you are done using any unusual settings.

46.__ Press the **l** [or the **L**] key to jump to the **Layers** menu and turn the **DOORS, WINDOWS**, and **ROOF** layers **on**.

47.__ Check a 3D view to see how things look.

••• **WARNING**••• Have you saved your drawing lately?

Adding a Roof

You might even want to do a rendered view with **Hide** or **Shader** to see how the wall openings have been filled in.

Adding a Roof

Next, a roof will be added to the playhouse. As a matter of fact, you can experiment with several different roof styles for the Playhouse. One way to construct roofs is to use inclined slabs (which are found in the **3D Entity** menu). An easier way is to use the **ROOFIT** macro.

The **ROOFIT** macro is similar to the macro used for doors and windows in that it automates the process of creating the different 3D entities required. It also provides you with a great deal of control over the various roof details.

48.___ Press the **Tab** key until the **ROOF** layer is active.

49.___ Select **ROOFIT** from the **Toolbox** selection.

The DataCAD 11 Project Book

The **RoofIt** menu is shown in Figure 10-7. The first option on the **RoofIt** menu allows you to toggle between rectangular and polygon shaped roofs. In this example, the **RectRoof** option should be selected. Note the different roof styles available. The **Settings** option provides control of the different roof details.

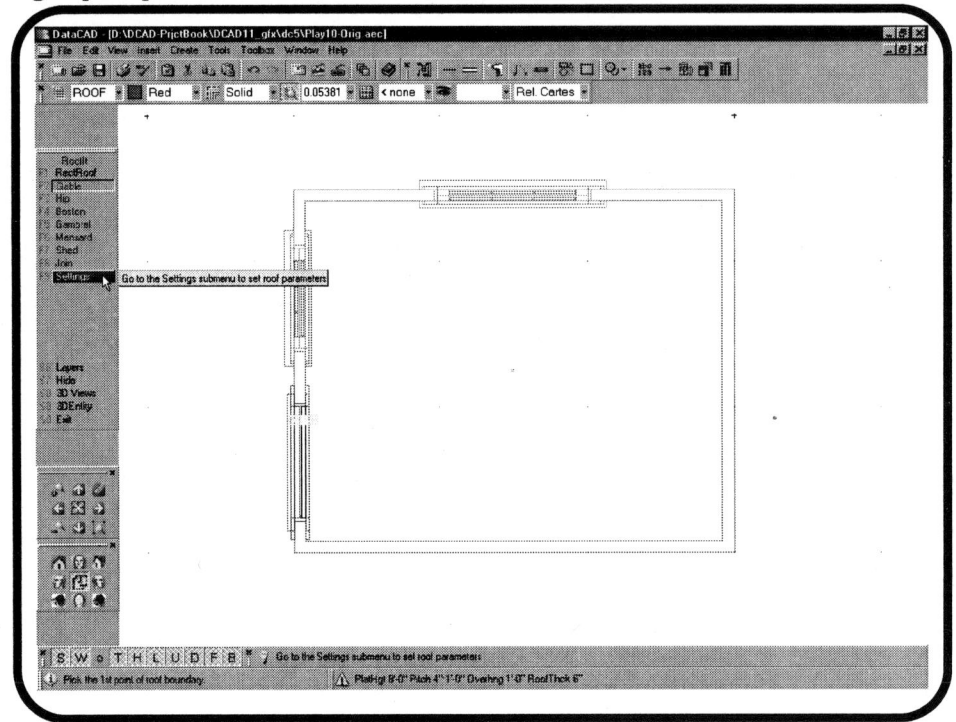

Figure 10-7, The RoofIt menu.

50. ___ Toggle the **Gable** option **on** in the **RoofIt** menu. Then select the **Settings** option.
51. ___ Select **PlatHgt** and enter **8.0** for the plate height.
52. ___ Select **Pitch** and then select a pitch of **4:12**.
53. ___ Select **RoofThck** and enter a roof thickness of **0.6**.
54. ___ Select **Overhng** and enter an overhang of **1.0**.
55. ___ Select **WallThck** and enter a wall thickness of **0.4**.
56. ___ Select **EndWall** (if it toggles off, select it again). Select the color that matches your wall color (this should be different from your roof color).

Adding a Roof

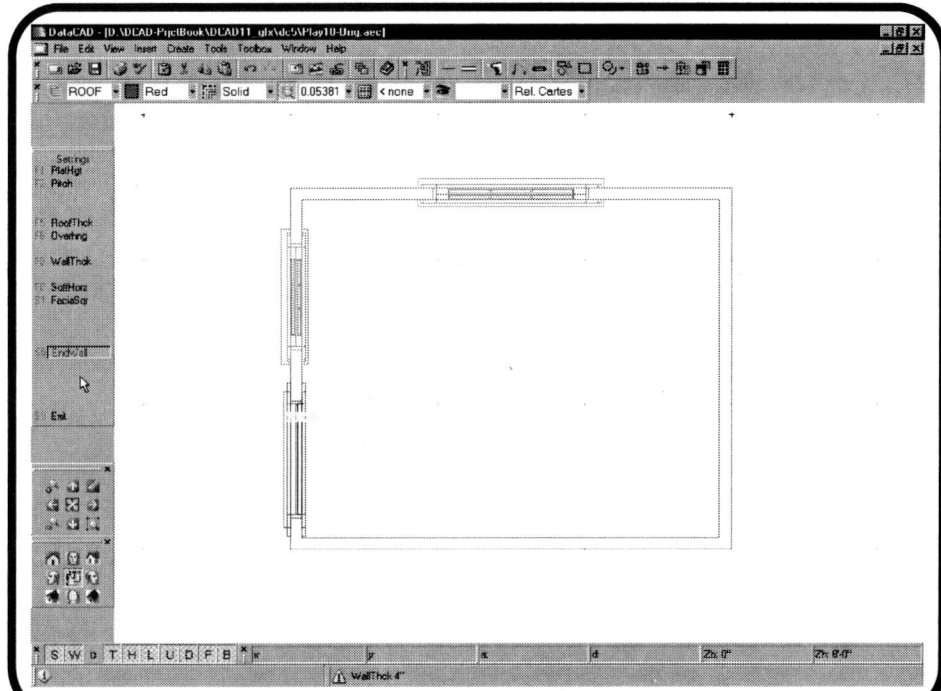

57.___ Set **SoffHorz, FaciaSqr,** and toggle **EndWall on** as shown in Figure 10-8.

Figure 10-8, Setting Roofit parameters.

58.___ Press the *right* mouse button *once* to return to the **Roofit** menu.

59.___ Use the *object snap button* to pick the *upper left* corner of the Playhouse as the *first* point of the roof boundary, and the *upper right* corner as the *second* point. Pick the *lower right* corner for the *second* point of the roof width.

Chapter 10, Page 191

The DataCAD 11 Project Book

After the roof calculations have been completed, take a look at the Playhouse in a pictorial view to see the result. Figure 10-9 shows an example pictorial view.

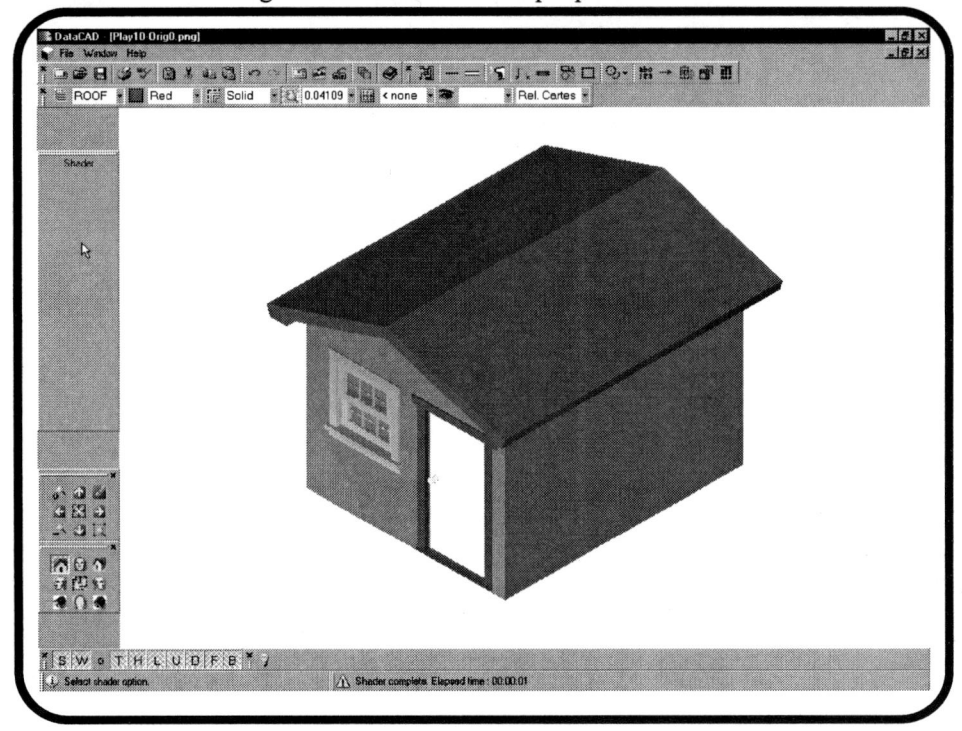

Figure 10-9, A gabled roof.

The **ROOFIT** macro puts all roof entities in a **Group**. You can use the **Group** erase function to erase a roof, and then try another with a different style, pitch, or other variations.

Adding a Roof

Once you have completed your Playhouse model, you will want to create different pictorial views. Use the **Hide** function to create line drawings of a pictorial view and save the image to a new layer. Then turn off all layers except the rendered image (remember that the rendered image is in 2D and must be displayed in the **Plan** view). Use the **Erase** function to remove unwanted lines. You can print the image and use felt markers or other media to generate professional looking results.

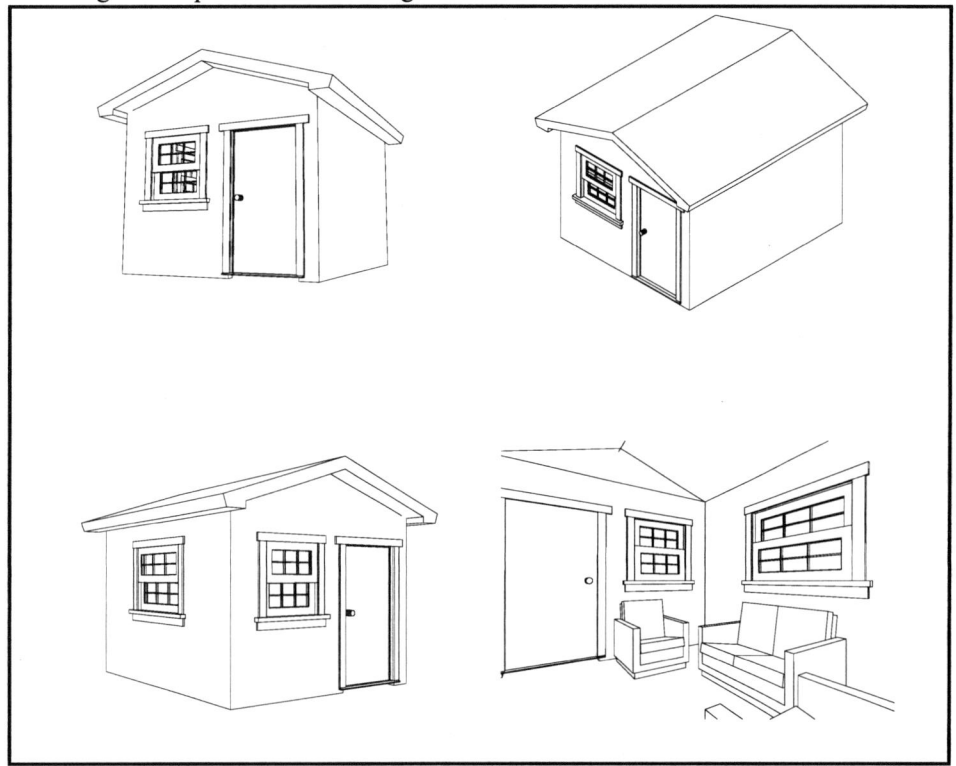

Figure 10-10, Sample rendered images.

This finishes your introduction to architectural CAD basics.

Chapter 10, Page 193

The DataCAD 11 Project Book

If you found this fun and interesting, and if you find that you like working with the computer and don't mind having to tell the computer all of the details of a design, you should try to take more classes in computer graphics. Architectural CAD is just one kind of computer graphics. Computer graphics is also used for mechanical design, animation, virtual reality, art, business graphics, and advertising. Computer graphics is an expanding field and more and more people in many professions are finding applications of computer graphics in their business. Of course, many people like to work with computer graphics just for the fun of it.

Adding a Roof

Review and Important Points

- To create a new drawing from an existing drawing, use **File, Save As**.

- To create good looking 3D doors and windows use the **Toolbox, AEC_MODL, Doors** and **Windows** macros.

- To change how much a door or window will be open, use the **% open** option found by clicking on **UnitType**.

- The **AEC_MODL, Doors** and **Windows** macros require that the exact wall width be specified (use the *object snap button*) so that the system can create the trim to fit the desired wall width.

- To fill in the gaps in the walls above the doors and windows and below the windows, set the **Z-Base** and **Z-Height** to the required sizes and create short filler walls.

- The **RoofIt** macro provides a variety of different roof types to pick from.

The DataCAD 11 Project Book

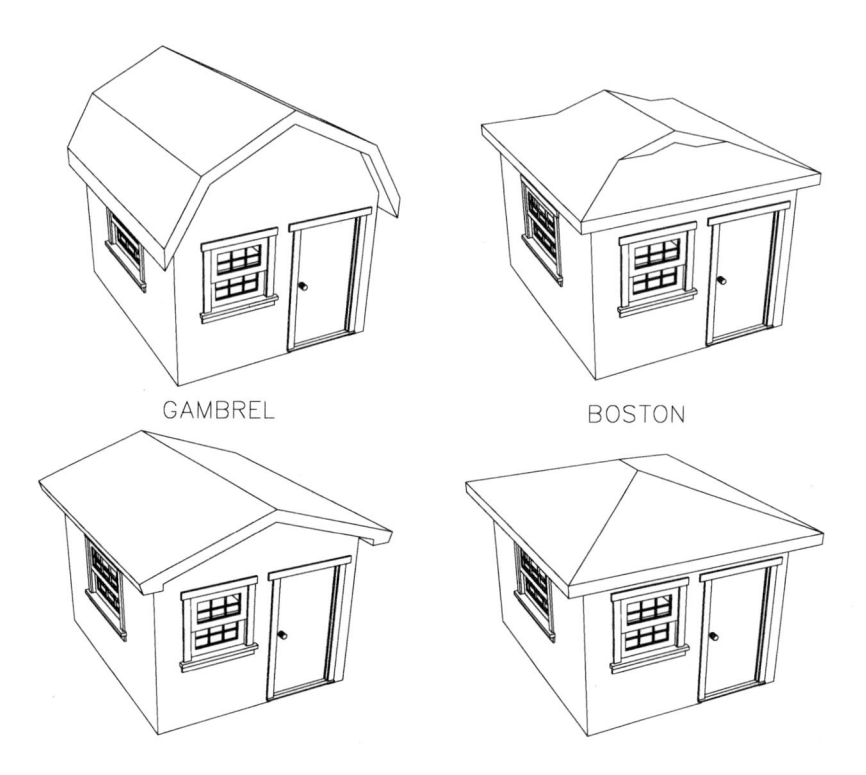

Practice Drawings

GAMBREL

BOSTON

Project Hints
Assignment 10-A, Roofs

- Create a drawing similar to that shown above to illustrate different roof styles.
- Keep each roof style on a different layer.
- Create hidden renderings for each roof and save the renderings on different layers.
- Use the **Enlarge** and **Move** functions to adjust the relative size and position of each example.
- Since the hidden renderings are 2D images, you can erase or add lines to the rendered views to make them look better.
- You will have to adjust the pitch and mid ridge height for the gambrel roof.

CHAPTER 11

Using the Framing Macro

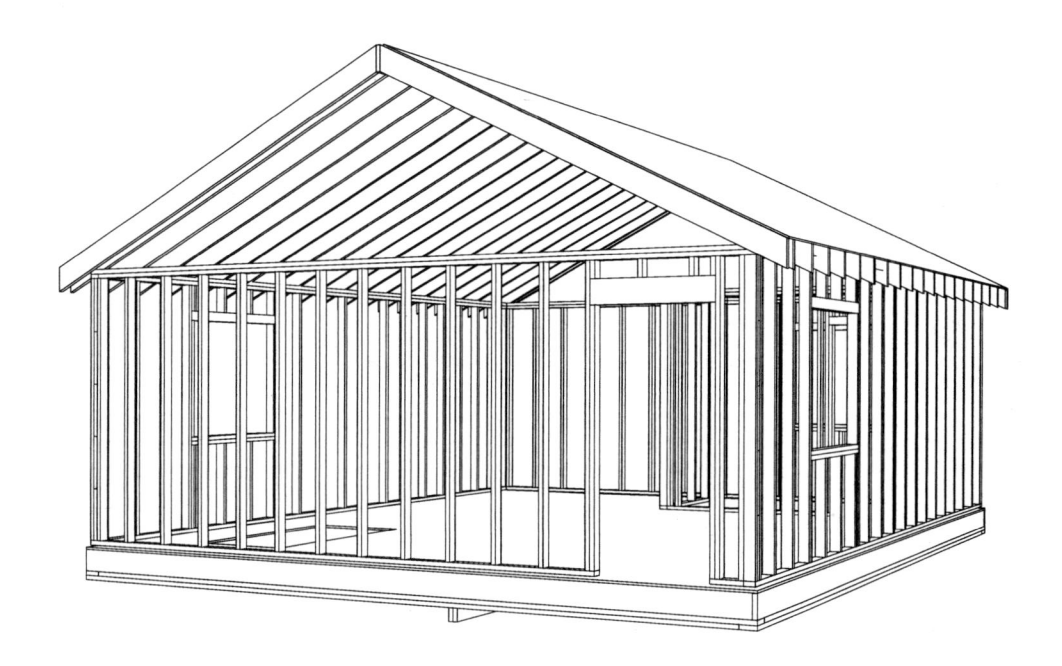

Project Goals

When you have completed this project, you will be able to:
1. Create exterior and interior walls using the **Framing** macro.
2. Frame door and window openings in walls.
3. Create floor framing using the **Framing** macro.
4. Frame floor openings.
5. Add beams to floor framing.
6. Create roof framing using the **Framing** macro.
7. Adjust roof framing parameters.

The DataCAD 11 Project Book

Prerequisite: DataCAD 11 interface experience

Introduction

The **Framing** macro was added in DataCAD release 6. It provides an easy means for creating 3D views of wall, floor, and roof framing, and provides for extracting information about the amount of framing material required. This project will start with a very simple rectangular building, and then move on to more complex framing designs as you gain experience with this powerful macro.

1. __ Start DataCAD with a new drawing file named **Frame**.
2. __ Select **Toolbox** from the **Edit** menu and **Open** the **FRAMEIT** macro.

Creating Wall Framing

There are three options in the **Frame-It** menu: **Floors, Walls,** and **Roofs**.

3. __ Select **Walls** from the **Frame-It** menu.

Creating Wall Framing

The **Walls** menu is shown in Figure 11-1.

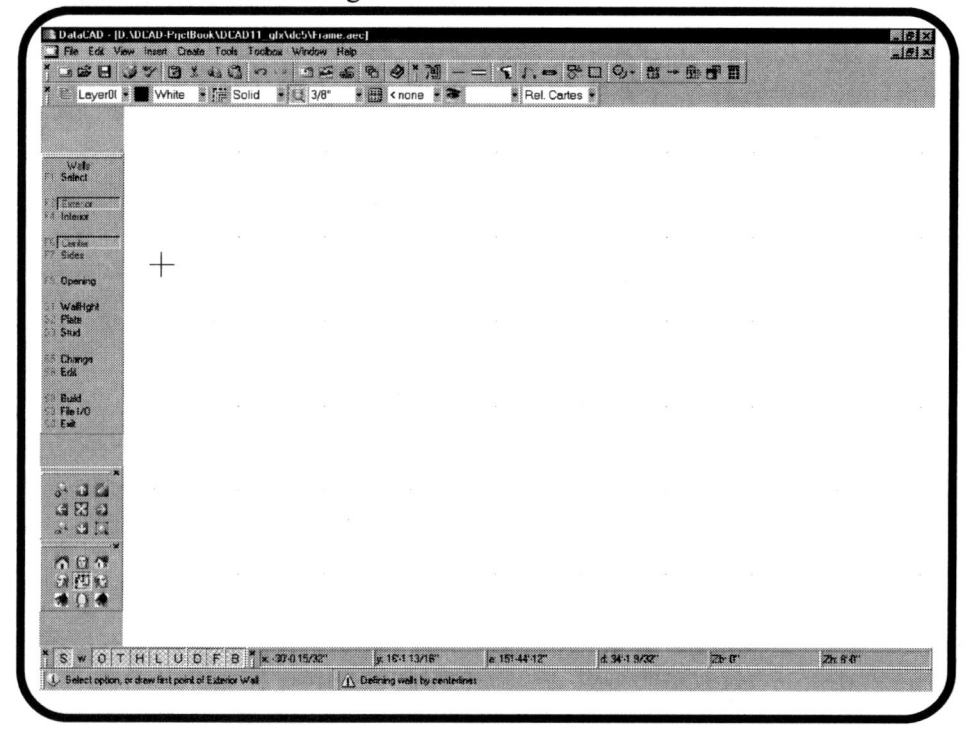

Figure 11-1, The Frame-It, Walls menu.

> The **Select** option on the **Walls** menu can be used to add walls to an *existing floor framing plan*. No, the **Framing** macro will *not* automatically select walls created using the **Archietct** menu. You can either trace your original plan (on a new layer) or you can create a new schematic outline for the walls to be framed. In this example, a new polyline will be created to define the outside walls without first creating a floor plan.

4. __ Toggle the **Exterior** and **Center** options **on**.
5. __ At the prompt *Select option, or draw first point of Exterior Wall,* click near the lower left corner of the drawing window to anchor the first corner of the wall.

Chapter 11, Page 199

The DataCAD 11 Project Book

You will be tracing around a 12' by 10' playhouse by selecting points on a polyline. (A polyline is a group of connected line segments.) This will be easier if coordinate entry is used.

6.__ Press the **Insert** key until *Relative Cartesian* coordinate entry mode is selected. Then press the **space bar**.

7.__ Enter an **X** distance of **12.0**, and a **Y** distance of **0.0**.

This opens the polyline menu. Polylines can either be closed or open. When the **Closed** option is selected, a line segment will be added from the starting point to the last segment drawn to create a closed polygon.

8.__ Toggle the **Closed** option **on** in the menu.

9.__ Press the **space bar** and enter values **X=0.0, Y=10.0**.

10.__ Press the **space bar** and (watch the minus sign) enter values **X=-12.0, Y=0.0**.

Your display should look similar to Figure 11-2.

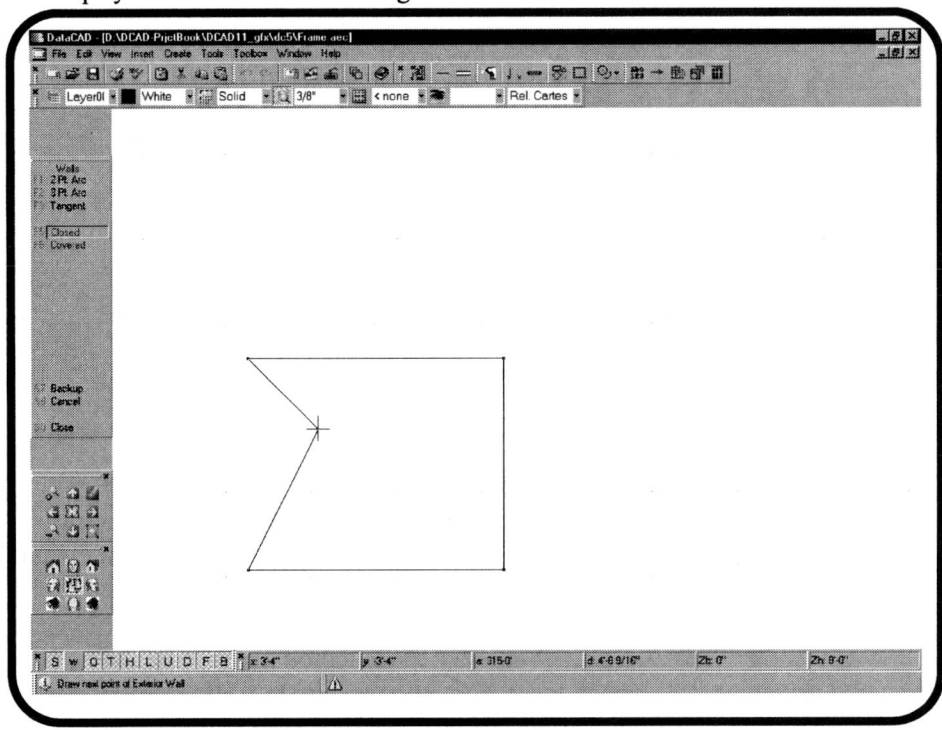

Figure 11-2, A framing schematic.

Chapter 11, Page 200

Creating Wall Framing

Three sides of a rectangular polygon have been created. Since the **Closed** option is **on**, the rectangle can be completed by pressing the *right* mouse button.

11.__ Press the *right* mouse button *once* to complete the wall boundary.

Now that a wall boundary schematic is in place, the system can build the framing. For now, the default settings will be used.

12.__ Select **Build** from the **Walls** menu. Toggle the **Group** option **on** and then click on a segment in the rectangular boundary.

The system will switch to a 3D view and build the wall framing. When it is done, your display should look similar to Figure 11-3.

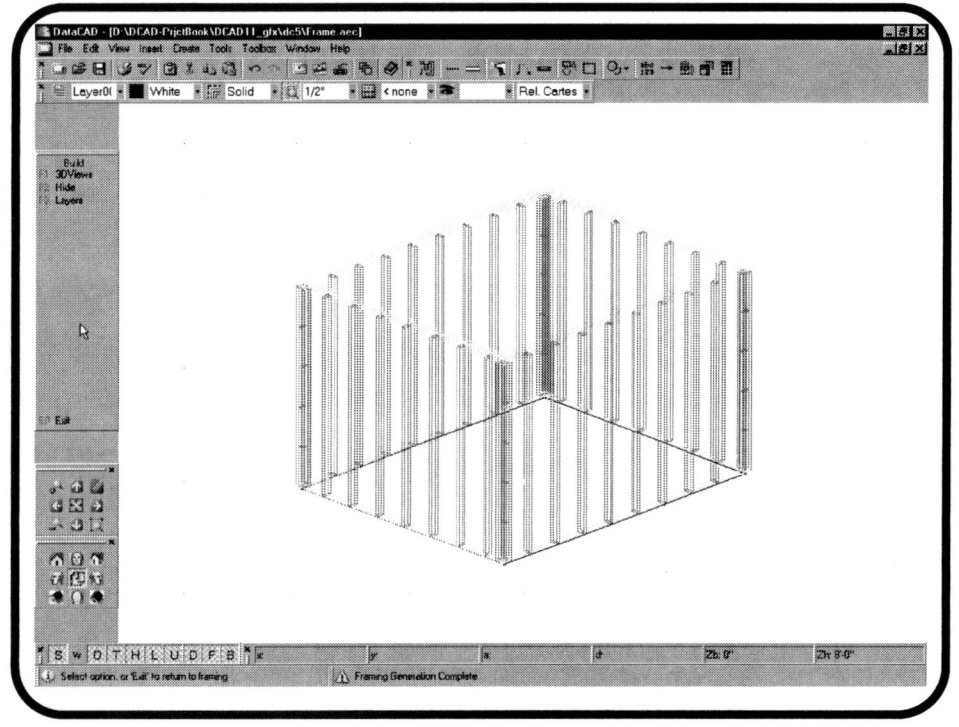

Figure 11-3, Framed walls.

Chapter 11, Page 201

The DataCAD 11 Project Book

A new menu appears that allows you to jump to the **3DViews, Hide**, or **Layers** menus. Access to these menus makes it easier to check the framing against your design. Of course, in this example there are no other layers or design geometry to work with, but when you are ready to add framing to your own designs, these options will be very helpful.

Adding Openings

Next, you will see how to add door and window openings to your framing plan. Each time you select **Build**, the old framing is removed and replaced, so it is not really necessary to delete the current frame walls. The framing will be erased in this example to make it easier to see where the doors and windows will be placed.

13. __ Press the *right* mouse button *twice* to return to the **Frame-It, Walls** menu. Press [CTRL] + [Z] to to delete the four framed walls. Press the **Esc** key to refresh the display.

14. __ From the **Frame-It, Walls** menu, select **Opening, Door, HeadStyl, Boxed**, and press the *right* mouse button *once*.

Adding Openings

15. __ Add a **3'0"** door opening **1'0"** above the lower left corner of the plan as shown in Figure 11-4.

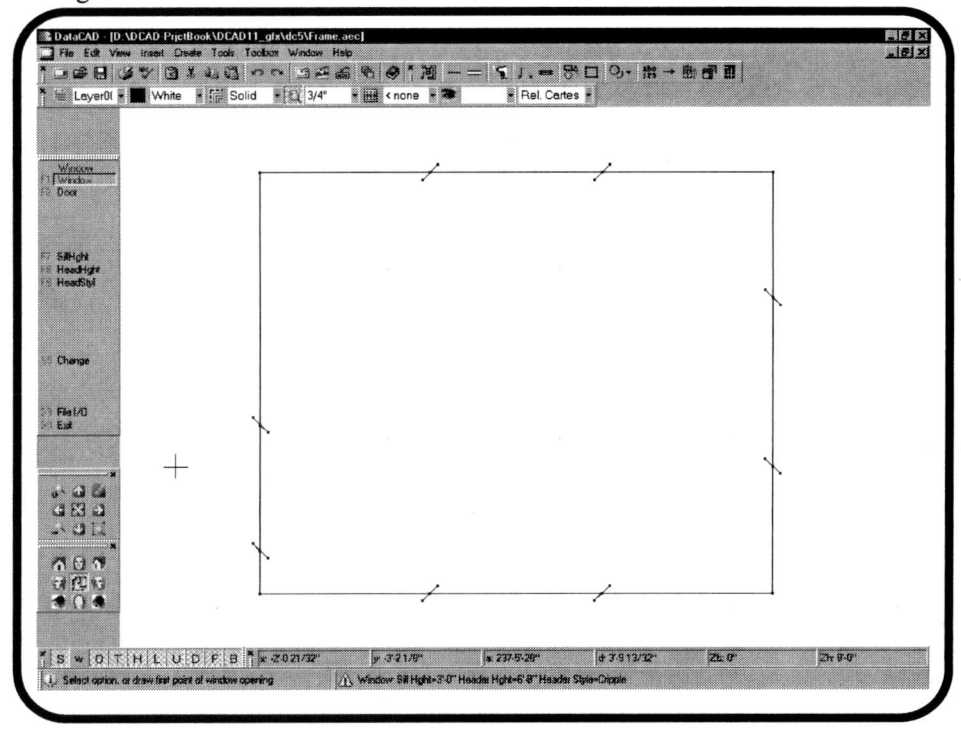

Figure 11-4, Adding openings.

16. __ From the **Frame-It, Walls, Opening** menu, select **Window, HeadStyl, Cripple**, and press the *right* mouse button *once*. Add three 4' wide windows to the center of three walls as shown in Figure 11-4.

17. __ Press the *right* mouse button *once* to return to the **Frame-It, Walls** menu.

18. __ Select **Build** and then *click* on a line segment in the plan.

The DataCAD 11 Project Book

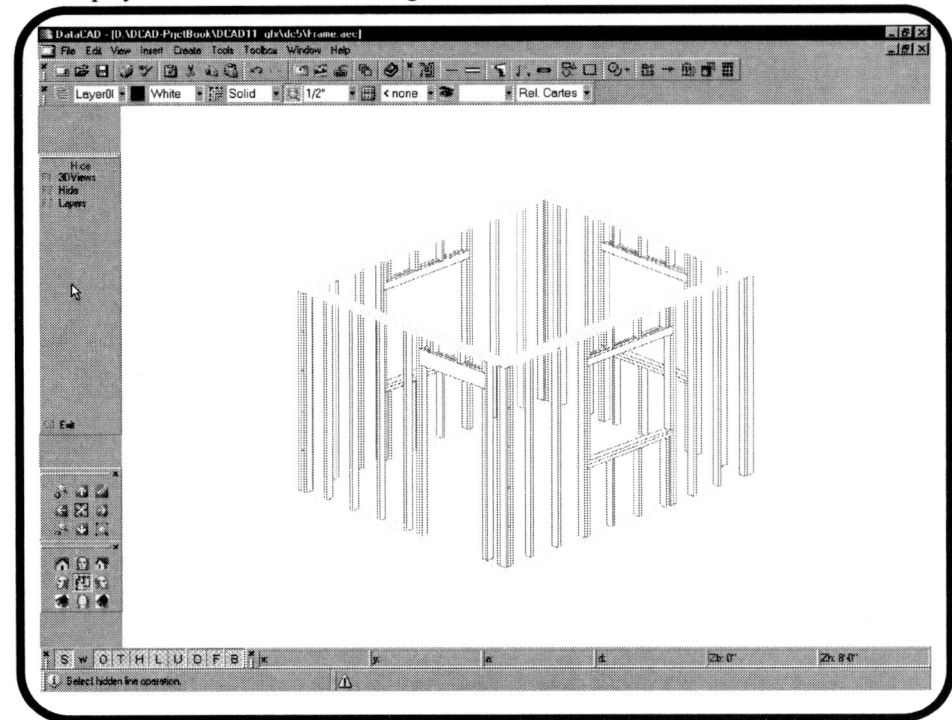

Your display should look similar to Figure 11-5.

Figure 11-5, Framed openings.

Actually, Figure 11-5 was captured after the **Hide** menu was used to create a hidden rendering of the framed walls.

Inspect the area above the doors and windows. There are two header styles available in the **Door** and **Window Opening** menus. Figure 11-5 was created by using the **Boxed** style for the door and the **Cripple** style for the windows.

It is also possible to adjust the parameters for the top and bottom plates and studs. Before exploring any further, it is time to save a copy of this framing plan.

19.___ Select **File, Save** to update the **Frame** file.

The above steps provided you with a quick look at how the **Framing** macro works. Now it is time to dig a little deeper into the various options available.

Adding Openings

Next, you will modify the design to include interior walls. To make things a little easier to manage, a new layer will be added and used for the framing. The schematic outline for the framing will be kept on a separate layer.

20.__ Press CTRL +Z to delete the four framed walls. Press the **Esc** key to refresh the display.

21.__ Press the **L** key [or the **l** key] to jump to the **Layers** menu.

22.__ Select **New Layer**, and add **1** layer.

23.__ Select **Name** and name Layer1 **WSCHEME**, and name Layer2 **WALLS**.

24.__ Press the **TAB** key to make the **WSCHEME** layer active.

25.__ Select **File, Save As**, and enter **Frame11** as the new file name.

26.__ Enter the **Framing** macro and select **Walls**.

Your display should look similar to Figure 11-6.

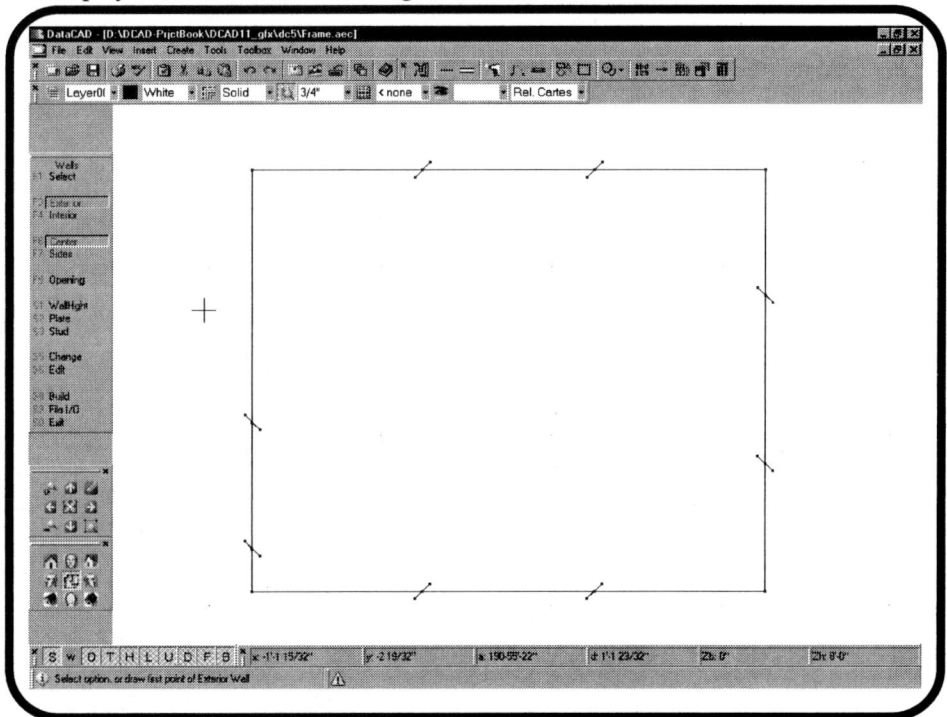

Figure 11-6, The Frame-It, Walls menu.

Chapter 11, Page 205

The DataCAD 11 Project Book

The **Framing** macro builds the frame based on a schematic outline that defines either a side or center line of the walls. As you have seen, door and window openings can be added to this schematic.

Editing a Frame

One of the options on the **Frame-It** menu is **Edit**. This option provides access to functions similar to those found in the regular DataCAD **Edit** menu. Next, you will expand the wall plan by using the **Stretch** option.

27.__ Select **Edit** from the **Frame-It, Walls** menu; then select **Stretch**. Position the cursor somewhere on the plan and click the *left* mouse button *once*.

28.__ For the second point of the distance to stretch, press the **space bar** and enter **X=12, Y=0**.

29.__ Select the **Area** option from the menu and capture the *right* vertical wall of the plan in an area box. *Click* on the **Zoom view** button in the center of the navigation pad to make the drawing fit the display.

30.__ Select **New Distanc** from the **Stretch** menu, click the *left* mouse button with the cursor somewhere in the plan, press the **space bar**, and enter **X=0, Y=10**.

31.__ Select the **Area** option from the menu and capture the *upper* horizontal wall of the plan in an area box. *Click* on the **Zoom view** button in the center of the navigation pad to make the drawing fit the display.

> **INFO STOP** You should spend some time reviewing the various **Edit** options. Each opening in the schematic is a **Group**. You can **Move** or **Erase** an opening by selecting it with the **Group** option. You can also use **Stretch** to change the size of an opening. Each time you **Build** the framing, the system tears down the old frame and builds it again with the new schematic.

Now that the perimeter of the plan has been expanded, some interior walls can be added.

32.__ Press the *right* mouse button *twice* to return to the **Frame-It, Walls** menu.

33.__ Toggle **Interior** and **Center on**.

Editing a Frame

34.__ Add a **10** foot long wall **6** feet to the *left* of the lower right corner of the plan and then add a **6** foot wall perpendicular to the **10** foot wall closing a new room as shown in Figure 11-7.

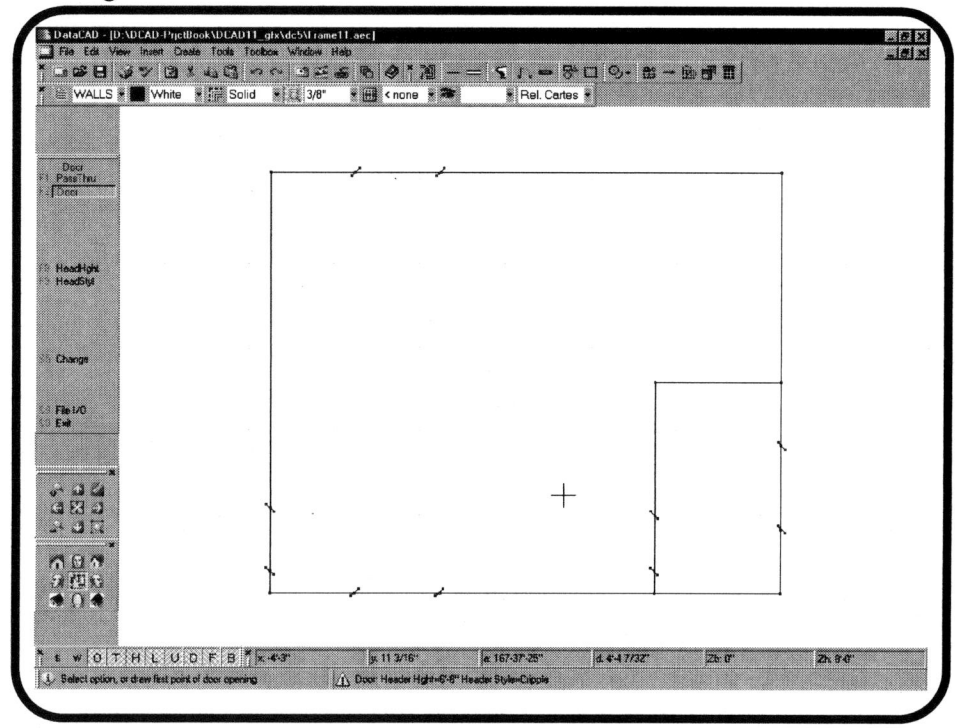

Figure 11-7, Adding interior walls.

35.__ Select **Opening, Door**, and add a **2'8"** door at the location shown in Figure 11-7.

INFO STOP
   When you select **Opening** while the **Interior** wall option is toggled **ON**, the choices are **PassThru** and **Door** rather than **Window** and **Door**.

••• **WARNING**••• Have you saved your drawing lately?

36.__ Press the **TAB** key to make the **WALLS** layer active.
37.__ Select **Build**, and toggle **ActvLyr on**.
38.__ Select **All** from the **Build** menu, then select **Yes**.

Chapter 11, Page 207

The DataCAD 11 Project Book

As soon as the frame is built, three options are available on the menu: **3D Views, Hide,** and **Layers**. These provide access to the standard **3D Views, Hide,** and **Layers** menus. When you use and then exit these menus, you will be returned to **Frame-It**. Figure 11-8 shows the results of using **3D Views** to define a new parallel projection, and then using **Hide** to do a hidden line rendering of the model.

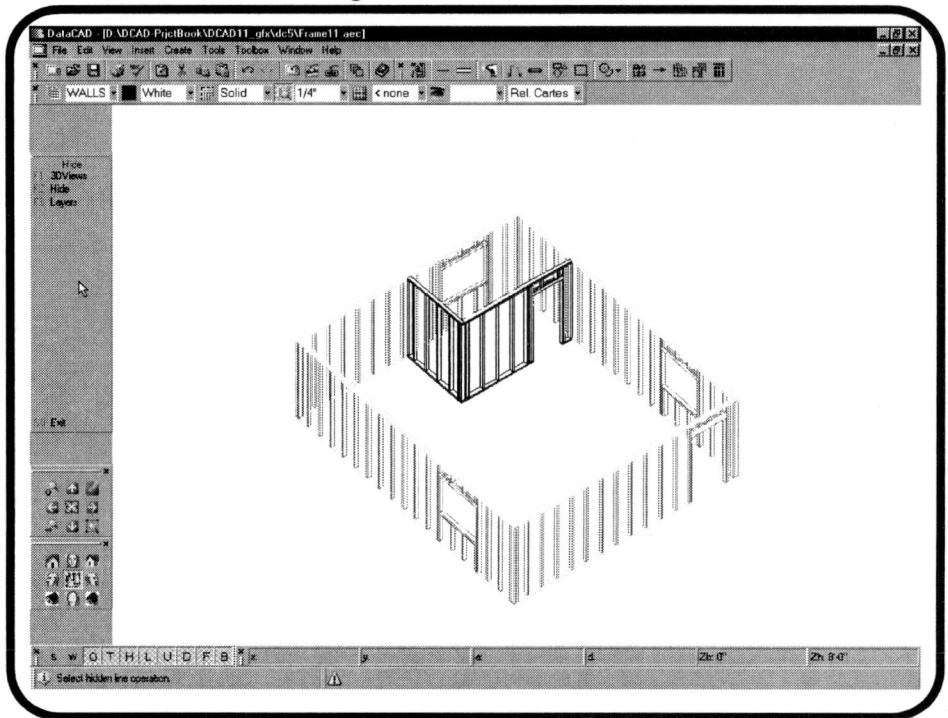

Figure 11-8, A rendered frame.

There are several other wall framing options you should explore. Watch the information line near the bottom of the display as you position the cursor on the various **Walls** menu options.

39.___ Select **Plate** from the **Walls** menu and review the various plate construction options.

Note that you can mess with the top and bottom plates separately.

40.___ Press the *right* mouse button *once* to return to the **Walls** menu; then select the **Stud** option.

Adding Floor Framing to Your Plan.

41. __ Note the different stud options. If you change the size of the stud and watch the information line, you will see that although nominal lumber sizes are given, the studs are actually constructed to dimension lumber sizes. You can also verify this by using the **Identify** button on the **Navigation Pad** (the one that looks like a pointing finger). It will show that each stud is created as a 3D slab using standard lumber sizes.

It might occur to you that as long as the system has to calculate how and where to put all this framing lumber, it could create a data file containing useful information.

42. __ Press the *right* mouse button until the **Frame-It** menu is displayed; then select **Extract, Yes**.

This option generates a framing member takeoff table that contains information about the lumber required for the frame. This table can be sent to the printer or to a text file by selecting the **ToPrint** option. If you have purchased the *DataCAD Estimator*, the data can be sent to a special file for use with the estimator.

43. __ Select **File, Save** and update your **Frame11** drawing file.

Adding Floor Framing to Your Plan.

1. __ If it is not on the display, **Open** the **Frame11** file created in the last section.
2. __ Select **File, Save As** and enter **Frame11F** as the file name.
3. __ Press the **L** [or the **l** key] to jump to the **Layers** menu. Select **New Layer**, and add **2** layers.
4. __ Select **Name** and name the new layers **FLOOR** and **FLSCHEME**.
5. __ Use the **Layers, On/Off** option to turn the **WALLS** layer **off**.
6. __ Make the **FLSCHEME** layer active, and press the **K** key until it is a color that you like.

To make the top of the floor even with the bottom plate of the wall, the **Z-Base** for the floor schematic will be adjusted.

7. __ Press the **z** [or the **Z**] key, and enter a new **Z-Base** (watch the minus sign) of **-1.0**. Do not worry about the **Z-Height**; just press enter to accept the default setting.

The DataCAD 11 Project Book

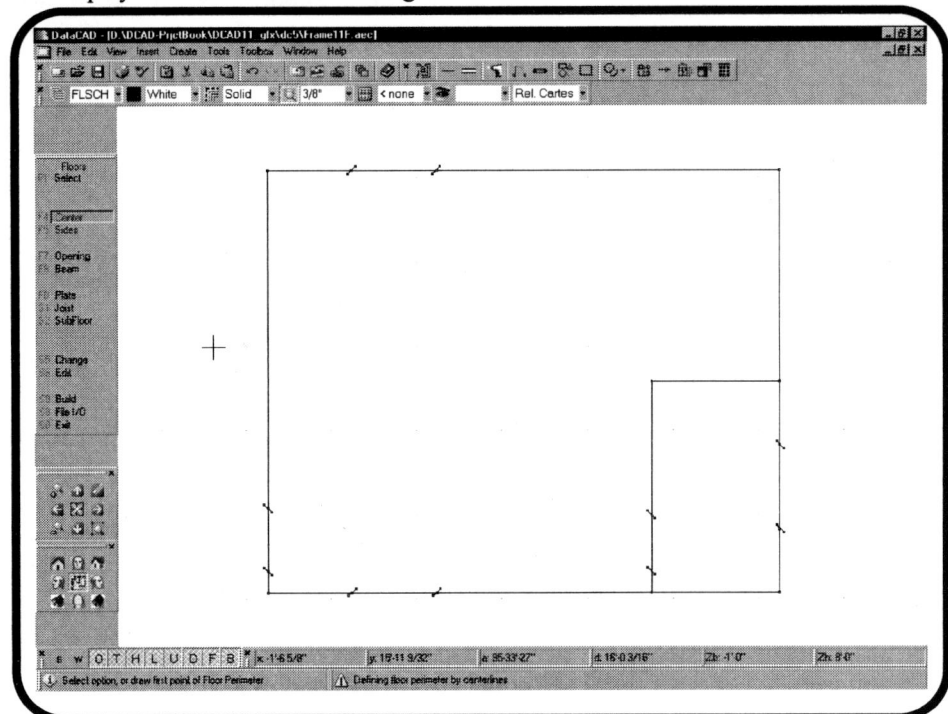

Your display should look similar to Figure 11-9.

Figure 11-9, The Frame-It Floors menu.

Figure 11-9 shows the schematic used for the wall framing in the last project. This same schematic will be traced to create a new schematic for the floor framing. Figure 11-9 also shows the **Frame-It, Floors** menu.

8.__ From the **Edit** menu, select **Toolbox, Frame-it, Floors**.

Next, you will create a floor framing schematic and build a basic floor frame. Then, you will explore adding beams and openings to the floor frame.

9.__ Draw a floor schematic by tracing the wall plan. When you have traced all *four* corners, press the *right* mouse button *once*.

10.__ Jump to the **Layers** menu and turn all layers off *except* **FLSCHEME** and **FLOOR**.

11.__ Press the **Esc** key to redraw the display.

12.__ From the **Frame-It, Floors** menu, select **Plate**. Make sure that **DoPlate** and **Double** are toggled **on**. Select **Size** and set the size for the sill plate to **2 X 6**.

Adding Floor Framing to Your Plan.

13.___ **Exit** back to the **Frame-it, Floors** menu and select **Joist**. Toggle **DoJoist on**, set the size to **2 X 8**, and the spacing to **1' 4"** (16 inches).

14.___ **Exit** back to the **Frame-it, Floors** menu and select **SubFloor**. Toggle **DoSubflr** and **Single on**, and set the thickness to **3/4** inch (**0.0.3/4**).

15.___ Exit back to the **Frame-it, Floors** menu and toggle **Center on**.

16.___ Press the **TAB** key to make the **FLOOR** layer active; then select **Build, All, Yes**.

Since the **Framing** macro is used to extract framing data, the elevation (Z-base) of the schematic is not normally a problem. If, however, you want to use the floor or wall framing together with other entities in your plan such as plumbing fixtures or furniture, you will have to adjust the elevation of the schematic used for framing to match the rest of the plan.

Figure 11-10 shows the results of using the **Hide** option after building the new frame and adjusting the 3D view to view the floor from below.

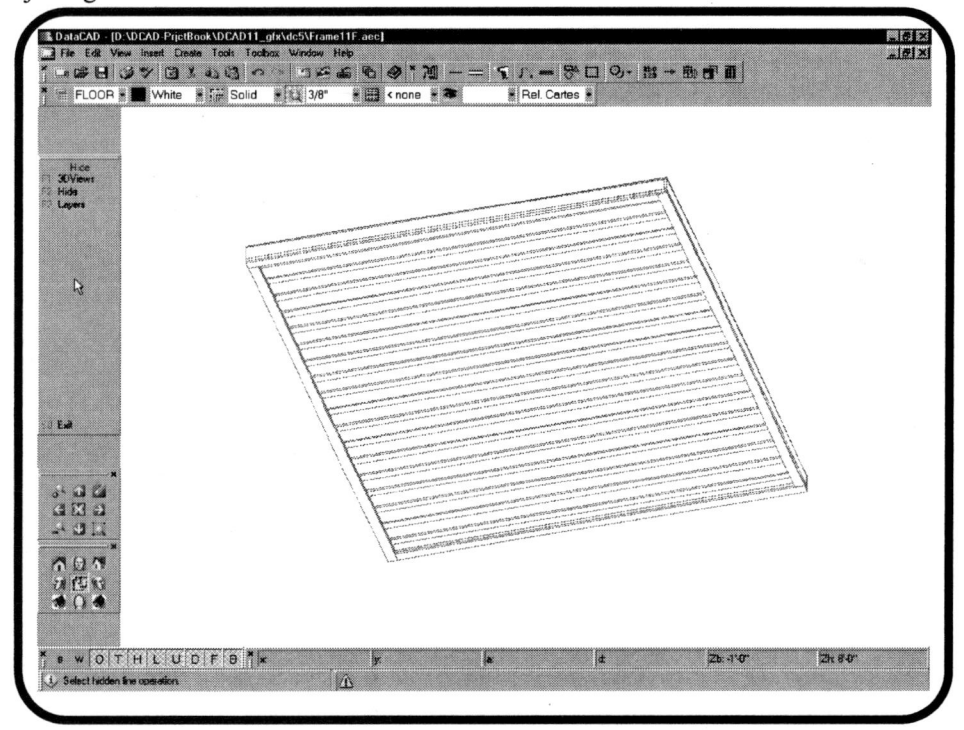

Figure 11-10, A framed floor.

Chapter 11, Page 211

The DataCAD 11 Project Book

Next, an opening will be created. First, the frame will be erased from the display using the **Layers** option in the **Floors, Edit** menu.

> •• **NOTE:** The **EraseLyr** option only erases entities on the selected layer. It does not delete the layer.

17.__ From the **Floors** menu, select **Edit, Layers, Clear Layer**; then select **FLOOR, Yes**.

18.__ **Exit** back to the **Frame-it, Floors** menu. Press the **Esc** key to redraw the display.

An opening will be added to the floor to simulate a place for stairs leading to the basement. This opening must be added to the schematic layer.

19.__ Press the **TAB** key until the **FLSCHEME** layer is active.

20.__ Select **Opening** from the **Floors** menu. Position the first corner of the opening **2** feet **below** and to the **right** of the upper **left** corner of the plan. Make the opening **6'** by **3'** as shown in Figure 11-11.

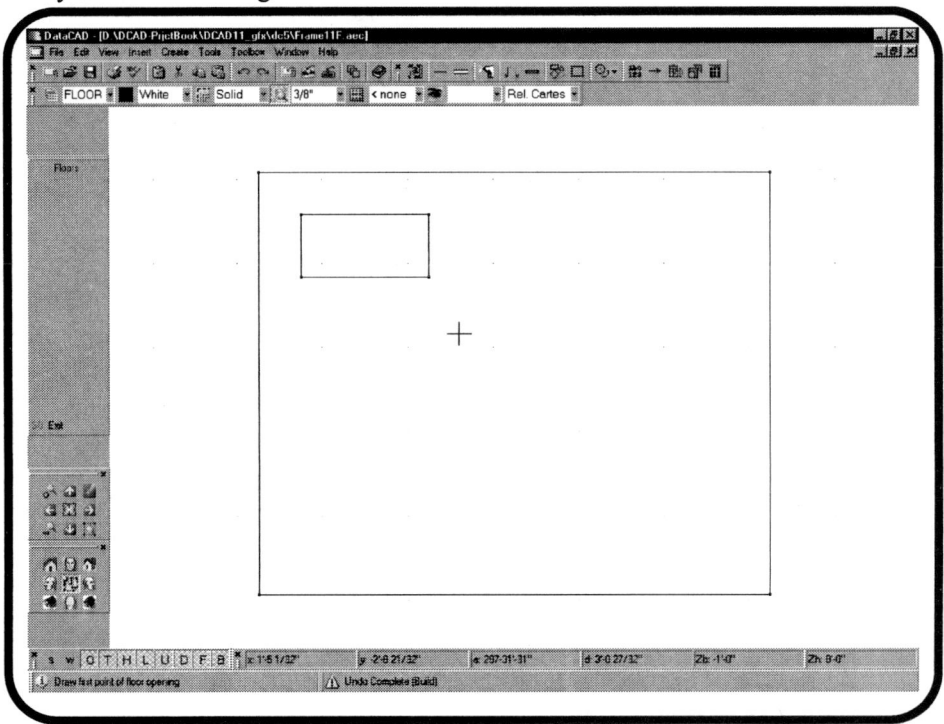

Figure 11-11, Adding an opening.

Chapter 11, Page 212

Adding Floor Framing to Your Plan.

21. __ Press the **TAB** key until the **FLOOR** layer is active.
22. __ **Exit** to the **Frame-it, Floors** menu, and select **Build**.
23. __ Select **All, Yes** to build the floor frame.

> •• **NOTE:** Remember that the **Frame-it, Floors** menu contains an **Edit** option. You can use the **Move** option in the **Edit** menu to move the opening after it has been placed. The **Edit** options allow you to fine tune your design without having to erase and recreate the schematic. If you want to change a parameter (such as size) of a framing member, you will have to remove the member and replace it after changing the parameter.

The new floor frame will have framing added to the opening. Next, a beam will be added across the center of the plan. Figure 11-12 shows the **Beam** menu.

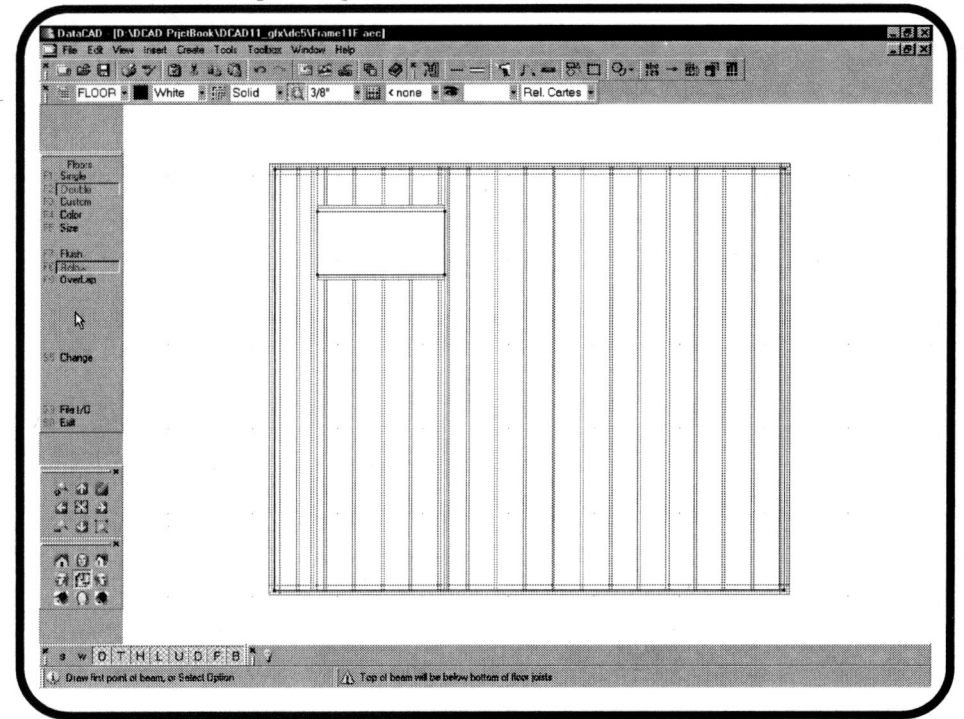

Figure 11-12, A floor frame with opening.

The DataCAD 11 Project Book

Beams can be placed either flush with, or below, the floor joists. When they are placed below, an overlap option appears that allows for specifying the amount the joists will overlap the beam. As with the opening example above, the beam should be placed on the schematic layer.

24.__ Press the **TAB** key until the **FLSCHEME** layer is active.

25.__ Select **Beam** from the **Floors** menu.

26.__ Toggle the **Double** and **Below** options **on**. Add a horizontal beam **12** feet in from the upper wall.

Actually, the beam could be centered on the plan. The spacing in this example is arbitrary.

27.__ Press the **TAB** key until the **FLOOR** layer is active.

28.__ **Exit** to the **Frame-it, Floors** menu, and select **Build**.

29.__ Select **All, Yes** to build the floor frame.

Roof Framing

The result of this construction is shown in Figure 11-13. Note that if you want to change the position of the beam from the **Below** to the **Flush** parameter, the beam must be removed and then replaced after the parameter has been changed.

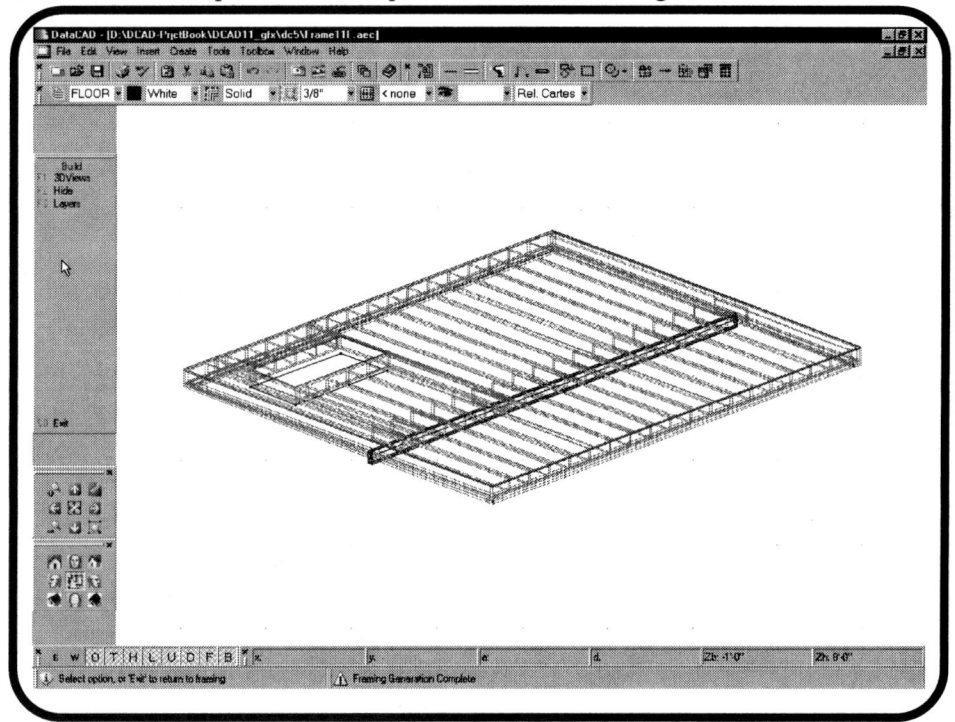

Figure 11-13, The floor frame.

You should spend some time exploring the various floor construction options until you feel comfortable with how they work.

Roof Framing

At this point you probably have the basic idea of how the **Framing** macro works. A schematic for the macro to use is created, parameters for the various components are adjusted, then the frame is built. The elevation of the roof is a function of the current **Z-Base** setting. If you want the roof to be placed on a frame's top plate that is set at 8 feet, you must set the Z-base to 8 feet before creating the roof schematic.

1.___ If it is not on the display, **Open** the **Frame11F** file created in the last section.
2.___ Select **File, Save As** and enter **Frame11R** as the file name.

The DataCAD 11 Project Book

3. __ Press the **L** [or the **l** key] to jump to the **Layers** menu. Select **New Layer**, and add **2** layers.
4. __ Select **Name** and name the new layers **ROOF** and **RFSCHEME**.
5. __ Make the **RFSCHEME** layer active, and press the **K** key until it is a color that you like.
6. __ Use the **Layers, On/Off** option to turn all layers off except **FLSCHEME, RFSCHEME,** and **ROOF**.

To make the base of the roof even with the top plate of the wall, the **Z-Base** for the roof schematic will be adjusted.

7. __ Press the **z** [or the **Z**] key, and enter a new **Z-Base** of **8.0**. Do not worry about the **Z-Height**; just press enter to accept the default setting.

When you trace the boundary in the next step be sure to select the points in the order specified. The first two points will specify the length of the roof, and the third will set the width. If you reverse these, the ridge will lie in a different direction. Also, since DataCAD uses a counterclockwise angle convention, it is a good idea to get into the habit of working counterclockwise when tracing perimeters.

8. __ From the **Frame-It, Roofs** menu, select *four* points (on top of the floor schematic) to define the roof schematic. Start at the *lower left* corner and select points in a *counterclockwise* direction.
9. __ Use the **Layers, On/Off** option to turn all layers off *except* **ROOF** and **RFSCHEME**.
10. __ From the **Frame-It, Roofs** menu, toggle **Gable on**.
11. __ Select **Settings** and set the pitch to **6/12**.
12. __ Press the **TAB** key to make the **ROOF** layer active.
13. __ **Build** the roof frame.

Roof Framing

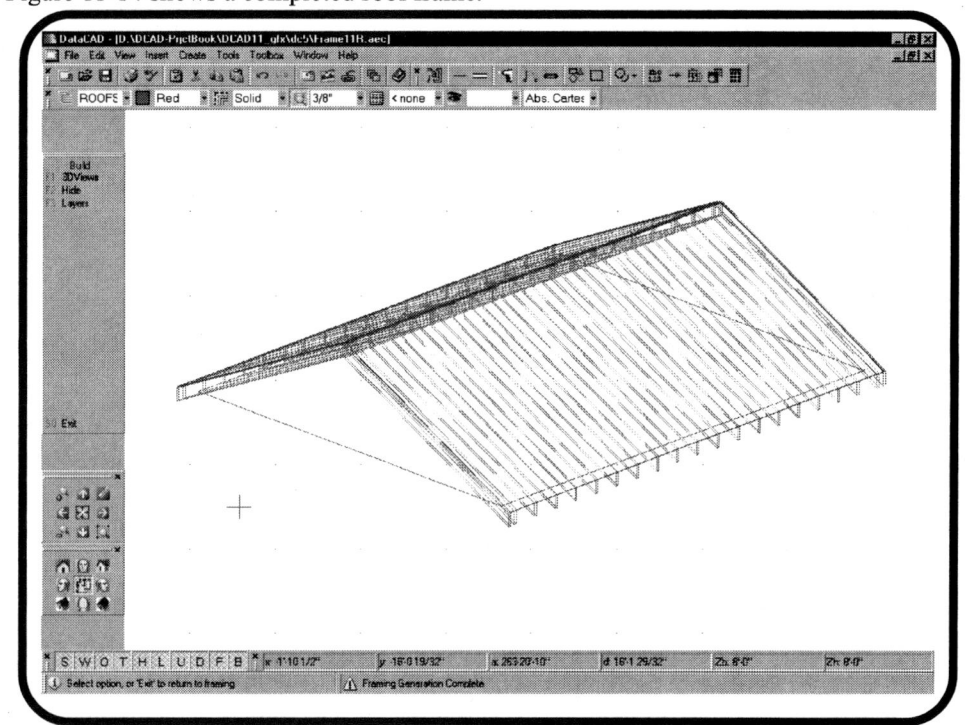

Figure 11-14 shows a completed roof frame.

Figure 11-14, A roof frame.

Now that you have created framing for the walls, floor, and roof, you can turn all the layers on and admire your work.

14.__ Turn on only the **ROOF, WALLS,** and **FLOOR** layers.

The DataCAD 11 Project Book

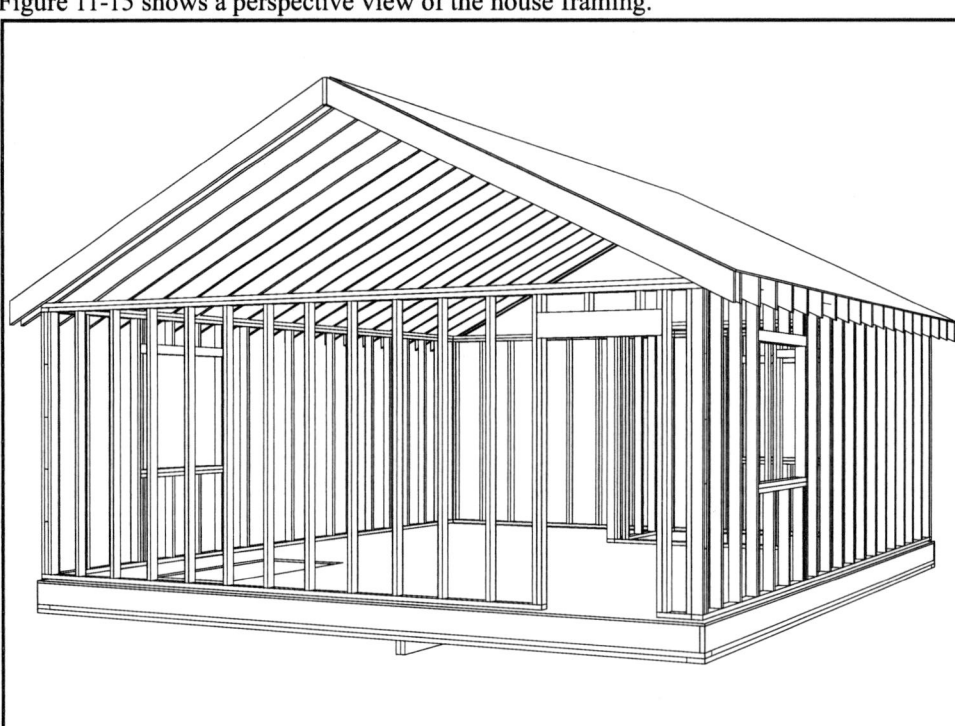

Figure 11-15 shows a perspective view of the house framing.

Figure 11-15, The completed house framing.

Review and Important Points

- When creating a schematic for floor, wall, or roof framing, be sure to set the **Z-Base** at the appropriate elevation.

- Establish a drawing layer for the framing that is different from that used for the framing schematic. This will allow the **Clear Layer** function to be used to remove frames without disturbing the schematic.

- Use the **Frame-It, Extract** function to create material take-offs.

Roof Framing

Practice Drawings

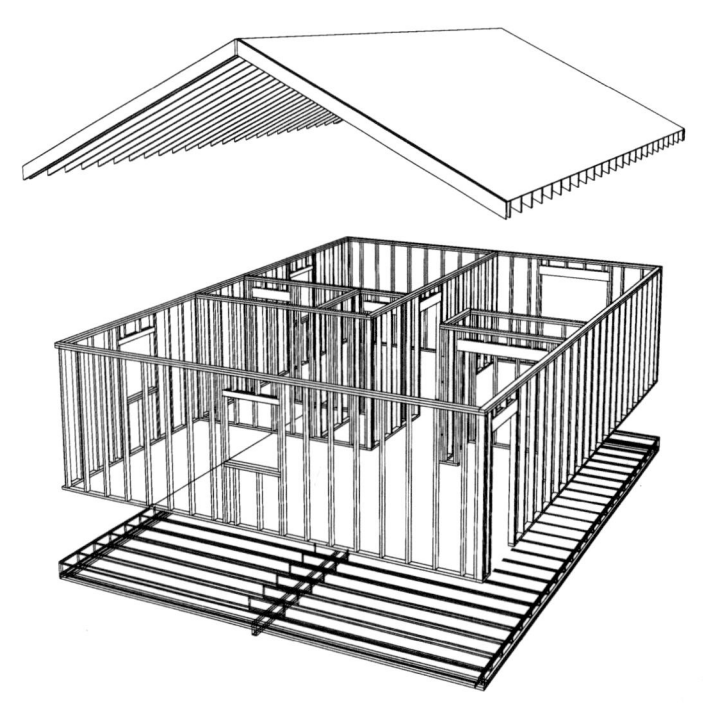

Assignment P11-A, Apartment Framing

- Use the dimensions shown in assignment 8-A and create a framing illustration for the apartment.
- Add layers for each schematic and frame.
- Change the Z-elevation for the roof and floor to allow for better viewing of the framing.
  - **•• NOTE:** The floor in the above illustration was rendered with **Draw Hidden on** in the **Hide, Options** menu, and then lines were trimmed around the wall framing.

The DataCAD 11 Project Book

Notes

CHAPTER 12

Using Shader and Object Viewer

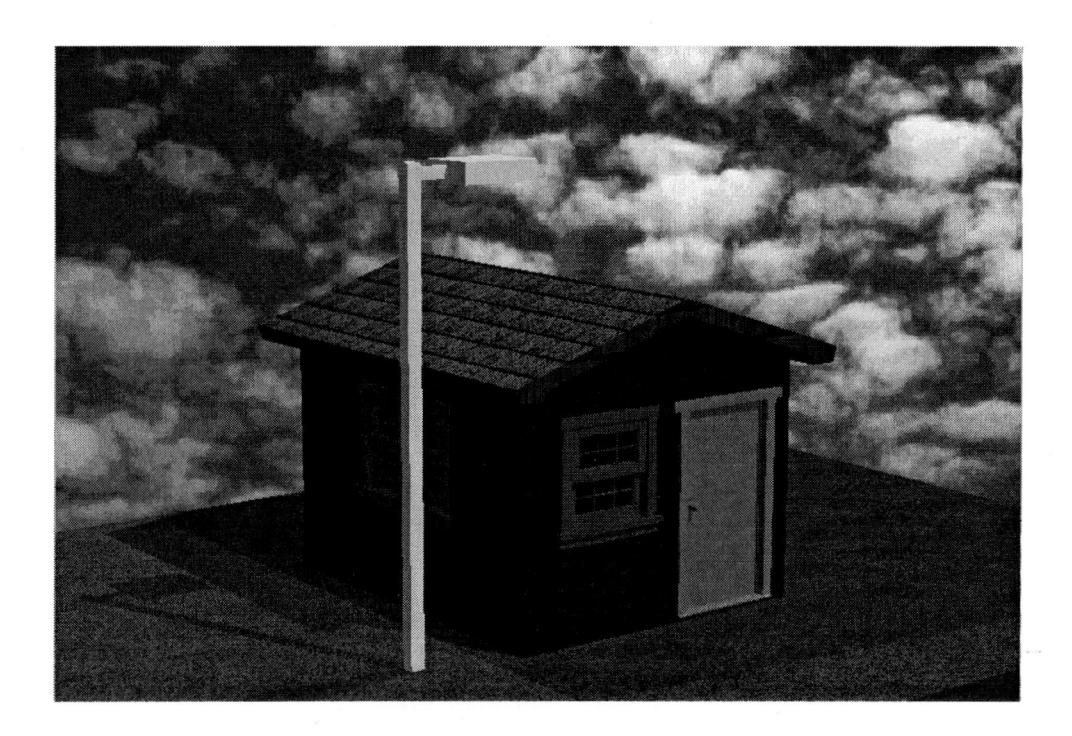

Project Goals

When you have completed this project, you will be able to:
1. Move objects to absolute coordinate 0,0.
2. Use the **GoTo View** function.
3. Display **Multi-View**s.
4. Create and save shaded images from your DataCAD models.
5. Add and adjust lights.
6. Adjust model colors.
7. Add background images to shaded scenes.
8. Use the **Object Viewer**.

The DataCAD 11 Project Book

Prerequisite: Chapter 1, 10

Introduction

In DataCAD release 10 the **Shader** function replaced the former **Quick Shader** option. The o2c **Object Viewer** was also added to DataCAD 10. These two features provide tools for creating three dimensional shaded images, and for dynamically viewing your designs by spinning and zooming around with easy mouse button controls. This project will introduce the basics of using the **Shader** and **Object Viewer** functions.

1.__ Start the **DataCAD** software and **Open** your **Play10** drawing file.

2.__ From the **MENU BAR**, select **File, Save As** and enter **Play12** as the file name.

Before getting into the *Shader* and *Object Viewer* functions, a few adjustments will be made to the model.

3.__ From the **MENU BAR**, select **View, Plan**.

Since the original playhouse plan was placed randomly on the drawing window, it will be moved so that it will be centered on the *Absolute Cartesian* coordinate system 0,0 location.

Moving the Playhouse to 0,0

4.__ Press **m** [not M] to jump to the **Move** menu.

5.__ From the **Move** menu, select **Drag**.

6.__ From the **Drag** menu, select **Area** and capture the entire playhouse in an area box (all lines in the playhouse should appear dashed after this step).

7.__ At the prompt *Select the point to drag from*, click at the **center** of the playhouse.

8.__ At the prompt *Select new position for these entities*, press the **Insert** key until the *Current input mode* is **Absolute Cartesian**.

9.__ Press the **space bar** and enter values of **0** for **X** and **Y**.

10.__ Select **Zoom view** at the center of the **Navigation Pad** to regenerate the display.

••• **WARNING**••• Have you saved your drawing lately?

The Playhouse is now centered at the origin of the absolute cartesian coordinate system. This will come in handy when you want to control the position of a light by entering a coordinate location.

Chapter 12, Page 222

Creating a Slab

Next, a slab will be added for the playhouse to rest on.

11. __ In the **Status Area**, *click* on the layer button to open the **Layer Manager** dialog box.
12. __ In the **Layer Manager** dialog box., select **New** and enter **1** new layer.
13. __ Enter **SLAB** as the new layer name.
14. __ Make the **SLAB** layer active and set the color to **Lt Green**.

Now the slab will be created using *Absolute Cartesian* coordinates.

15. __ Press the **J** [not the j] key to jump to the **3D Entity** menu.
16. __ From the **3D Entity** menu, select **Slabs**. Then select **Horizontal** from the **Slabs** menu.
17. __ From the **Slabs** menu, toggle **Base+Thk on**. Select **Thickness** and enter a value (watch the minus sign) of **-1.0**.
18. __ At the prompt *Select the 1st point of the slab,* press the **space bar** (you should still be in *Absolute Cartesian* mode) and enter values of **15** for **X** and **20** for **Y**.
19. __ At the prompt *Select the 2nd point of the slab,* press the **space bar** and (watch the minus sign) enter values of **-15** for **X** and **20** for **Y**.
20. __ At the prompt *Select the 3rd point of the slab,* press the **space bar** and (watch the minus *signs*) enter values of **-15** for **X** and **-20** for **Y**.
21. __ At the prompt *Select the 4th point of the slab,* press the **space bar** and (watch the minus sign) enter values of **15** for **X** and **-20** for **Y**.
22. __ Press the *right* mouse button to complete the slab.
23. __ From the **MENU BAR**, select **View, Isometric**.
24. __ Click on the **Zoom view** button at the center of the **Navigation Pad** to regenerate the display.

••• **WARNING**••• Have you saved your drawing lately?

The DataCAD 11 Project Book

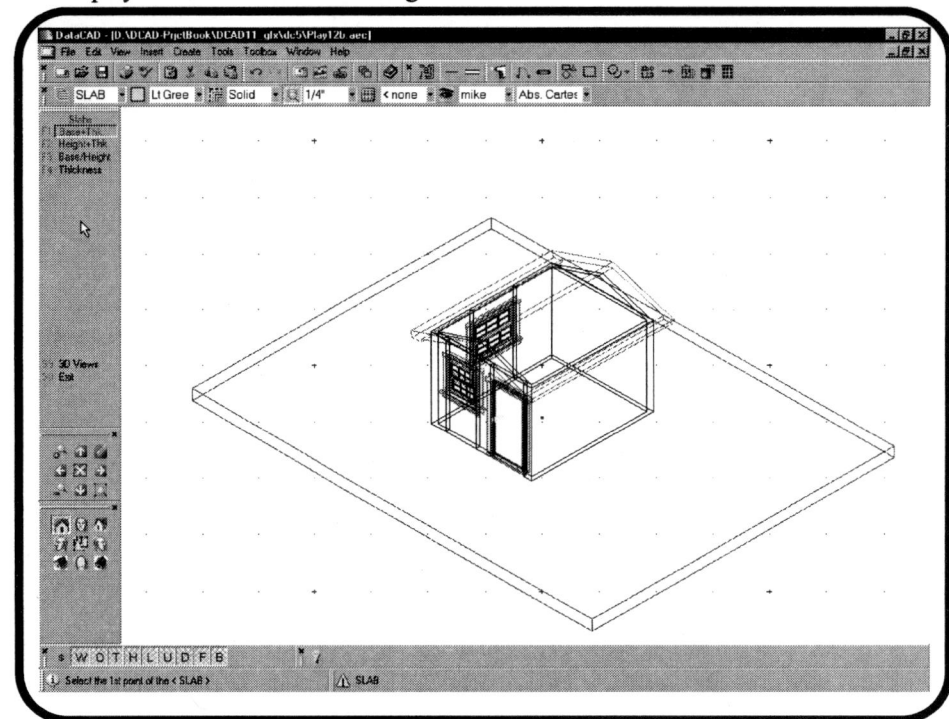

Your display should look similar to Figure 12-1.

Figure 12-1, The playhouse model.

It would look a little better if we could see all of the windows. A perspective view will be created using a normal camera lens and with the observer standing 10 feet above the ground.

25.___ Press the **y** [not the Y] key to jump to the **3D Views** menu.

26.___ From the **3D Views** menu, select **Set Persp**.

27.___ Select **Camera** and then set the camera focal length to **55mm**.

28.___ Select **Eye Point Z** and then set the Z coordinate value to **10.0**.

29.___ Select **Focal Pnt. Z** and then set the Z coordinate value to **5.0**.

30.___ Toggle **Fixed Focal off**, and toggle **Fixed Dist.** and **Pan View on**.

31.___ Press the **o** {or the O] key and toggle **Ortho** mode **off**.

32.___ At the prompt *Select position of eye point for perspective view*, press the **space bar** and (watch the minus sign) enter values of **-30** for **X** and **15** for **Y**.

Using GoToView

33.___ At the prompt *Select position of center of perspective view*, click at the *center* of the playhouse.

Your display should look similar to Figure 12-2.

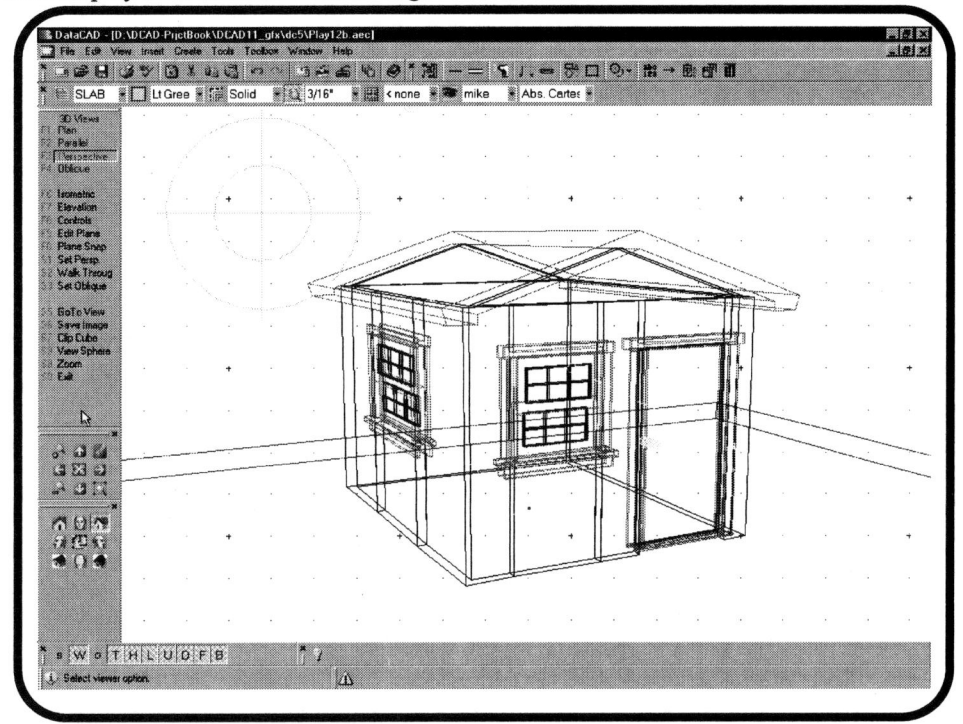

Figure 12-2, A perspective view.

••• **WARNING**••• Have you saved your drawing lately?

Using GoToView

DataCAD has a very powerful feature called **GoTo View** that allows you to define a number of different views of your drawing (2D or 3D) and to return to a particular view whenever you like. You don't return to a saved image. You return to the saved view of your current drawing or model. Next, the current perspective view will be added to the **GoTo View** list.

34.___ From the **Utility** menu, select **GoToView**.

Chapter 12, Page 225

The DataCAD 11 Project Book

35.__ From the **GoTo View** menu, select **Add View**. Enter **Persp_1** as the new view name (you are limited to a maximum of 8 characters).

To demonstrate another DataCAD feature, some new views will be added.

36.__ Press the **y** [not the Y] key to jump to the **3D Views** menu.

37.__ Select **Isometric**; then click on the viewing sphere near the location shown in Figure 12-3.

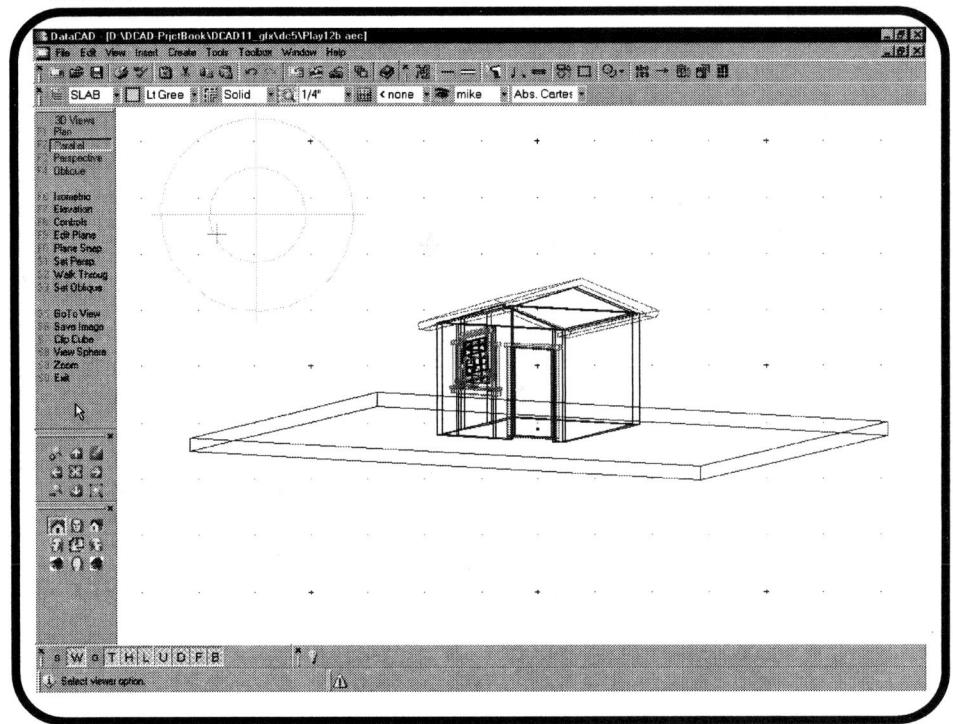

Figure 12-3, Creating a parallel projection.

Now, save this new view in the **GoTo Views** list. Notice that the same **Goto View** option that was available in the **Utility** menu is also available in the **3D Views** menu.

38.__ From the **3D Views** menu, select **GoTo View**.

39.__ From the **GoTo View** menu, select **Add View**. Enter **Para_1** as the new view name.

Next, you will see how handy the **GoTo View** list can be.

Using GoToView

40.__ Press the `:` key to jump to the 2D **Edit** menu. From the **MENU BAR**, select **View, Plan**. *Click* on the **Zoom view** button at the center of the **Navigation Pad** to regenerate the display.

41.__ Press the **v** [not the V] key. Press the function key for the **Persp_1** view.

42.__ Press the function key for the **Para_1** view.

43.__ *Click* on the **Front** button in the **Navigation Pad** to display an elevation view.

But wait. It gets even better!

44.__ Press **Ctrl + W** (or select **View, Multi-View**) to open the multi view window.

Your display will look similar to Figure 12-4 (your view windows will be empty until you add some views to them).

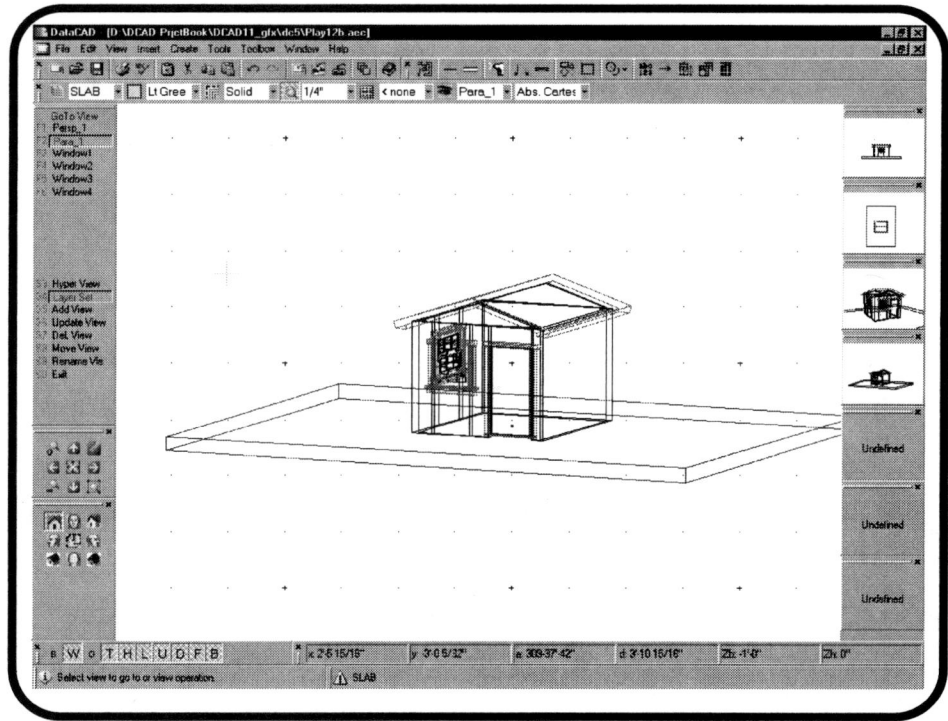

Figure 12-4, Displaying multiple views.

45.__ Place your cursor in the top *multi view window*, then *right-click* your mouse. A new menu will appear under your cursor.

Chapter 12, Page 227

The DataCAD 11 Project Book

46.__ *Click* on the **View In** option to display the current *drawing window* view in the top multi-view window.

47.__ Click on the **Plan view** button in the **Navigation Pad** to display the plan view of your model in the main *drawing window*.

48.__ Place your cursor in the second from the top *multi view window*, then *right-click* your mouse. *Click* on the **View In** option.

49.__ Now press the **v** [not the V] key to jump to the **GoTo View** menu.

50.__ Select the **Persp_1** view. Place your cursor in the third from the top *multi view window*, then *right-click* your mouse. *Click* on the **View In** option.

51.__ In the **GoTo View** menu, select the **Para_1** view. Then in the fourth from the top *multi view window*, place your cursor there, then *right-click* your mouse. *Click* on the **View In** option.

Your display should look similar to Figure 12-5. Note that Figure 12-5 shows one of the views shaded.

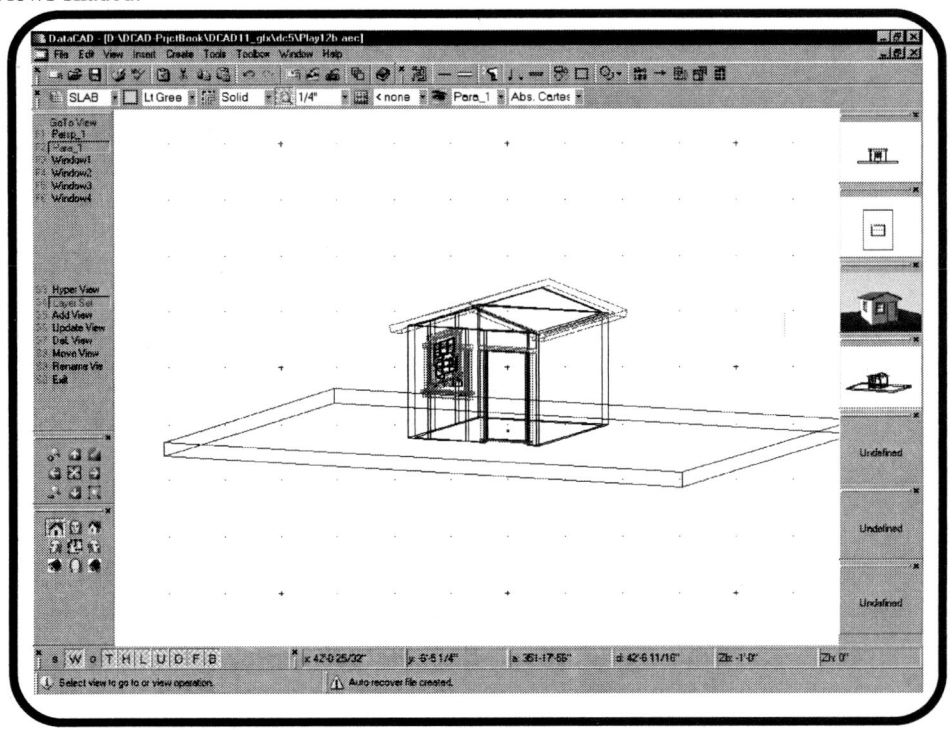

Figure 12-5, The multi view window display.

Using the Shader Function

INFO STOP

Each of the small views has several options which are available by placing the cursor over one of the windows, then *clicking* on the right mouse button. The **View In** option moves the contents of the current drawing window into that view window. The **View Out** option moves the contents of that view window into the drawing window. **Clear** will remove the contents of the window and display the default *Undefined* text. When you select the **GoTo View** option, you can select an existing GoTo View to be displayed in the window. The next tool, **Shade**, allows for displaying the image in the view window as a shaded image. The **Refresh** option refreshes the contents of the multi view window, while **Refresh All** will refresh all of the multi view windows. Remember that the multi view window is toggled on and off by pressing **Ctrl+W**

The multi view window can improve productivity by allowing you to quickly switch between different views as you create your designs. The **GoTo View** function has the added advantage of allowing quick access to a long list of preferred user views.

Using the Shader Function

Now it is time to explore the various options in the Shader tool.

52.__ Use either the multi view window or the **GoTo View** menu to display the **Persp_1** view. Then press **Ctrl+W** to close the multi view window.

53.__ Press the **j** [not the J] key to jump to the **3D Menus** menu. Then select **Shader**.

Chapter 12, Page 229

The DataCAD 11 Project Book

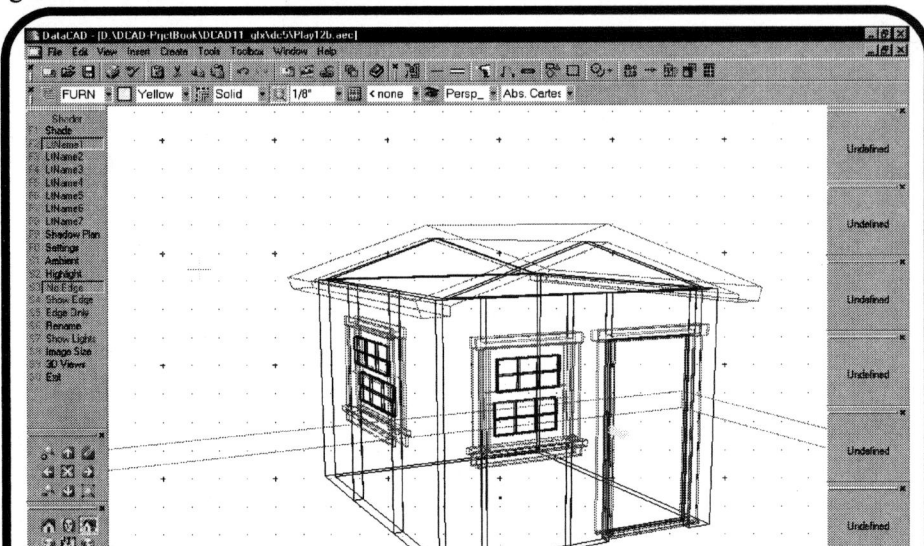

Figure 12-6 shows the **Shader** menu.

Figure 12-6, The Shader menu.

As is true in photography, control of lighting will make or break a shaded image. The **Shader** tool provides for up to seven different lights in addition to ambient light. Since messing with a number of different lights at the same time will most likely also mess with your brain, it is best to work with one light at a time until each light used is under control.

54.___ Toggle all lights off. (If any light button is depressed, select that light and then click on the **On/Off** button to turn it off. Then click the *right* mouse button to return to the **Shader** menu.)

55.___ From the **Shader** menu, select the **Settings** button.

Using the Shader Function

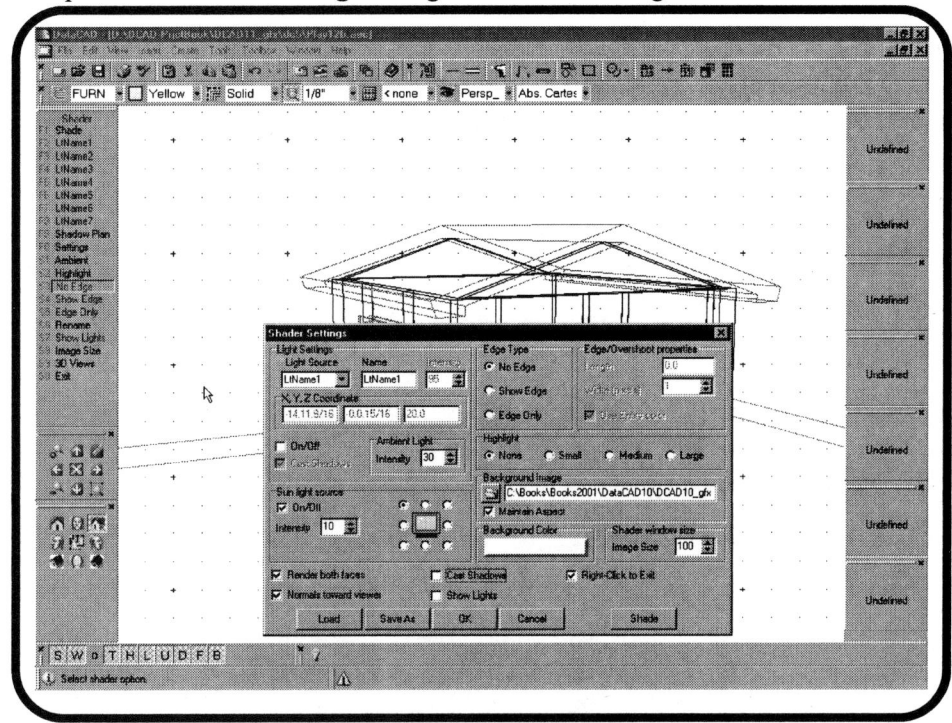

This opens the **Shader Settings** dialog box as shown in Figure 12-7.

Figure 12-7, The Shader Settings dialog box.

The **Shader Settings** dialog box provides one-stop shopping for all of the **Shader** options. Many of these are duplicated by separate buttons on the **Shader** menu. For example, some of the settings for each of the lights can be controlled by picking the **LtName** in the **Shader** text window, then picking **On/Off**. From there you can set light's intensity, shadows and position.

In the illustrations used here, the background color has been changed from the default black to white. (This makes the print shop folks happy.)

56.___ Set the **Ambient Light Intensity** value to **70**. (Note that ambient light values can range from 0 to 100.)

57.___ Toggle the **Right-Click to Exit** option **ON**.

The DataCAD 11 Project Book

INFO STOP

The **Right-Click to Exit** option will automatically close the shaded image window when the right mouse button is clicked. If this option is on, clicking the right mouse button in a shaded image window will provide several options including saving the image to a file. When you are experimenting with shading, it is best to toggle the **Right-Click to Exit** option **ON**.

58.___ In the **Shader Settings** dialog box, click the **Shade** button.

Figure 12-8 shows the shaded image.

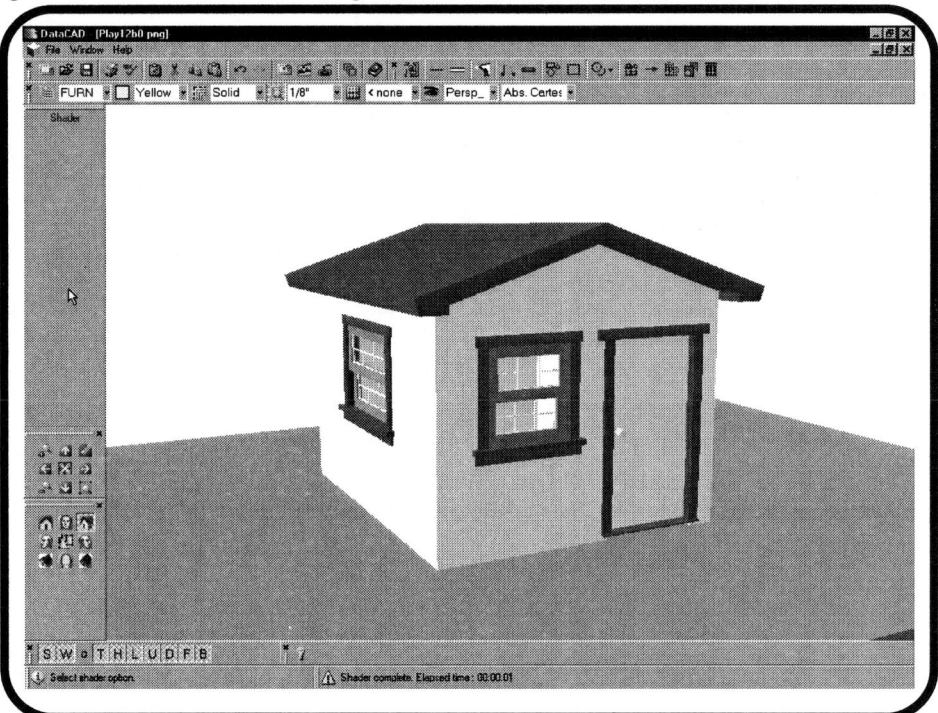

Figure 12-8, A shaded perspective view.

This image is very flat and not too interesting. Let's reduce the ambient light level, and then add a new light.

59.___ Right click on the shaded image to return to the **Shader** menu.

Using the Shader Function

60.___ From the **Shader** menu, select the **Ambient** button and then enter a new value of **30**.

61.___ From the **Shader** menu select **Settings**. The **Shader Settings** dialog will appear.

62.___ Tturn **OFF** the **Sun light source** (more on this later).

63.___ From the **Shader** menu, select the **Shade** button and note the change caused by the different ambient light setting. Then right click on the shaded image to return to the **Shader** menu.

Next, a light will be added. It will be positioned 20 feet above the ground right over the upper left corner of the slab.

64.___ From the **Shader** menu, select the **LtName1** button. Toggle **LtName1 on**.

65.___ In the **LtName1** menu, click on the **Position** button.

66.___ If it is ON, toggle the **3DCursor** button **OFF**.

67.___ Select the **Set Z User 1** button and enter a value of **20**.

68.___ Toggle the **Z-User 1** button **ON**. Move the cursor directly over the *upper left* corner of the slab and click the *left* mouse button to position the light.

69.___ Click the **Shade** button.

Well, that's not too exciting because the **Position** menu automatically switched to the **Ortho** view. However, if you examine the shaded view carefully, you should observe the slight difference in the color of the slab further from the light.

70.___ Right click to close the shaded image window. Right click once more to return to the **Shader** menu. Then click the shade button.

This image should have a little more definition than the ambient light alone. Let's experiment a little with the intensity of light 1.

71.___ From the **Shader** menu, select the **LtName1** button. Select the **Intensity** button and set the intensity to **90**. Click the **Shade** button and inspect the result.

The reason for this fooling around is for you to get a feel for how the **Shader** lights work. Next, let's try a different position for light 1.

Chapter 12, Page 233

The DataCAD 11 Project Book

> **INFO STOP**
> The **Sun light source** option in the **Shader Settings** dialog allows for a uniform light source (the sun), and allows you to define one of 8 general directions from which the light comes. It adds one more level of lighting realism.

> **INFO STOP**
> The **Show Lights** toggle in the **Shader** menu will add a yellow (tough to see on a white screen) symbol to show the position of each light. Each light symbol has a number in it corresponding to the **LtName** number. This is especially useful when you are manipulating a number of different lights.

72.__ From the **Shader** menu, select the **LtName1** button. Click on the **Position** button, make sure that **Z-User 1** is still toggled **ON**, and click the left mouse button once with the cursor directly over the *lower left* corner of the playhouse roof.

73.__ Right click back one menu and then select the **Shade** button.

At this point you should see a difference in brightness across the front of the playhouse. Next, shadows will be added.

74.__ Right click to return to the **Shader** menu. From the **Shader** menu, select the **Settings** button.

75.__ In the **Light Settings** area of the **Shader Settings** dialog box, make sure that the **Cast Shadows** option is checked **ON**.

76.__ Near the lower right corner of the **Shader Settings** dialog box, make sure that the **Cast Shadows** option is checked **ON**.

77.__ For now, make sure the **On/Off** option in the **Sun light source** section is <u>un</u>checked.

78.__ Select the **Shade** button.

Hmm... still no shadows. There is still one more thing to do before the shadows will appear. That is to tell the system which objects will receive shadows.

Using the Shader Function

INFO STOP

An object can either receive shadows, or cast shadows. It cannot do both. If it has been set to receive shadows, it cannot cast shadows.

79.__ Right click to return to the **Shader** menu. From the **Shader** menu, select the **Shadow Plan** button.

80.__ In the menu, select **Group** and **Recv. Shado**. Then click on one of the lines in the slab. (All lines in the slab should appear dashed after this step.)

81.__ Right click back to the **Shader** menu and **Shade** the view.

Now you should see a shadow appear on the slab. Figure 12-9 shows shadows cast with the light at 20 feet above the edge of the slab directly in front of the playhouse.

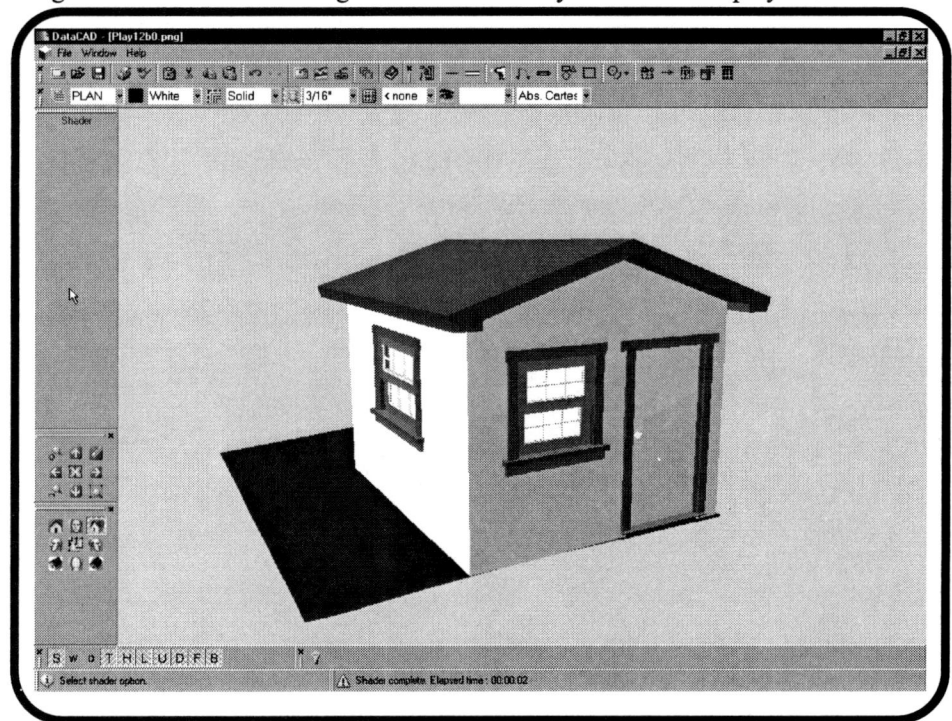

Figure 12-9, A shaded image showing shadows.

You may want to experiment with different positions for light 1 to see what affect it has on the shadow. Before you do, you should learn how to save your light settings.

Chapter 12, Page 235

The DataCAD 11 Project Book

82.___ Right click to return to the **Shader** menu. From the **Shader** menu, select the **Settings** button.

83.___ In the **Shader Settings** dialog box, click the **Save As** button. Adjust the **Save in:** folder as desired, and enter a name for your light settings.

Adding a Street Lamp

To demonstrate a second lighting technique, a street lamp model will be added to the design.

84.___ Click on the **Plan** view button on the **PROJECTION PAD** to display the plan view.

85.___ In the **STATUS AREA**, click on the layer reminder to open the **Layer Manager** dialog box.

86.___ In the **Layer Manager** dialog box., select **New** and enter **1** new layer.

87.___ Enter **StrLgt** as the new layer name, and press **OK**.

88.___ Press the **TAB** key until the **StrLgt** layer is active.

89.___ Press the **Ctrl+T** key to open the **Symbol Browser** menu.

90.___ Click on the yellow folder button, and pick the **Folder** option.

91.___ Navigate to the **Symbols, Site, 3D Parking Lot Lights** folder.

92.___ Select **PLL1S** light fixture (at top center).

INFO STOP

To see what a 3D symbol will look like, before you place it, locate the cursor over the symbol, *right-click* the mouse, then pick the **o2c Viewer** option. An **Object Viewer** window will open, allowing you to see, and rotate, the symbol in 3D. More on this later.

93.___ From the **Ins Symbol** menu, select **Dyn. Rotate** *twice* and enter an angle of **270**.

94.___ From the **Ins Symbol** menu, select **Z Enlargmt.**, and enter a value of **0.75**.

95.___ Press the **Space Bar** and (watch the minus sign) enter the absolute coordinates of **X=-5** and **Y=8**.

Adding a Street Lamp

This should have placed a light near the upper front corner of the playhouse in the plan view. Next, a new **Shader** light will be added so that it is right where the model lamp has been placed.

96.___ *Right-click* to stop placing symbols, then turn the **Symbol Browser** off.

97.___ Press the **y** [not the Y] key to jump to the **3D Views** menu.

98.___ From the **3D Views** menu, select **Set Persp**.

99.___ Toggle **Fixed Focal OFF**, and toggle **Fixed Cone, Fixed Dist.** and **Pan View ON**.

100.___ Select **Eye Point Z** and then set the Z coordinate value to **20.0**.

101.___ Select **Focal Pnt. Z** and then set the Z coordinate value to **5.0**.

102.___ At the prompt *Select position of eye point for perspective view*, place the cursor at the Eye Point shown in Figure 12-10, and *click* the left mouse button.

103.___ At the prompt *Select position of center of perspective view*, place the cursor at the Center Point shown in Figure 12-10, and *click* the left mouse button.

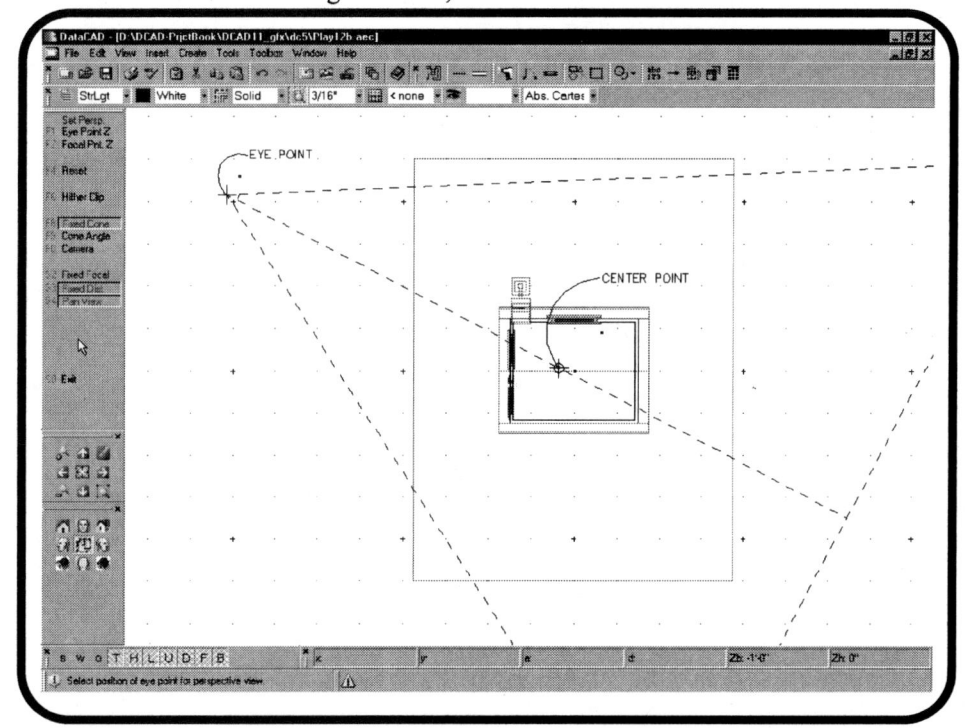

Figure 12-10, Perspective Eye Point and Center Point.

Chapter 12, Page 237

The DataCAD 11 Project Book

The Drawing Window will change to show the new perspective view that you just defined.

104.__ From the **3D Views** menu, select **GoTo View**.

105.__ From the **GoTo View** menu, select **Add View**. Enter **Persp_2** as the new view name.

106.__ Press the **Identify** button in the **Navigation Pad** (looks like a pointing finger) and *click* on the street lamp. Note that **Z-Max** is a little under 15 feet. Also note that the head of the light fixture is off to the side of the light pole.

107.__ Press the **j** [not the J] key to jump to the **3D Edit** menu and select **Shader**.

108.__ Toggle **LtName1 OFF**.

109.__ Toggle **LtName2 ON**. Make sure **3D Cursor** is **OFF**, then select **Position**. Set **Set Z User 1** to **13.6**, toggle **Z-User 1 ON**; then press the **Space Bar** and (watch the minus sign) enter values **X=-5, Y=6**.

110.__ Right click and **Shade** the view.

We just placed the light source 2 where the light bulb would be in the light fixture. See how the shaded image shows more intense light close to the light, and weaker light and shadows farther away.

111.__ Experiment with the **Intensity** of **LtName2**, and with the **Ambient** option.

There is one more light option to explore.

112.__ From the **Shader** menu, select **Highlight**. Select **Small** and **Shade** the model.

113.__ Experiment with the other **Highlight** options and observe the effects.

Adding Background Images

There is one more **Shader** feature that will be explored. It is possible to replace the colored background with a graphic image. For this you will need a bmp format graphics file.

INFO STOP
Most *Windows* installations include a file named **clouds.bmp**. If your system does not have this file available, you will have to substitute another file for the background.

114.__ From the **Shader** menu, select the **Settings** button.

Adding Background Images

115. ___ Click on the **Background Image** folder icon, Set the **Look in:** folder to *c:\DATACAD\Textures\Backgrounds* and then open the **Sky.bmp** file. (If you prefer, you can substitute any other bmp format file.)

116. ___ Click on the **Shade** button, and inspect the results.

•• **NOTE:** To remove a background image, delete all text in the **Background Image** text box in the **Shader Settings** dialog box.

Figure 12-11 shows a shaded image with background clouds.

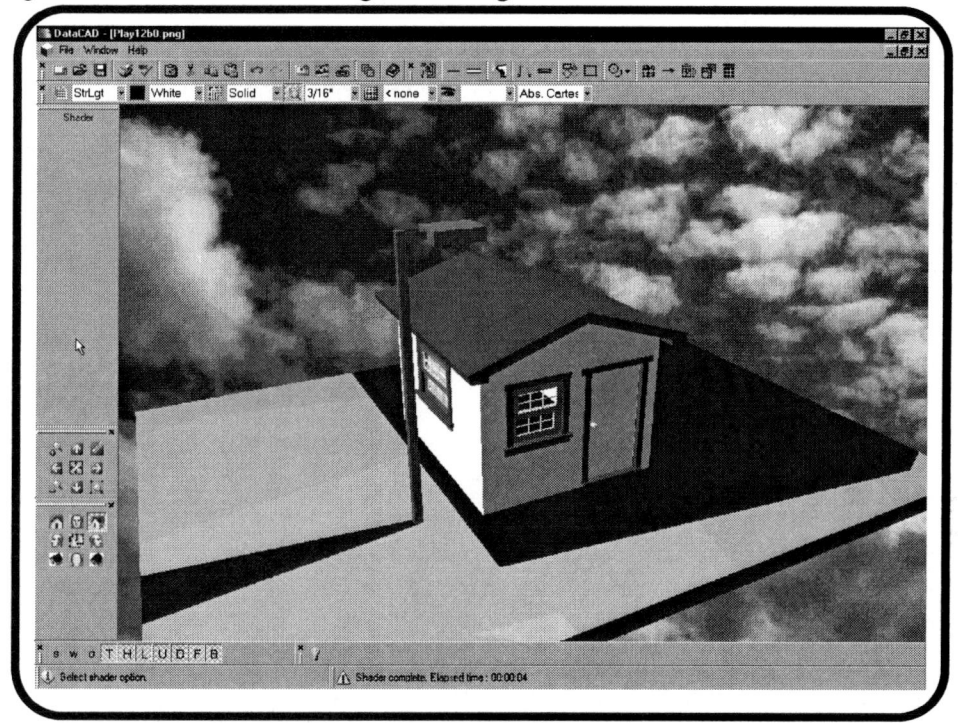

Figure 12-11, A shaded image with added background.

The DataCAD 11 Project Book

> **INFO STOP**
>
> For a more realistic image, the horizon should appear as a horizontal line. You can simulate this by adjusting the border of the slab. (The **Stretch** function can be used to move one corner of the slab at a time.) The problem with this is that you will have to adjust the slab for every different pictorial view.

Adjusting Model Colors

The color of any object in the DataCAD model can be changed to any color in the DataCAD palette. This provides a lot of flexibility for shading models.

1.___ With the **Play12** file open in DataCAD, press **Alt+c** to jump to the **Change** menu.

The **Change** menu is shown in Figure 12-12.

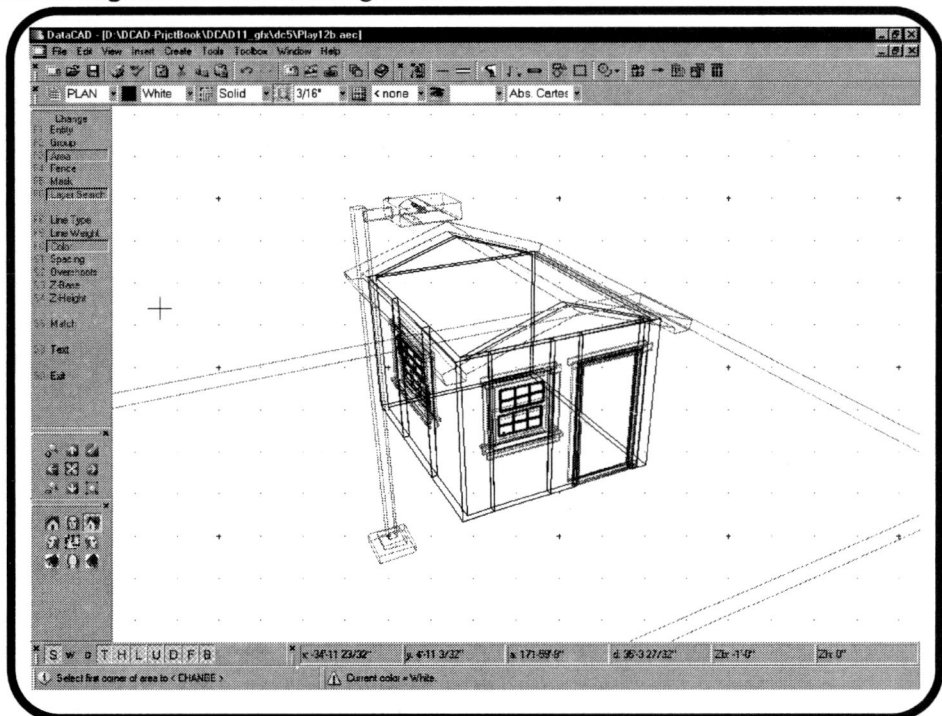

Figure 12-12, The Change menu.

Chapter 12, Page 240

Adjusting Model Colors

The **Color** option on the **Change** menu can be used either to select one of the 15 standard colors, or to open the **Color Palette** dialog box.

2. ___ From the **Change** menu, select **Color**. Then select the **Custom** option. The **Color Palette** dialog will appear.

From here you can select any one of the 255 colors in the palette by simply clicking on a color. But let's first see what other color options DataCAD has.

3. ___ Close the **Color Palette** dialog by clicking **Cancel**.

DataCAD comes with several different Color Palettes. It is also possible to create and save your own custom palettes. DataCAD will use whichever color palette is current. To change the color palette:

4. ___ From the **Menu Bar** select the **Tools** option, and then the **Color Palettes** option, near the bottom of the list. In the **Color Palette** dialog box, click on the down arrow button to the left of the **Save As** button, and select the **Prizma** palette from the list. Notice how the available colors have changed.

5. ___ Click **OK**. The **Prizma** palette is now the active palette.

The **Prizma** palette contains a selection of colors popular with many illustrators.

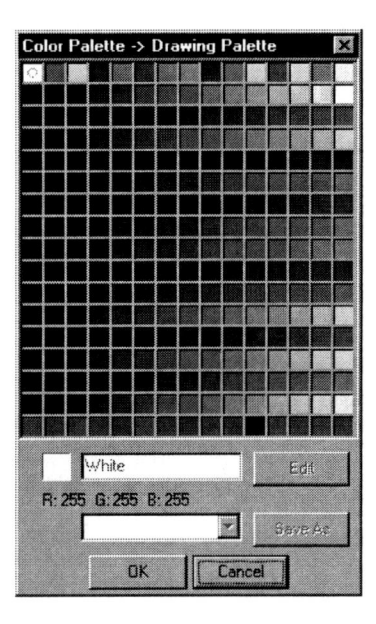

Chapter 12, Page 241

The DataCAD 11 Project Book

6.__ Press the **U** key and note that all of your objects have changed to new colors (in fact, if you have a white Drawing Window, some may have seemed to disappear, since many of the colors blend into the white screen background).

> **INFO STOP**
> Each color in the palette has a particular location in the palette. If you load a different **Color Palette** and then press the **U** key to regenerate the display, you will see that the system substitutes the original colors of objects with the colors used in that palette location. This means that you can create a custom palette that, for example, replaces every object drawn with standard red with a color from a custom palette.

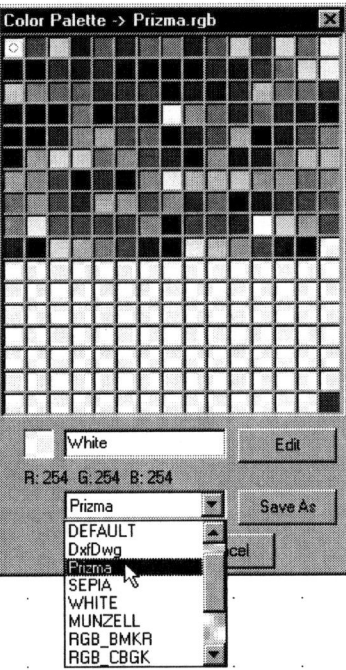

But if you now **Shade** the model you will notice that none of the new colors show up in the shaded image. Why not? Shader colors are controlled separately from the model colors using a DMF (DataCAD Materials File) file. When you load a palette file, the DMF file assigned to the drawing does not change. To get the **Shader** colors to match the palette colors, you need to change the DMF file to match.

7.__ With the **Prizma** palette already selected, go to the **Menu Bar** select the **View, Rendering Settings** option, find the area labeled **Rendering Settings File**, and click on the **Load Defaults** button (more on the **Rendering Settings File** later).

You have now told DataCAD to use the current palette as the default color palette for everything in the drawing file, including the **Shader** images.

Using the Object Viewer

> **INFO STOP**
> In the **Default** palette, the 15 colors across the top of the palette are the 15 standard DataCAD colors. When you click on one its name (Red, LtGreen, etc.) is displayed in the white box near the bottom of the dialog box. All the other colors' names are represented by their numerical position in the palette.

8. ___ Press **Alt+c** to jump to the **Change** menu. Select **Color, Custom**, then pick a nice version of Red from the **Color Palette**. Click **OK**.
9. ___ From the **Change** menu select **Group**. Then *click* on a line on the slab.
10. ___ Press the **j** [not the J] key to jump to the **3D Menus** menu. Then use the **Shader** function to shade your model.
11. ___ Press the **U** key and note that the slab color has changed to the new color.

> **INFO STOP**
> The **Toolbox** macros used for creating the door and windows automatically assigned colors to the different components of the model, and also made the entire door or window a group. If you want to change the color of part of a group, you can use the **Selection Set** tool to mask on a particular color. Using **Selection Sets** is discussed in *Appendix B*.

12. ___ Adjust the colors of your model as desired.

Using the Object Viewer

DataCAD comes with another built-in package called the **o2c** (Objects to See) **Object Viewer**. This program allows for real-time zooming, rotating and panning of a shaded view.

The DataCAD 11 Project Book

INFO STOP

Although the images generated by the **Object Viewer** look similar to those created by the DataCAD **Shader** tool, keep in mind that the **Object Viewer** window is running a new program. Although there are hooks between DataCAD and **Object Viewer**, they are actually two separate programs.

1. ___ With the **Play12** file open in DataCAD, from the **MENU BAR** select **View, Object Viewer**.

This opens a new window as shown in Figure 12-13.

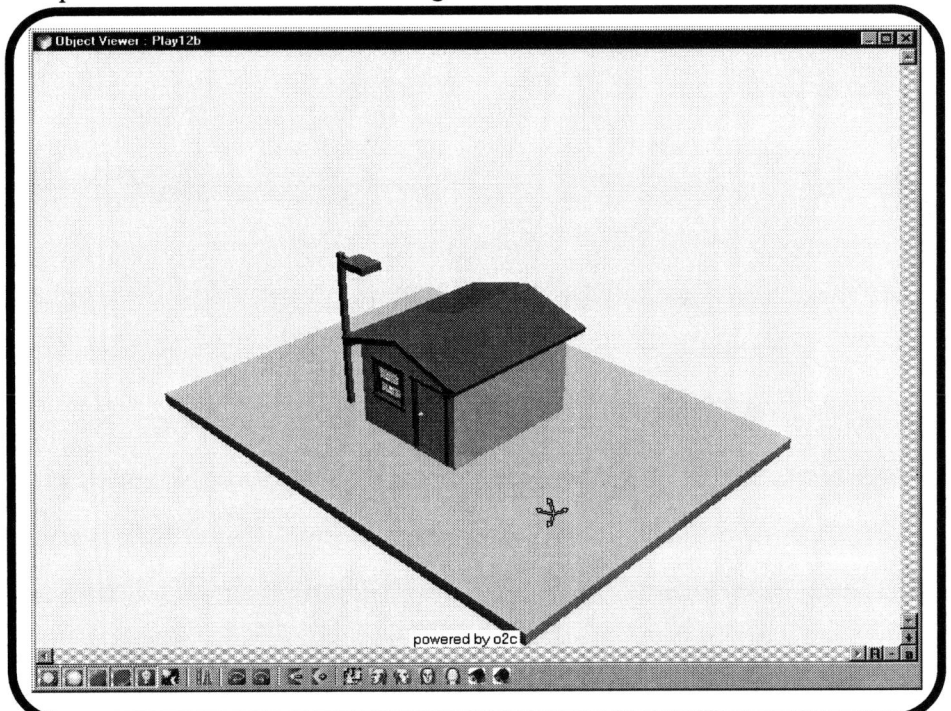

Figure 12-13, The Object Viewer window.

The **Object Viewer** uses the mouse to dynamically control rotate and zoom.

2. ___ Click and hold the *left* mouse button down and (while holding the left mouse button down) move the mouse left and right and up and down.

Using the Object Viewer

This rotates the display around horizontal and vertical axes.

3.__ Click and hold the *right* mouse button down and (while holding the *right* mouse button down) move the mouse up and down.

This zooms the display in and out. With just a little practice you will be able to travel anywhere around your model in a hurry. This is a great tool for previewing your design and can increase your design productivity.

4.__ Try to adjust the display to show a view similar to that shown in Figure 12-14.

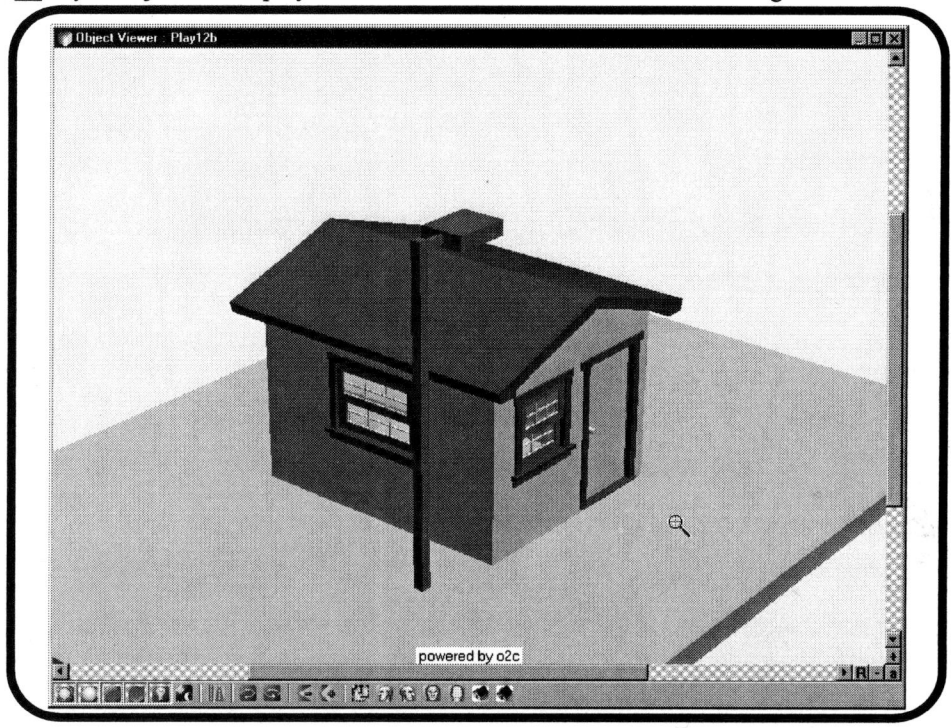

Figure 12-14, Dynamically adjusting the view.

Observe that the **Object Viewer Toolbar** appears along the bottom edge of the display. These tools allow for adjusting various **Object Viewer** functions. Holding the cursor on a tool will open a **Tool Tip** that will identify each function. **Scroll Bars** along the edge of the display can be used to adjust the position of the image.

There are a few more **Object Viewer** features to review.

Chapter 12, Page 245

The DataCAD 11 Project Book

5.__ Position the cursor in the **Object Viewer** and click the right mouse button once.

This opens the **Object Viewer** dialog box.

The **Show all** option returns the display to the initial view position.

The **Save image** option allows you to save the current image as a .BMP format graphics file.

The **Save object** option allows you to save the current object as an o2c format graphics file. This file can be opened by anyone who has the free o2c viewer program installed on their system (check the DataCAD web site, www.datacad.com). It is even possible to install an o2c viewer plug in to programs such as Microsoft Word, and use the viewer directly in the Word document.

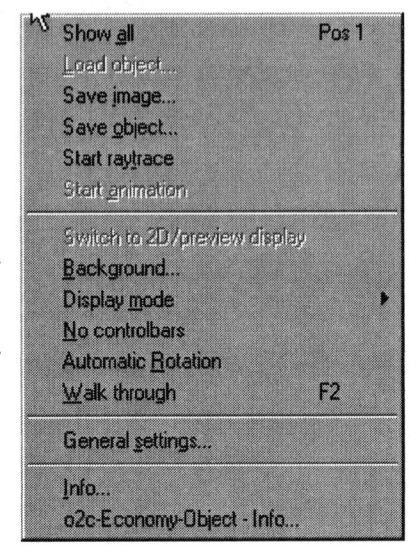

The **Start raytrace** option will generate a high quality ray traced image of your view. This image will include shadows from a default light source.

The **Background** option will allow you to add a background image similar to the Shader function.

The **Display mode** option opens a pop out menu with additional viewing options.

The **No controlbars** option toggles scroll bars along the bottom and right edge of the **Object Viewer** window on and off.

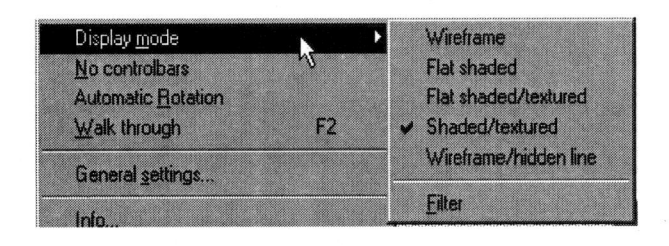

The **Automatic Rotation** option can be toggled on and off. It allows you to start rotating the object with the mouse, and if you release the left mouse button while the mouse is moving, the object will continue to spin with the direction and speed the mouse was moving. Click the left mouse button to stop the rotation.

Chapter 12, Page 246

Using the Object Viewer

The **Walk through** option is also an on off toggle. When it is on, you click and hold the left mouse button to pan the display. When it is off, you can use the arrow keys on the numeric keypad (the 4,8,6, and 2 keys) to pan the display.

The **General settings** option opens a dialog box that has options of interest to those who are communicating o2c objects over the internet, or embedding them in other applications.

6.__ Experiment with the various **Object Viewer** options.

There is another feature of the connection between DataCAD and the **Object Viewer** to be discussed. This involves setting DataCAD colors to be viewed as different colors in the **Object Viewer**. For example, the image shown in Figure 12-14 shows black walls because this was the wall color used in DataCAD. The color used for walls by the **Object Viewer** can be modified without changing the color used in DataCAD.

7.__ Close the **Object Viewer** window.

8.__ Hold the **Ctrl** key down and (while holding the **Ctrl** key down) click the *right* mouse button while the cursor is over one of the lines on the walls.

9.__ Select the **Rendering Settings** option in the pop-up box.

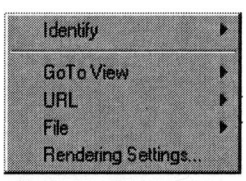

This opens the **Rendering Settings** dialog box. This dialog box can be used to set the relationship between the DataCAD palette and the o2c palette.

10.__ Click on the vertical colored box in the **Rendered Color** section and select a new rendering color for the walls. Then select **OK**.

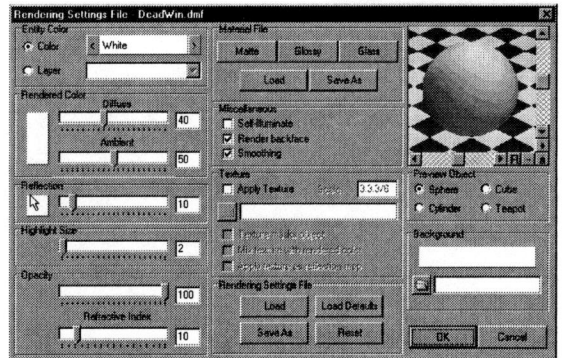

11.__ From the **MENU BAR**, select **View, Object Viewer** and inspect the results.

12.__ Again, hold the **Ctrl** key down and click the *right* mouse button while the cursor is over one of the lines on the walls of the playhouse. Select **Rendering Settings**.

13.__ Now go to the **Texture** section of the dialog, and check the **Apply Texture** box. The folder icon below will turn yellow, allowing you to click on it.

Chapter 12, Page 247

The DataCAD 11 Project Book

14.__ Click on the yellow folder. The **Select Texture** dialog will appear.

15.__ Open the **Brick** folder and select the **RedBrick.bmp** file. Click **Open** to select it.

16.__ Click **OK** to close the **Rendering Settings** dialog.

17.__ From the **MENU BAR** select **View, Object Viewer**. Notice that all the White objects, including the door, now look like brick.

In the **Rendering Settings** dialog, under the **Entity Color** section, you can choose to assign a color or bitmap texture to all objects of a particular color (as we did with all the White objects in the above example), or to all objects on a particular layer.

In the illustration below the **Layer** option was used to add a grass texture for the slab, a brick texture for for the walls, and a shingle texture for the roof. The image was created using the **Raytrace** option. As you can see, the **Object Viewer** adds another powerful set of tools for creating and displaying your designs.

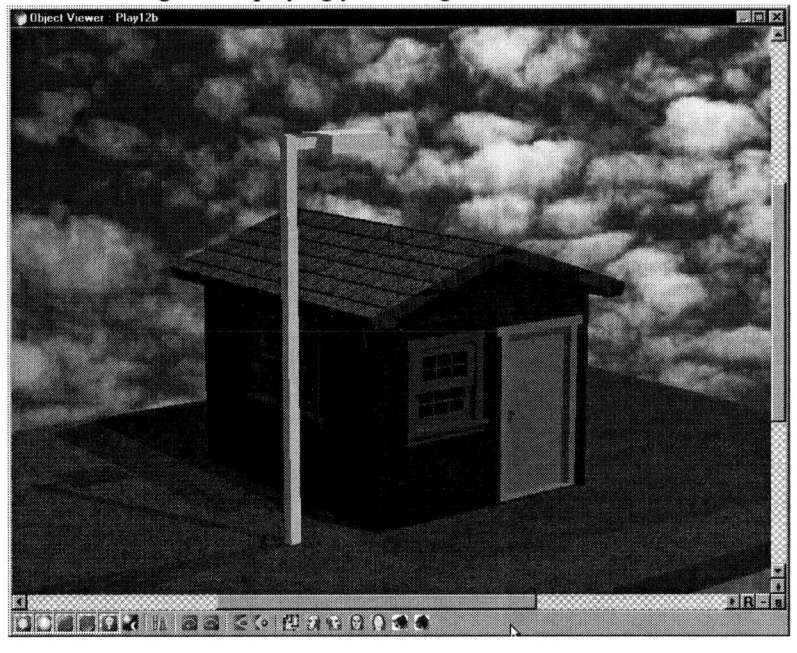

You can use *either* the **Color** or the **Layer** option in any drawing file, but not both.

Using the Object Viewer

Review and Important Points

- Objects can be moved to absolute coordinate positions.
- The **GoTo View** function can be used to save or recall views.
- The multi view display can be used to quickly switch between several different views.
- Many **Shader** settings can be adjusted either through the **Shader** menu, or through the **Shader Settings** dialog box.
- If the **Right-Click to Exit** option is toggled **OFF** in the **Shader Settings** dialog box, right clicking on a shaded image will open a dialog box that provides image saving or window adjusting options.
- If the **Right-Click to Exit** option is toggled **ON** in the **Shader Settings** dialog box, right clicking on a shaded image will close the image without saving. If you want to save a shaded image when this option is **ON**, use **File, Save As** from the **MENU BAR**.
- To show light placement symbols on the display, toggle **Show Lights ON** in the **Shader Settings** dialog box.
- Objects can either cast shadows, or receive shadows; they cannot do both.
- Only one light can cast shadows in any rendering. If more than one light has the **Cast Shadows** toggled **ON**, no shadows will be shown.
- The DataCAD color palette contains 255 colors that contain 15 initial DataCAD colors. The remaining are numbered from 16 to 255.
- The color of any entity or group can be changed to any color in the DataCAD palette. If a different DataCAD palette is loaded, all entities will take on the colors (that have the same color number) of the new palette. New palettes can be created, saved, and loaded.
- The **Object Viewer** either uses the DataCAD color palette, or new colors that are assigned by **Ctrl + right clicking** on a entity and selecting the **Rendering Settings** option in the pop up menu dialog.
- Right clicking in the **Object Viewer** window opens the **Object Viewer** dialog box.

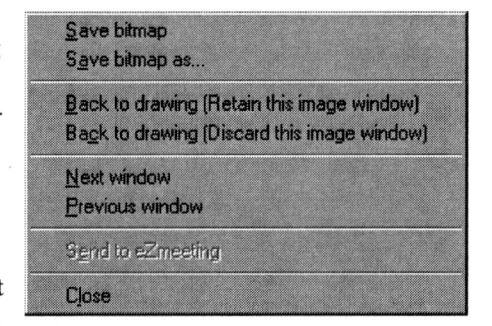

Chapter 12, Page 249

The DataCAD 11 Project Book

Practice Drawings

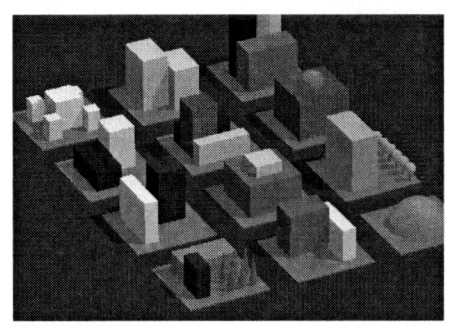

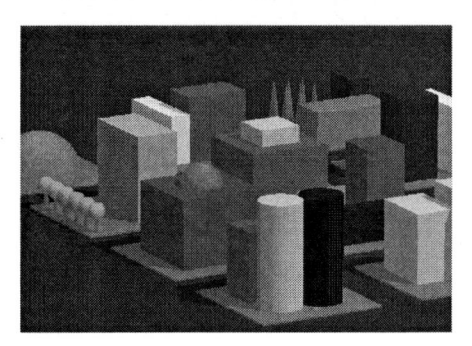

Project Hints
Assignment 12-A, Ferd City

- Open the **Ferd City** drawing file and change the colors of different buildings.
- Use **GoTo View** to save different parallel and perspective views of the city.
- Use the **Shader** options to create and save shaded views of the city.
- Use the **Object Viewer** to tour, and save images of **Ferd City**.

Notes

The DataCAD 11 Project Book

CHAPTER 13

Detail Drawings

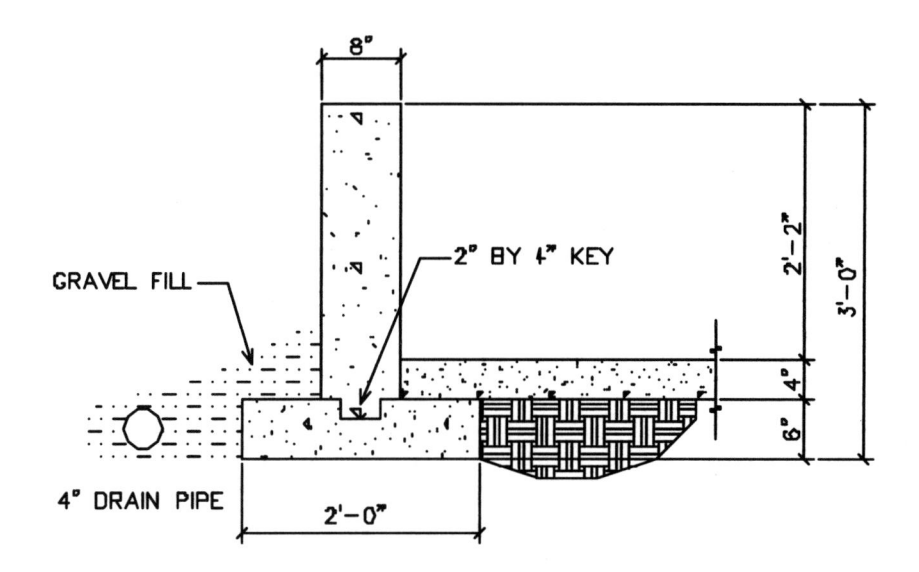

Project Goals

When you have completed this project, you will be able to:
1. Use the **Layer Manager** to create and control drawing layers.
2. Use DataCAD to create a typical foundation detail.
3. Use different hatch patterns to show different materials used in construction.
4. Adjust hatch pattern scale.
5. Hatch around a non-hatched area.

The DataCAD 11 Project Book

Prerequisite: DataCAD 11 interface experience

Introduction

In this session, you will begin to explore using DataCAD to create detail drawings. One of the common features of architectural details is the use of hatching to indicate different materials. DataCAD provides a large selection of different hatch patterns. The illustration on the previous page shows a typical foundation detail that will be created in this session. Observe the different hatch patterns used. Also, note that DataCAD is smart enough to hatch around areas such as the drain pipe.

1.__ Start DataCAD and create a new drawing file named **Detail13**.

Using the Layer Manager

The detail drawing for the foundation will have three drawing layers: **DETAIL, DIM, and HATCH**.

2.__ Press **Ctrl+L** to open the **Layer Manager** dialog box [or click the *left* mouse button when the cursor is over the button to the left of the layer name in the **STATUS AREA**].

INFO STOP   Although Chapter 2 covered the use of the **Layer Manager**, let's just take another quick look at its various options.

- The **ON** column shows a light bulb that is *green* for ON and active, *white* for off, and *yellow* for on.

- The **Lock** option prevents any changes to a locked layer (the padlock will alternate between locked and unlocked as you click on it).

- The **Count** column shows the number of entities on the layer.

- The three buttons near the top of the **Layer Manager** dialog box can be used to toggle all layers on, set a layer to be active, or turn all layers off except for the active layer.

Using the Layer Manager

- The six buttons on the right side of the **Layer Manager** dialog box can be used to add **New** layers, **Erase** the entities on a layer (without removing the layer from the drawing), **Delete** a layer from the drawing, **Delete Empty** layers (layers with 0 entities on them) from the drawing, or **Move** a layer up or down the layer list.

    ••• **WARNING**••• Have you saved your drawing lately?

3. ___ In the **Layer Manager** dialog box, select **New**. Then enter **2** in the pop up window and click **OK**.

4. ___ Edit the layer names (by double clicking on the layer name column) so that they read as follows: layer 1 - **DETAIL**, layer 2 - **DIM**, layer 3 - **HATCH** (see Figure 13-1).

After creating and naming new layers, your display will look similar to Figure 13-1.

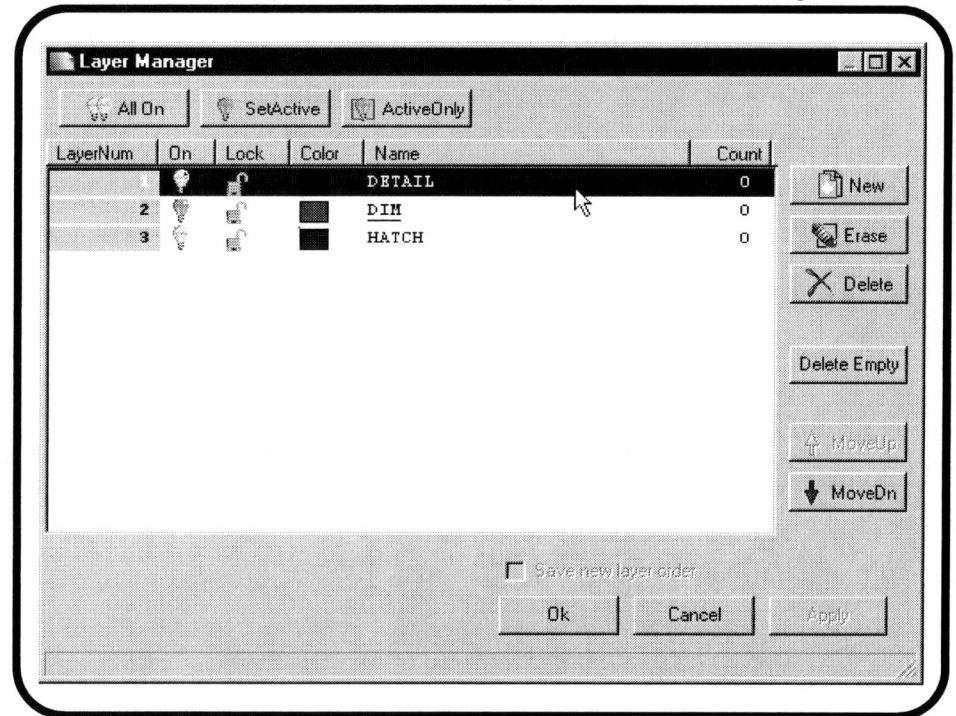

Figure 13-1, The Layer Manager dialog box.

5. ___ Set the desired color for each layer by clicking on the **COLOR** column and then selecting a color from the **Color Palette** dialog box.

Chapter 13, Page 255

The DataCAD 11 Project Book

6.___ Turn all layers **ON**, and make the **DETAIL** layer active. Then select **OK** to close the **Layer Manager**

Creating a 2D Detail Drawing

Now that the layer business has been taken care of, it is almost time to start drawing the foundation detail. Since this drawing will be completely in 2D, the **Z-base** and **Z-height** will both be set to 0.

7.___ Press the **z** [or the **Z**] key, and enter **0** for the **Z-base** and **Z-height**.

8.___ Press the **s** [not the **S**] key and set the snap to **0.2** for both X and Y.

9.___ Check the **w** in **SwOTHLUDFB** to be sure that walls are toggled **OFF**. (Walls can also be toggled on and off from the **Architect** menu, or by pressing the = key on the keyboard.)

When you first start drawing, the display scale might not be appropriate. This can be quickly fixed as soon as you have created a couple of lines.

10.___ At the prompt *Select first end point of new line/wall*, click the *left* mouse button with the cursor near the *lower left* corner of the display.

11.___ If the information line does not show **Relative Cartesian**, press the **Insert** key until **Relative Cartesian** entry mode is selected. Press the **Space Bar** and enter values of **X=2, Y=0**.

12.___ Press the **Space Bar** and enter values of **X=0, Y=0.6**.

Now the display scale can be adjusted.

13.___ Press the **Zoom to area** button in the **NAVIGATION PAD** [or press the / key] and capture the two lines in a zoom window.

14.___ Press the *right* mouse button *once* to exit the **Zoom** menu.

15.___ Complete the footing outline using the dimensions show here.

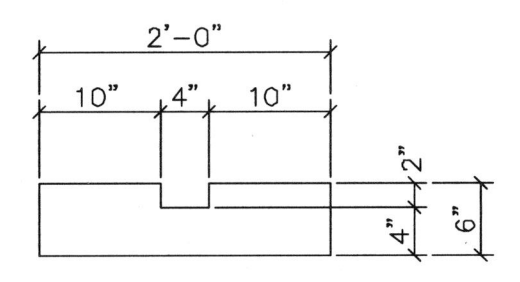

Creating a 2D Detail Drawing

16.__ When the footing has been completed, add the **8** inch foundation wall **2** feet **6** inches above the footing as shown in the illustration.

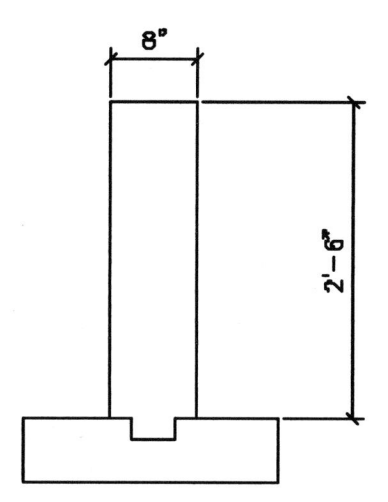

The next step will be to add a 4 inch slab to the right of the wall. This slab will be shown with a conventional break line.

17.__ Draw two horizontal lines 4 inches apart that extend 2 feet to the right of the foundation wall. (See Figure 13-2)

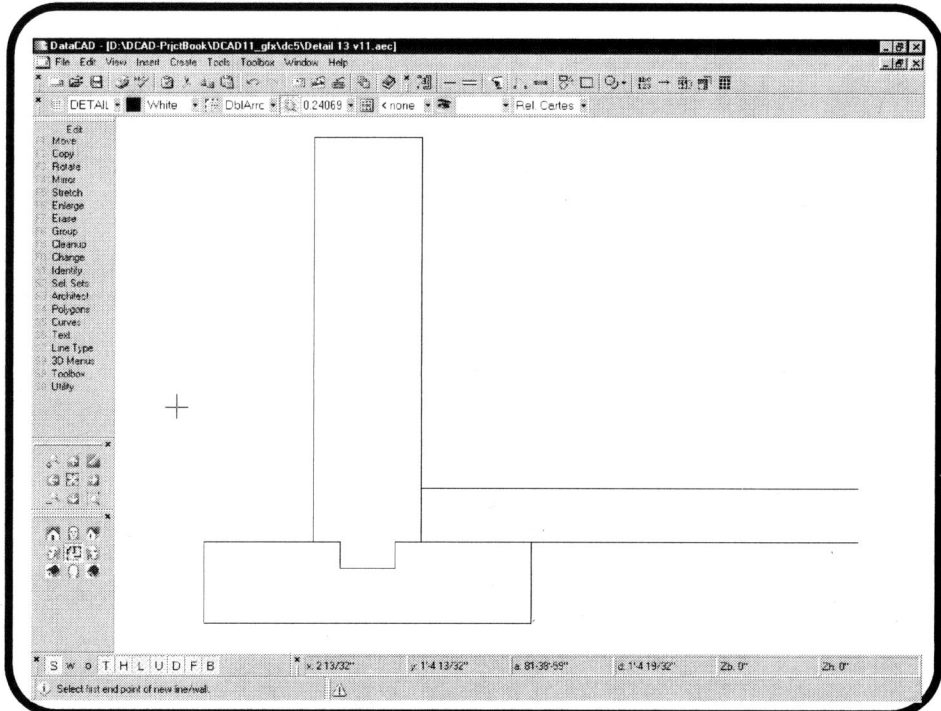

Figure 13-2, Adding a floor slab.

Chapter 13, Page 257

The DataCAD 11 Project Book

Changing Line Type

A different line style will be used to show that the slab has been broken. Observe that the **STATUS AREA** shows the current line type. A new line type can be selected either by pressing the **q** or the **Q** key to scroll forward or backward through the list of lines, or from the **Line Type** menu.

18.__ From the **Edit** menu, select **Line Type**. Select **Scroll Fwrd** to the next page of line types and then select **NewSectn** as shown in Figure 13-3.

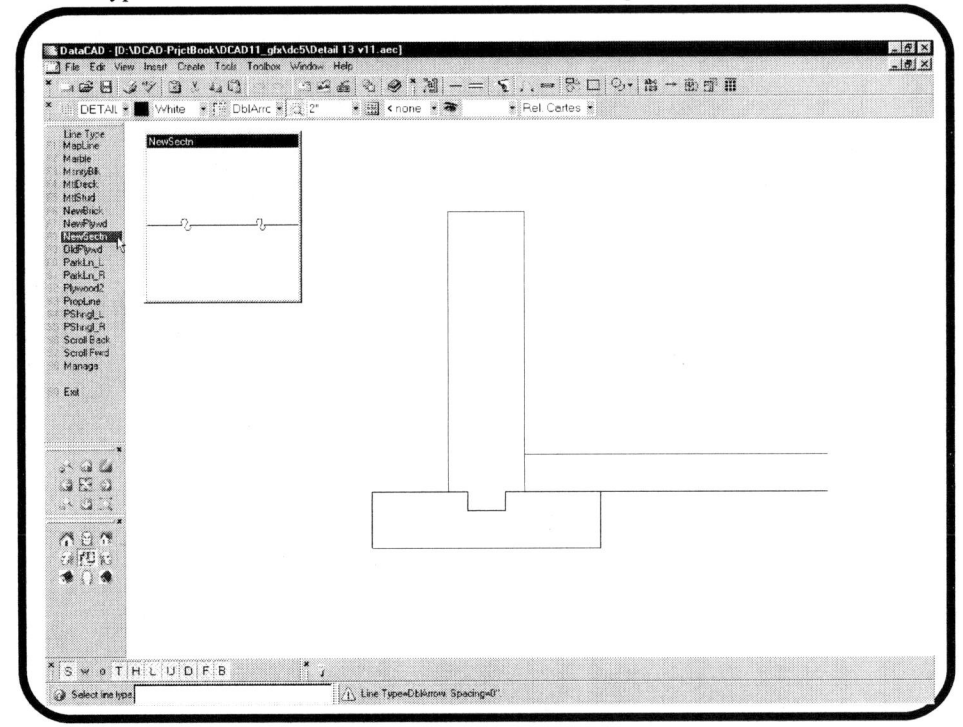

Figure 13-3, Selecting a new line type.

19.__ From the **Line Type** menu, select **Spacing** and enter a spacing of **0.5**.
20.__ Add a vertical line at the right end of the floor slab as shown in Figure 13-4.
21.__ Press the **Q** [or the **q**] key until the line type shows as **Solid** in the **STATUS AREA**.

••• **WARNING**••• Have you saved your drawing lately?

Adding a Circle

Adding a Circle

Next, a 4 inch diameter drain pipe will be added to the left side of the foundation. Press the unshifted ~ key and then click the *object snap* button to reset the relative 0,0 coordinate at the center of the *left vertical* line in the *footing*.

22.__ From the **Edit** menu, select **Curves, Dia. Circ.** (circle by diameter). Make sure **Dynamic** is ON.

23.__ *Left-click* in the drawing window, to the left of the footing, to define the left side of the circle to represent the drain pipe. Move the mouse to the right and you will see a circle being dynamically drawn, attached to the starting point. Relative coordinates will be used to set the other point on the diameter of the circle.

24.__ Press the **Space Bar** and enter values of **X=0.4, Y=0**.

Your display should look similar to Figure 13-4 (after adjusting the display scale).

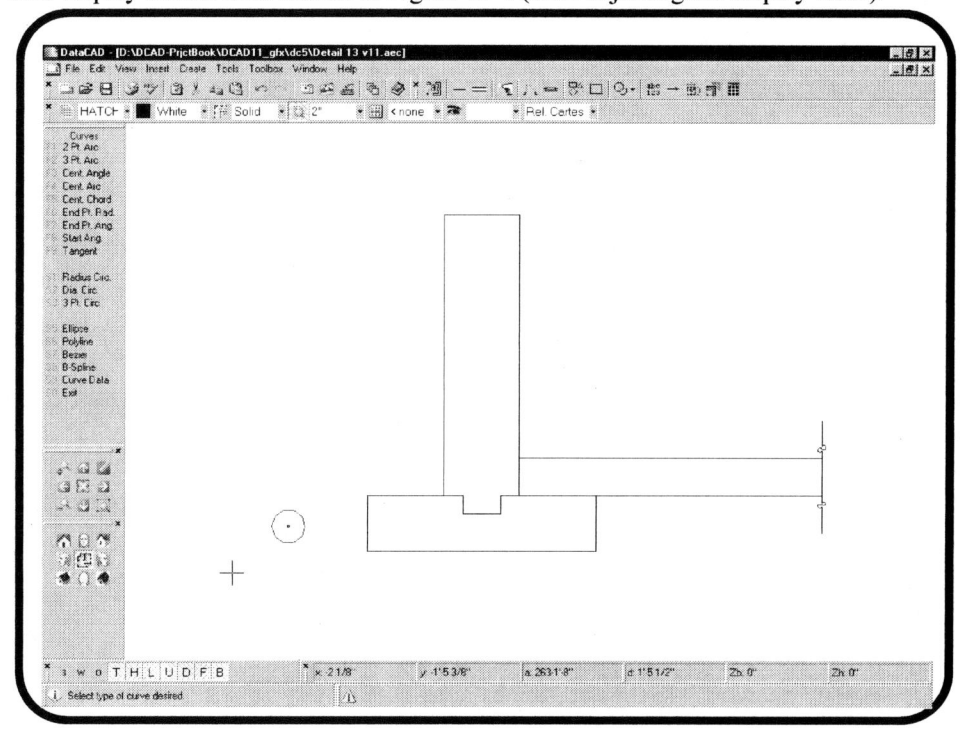

Figure 13-4, Adding a drain pipe.

Chapter 13, Page 259

The DataCAD 11 Project Book

Adding Hatching

This completes the geometry for the footing detail. Next, some hatching will be added. Although DataCAD can handle the hatching of complex areas, the boundary for the hatching must be clearly defined. For simple boundaries like the footing, floor slab, and foundation wall, a search option can identify a closed area for you.

Associative Hatching

There are two kinds of hatching in DataCAD: *Associative* and *non-Associative*. We'll start with Associative hatching.

25.__ Press the **TAB** key until the **HATCH** layer is active.

26.__ Press the **h** [or the **H**] key to jump to the **Hatch** menu.

27.__ From the **Hatch** menu, toggle **Associative ON**.

28.__ From the **Hatch** menu, select **Pattern**. Then select the **concrete** hatch pattern.

29.__ From the **Hatch** menu, select **Scale**. Enter a hatch scale of **15**.

30.__ From the **Hatch** menu, select **Boundary**. Then select **Cont. Search**.

> **INFO STOP**
>
> **Cont. Search** stands for *Contour Search*, so named because DataCAD will search for the boundaries, or contours, of an area to be hatched. However, **Cont. Search** will only work if the boundaries are completely closed (no openings or gaps). In this example the footing and the foundation wall are completely closed. The slab would only be enclosed if the cut line were touching the slab lines, forming a closed rectangle.

31.__ Move the cursor somewhere inside of the *floor slab* area and click the *left* mouse button *once*.

The perimeter of the floor slab should now be highlighted with the current drawing color.

32.__ Press the *right* mouse button once to exit the **Boundary** menu and the concrete hatch pattern will be drawn.

33.__ The floor slab should now be hatched. Next, the scale will be changed and the footing will be hatched.

34.__ From the **Hatch** menu, select **Scale**. Enter a hatch scale of **25**.

Adding Hatching

35. __ To provide a different look, select **Angle** and enter **45**.
36. __ From the **Hatch** menu, select **Boundary**. Then select **Cont. Search**. Move the cursor somewhere inside the *footing* and click the *left* mouse button *once*.
37. __ Press the *right* mouse button once to exit the **Boundary** menu and the concrete hatch pattern will be drawn.
38. __ Change the **Scale** to **30** and the **Angle** to **90**; then add hatch to the foundation wall.

Your display should look similar to Figure 13-5.

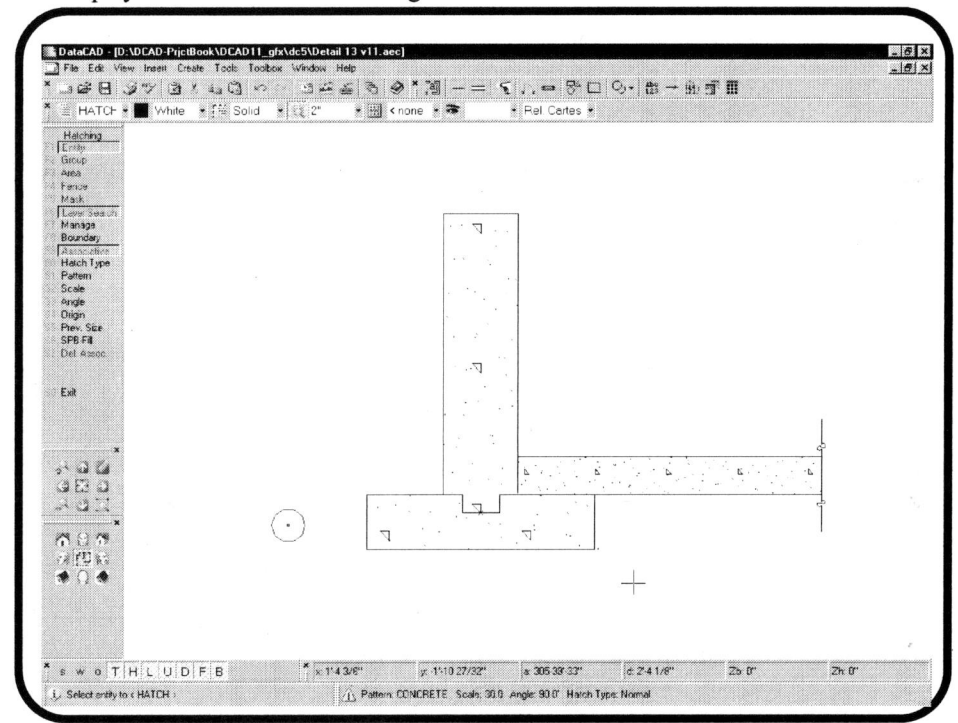

Figure 13-5, Adding hatching.

Next, some earth will be shown under the floor slab. There is no closed area for this, so a hatch boundary will be constructed.

39. __ From the **Hatch** menu, select **Pattern**. Then select **Scroll Fwrd** and select **earth**.
40. __ Change the **Scale** to **100** and the **Angle** to **0**.
41. __ Toggle **Associative ON**.

Chapter 13, Page 261

The DataCAD 11 Project Book

42.__ Select **Hatch Type** and toggle **Normal** and **Outline ON**.

43.__ From the **Hatch** menu, select **Boundary**. Use the mouse to sketch out an area as shown in Figure 13-6. Press the *right* mouse button **twice** to exit the **Boundary** menu and the concrete hatch pattern will be drawn.

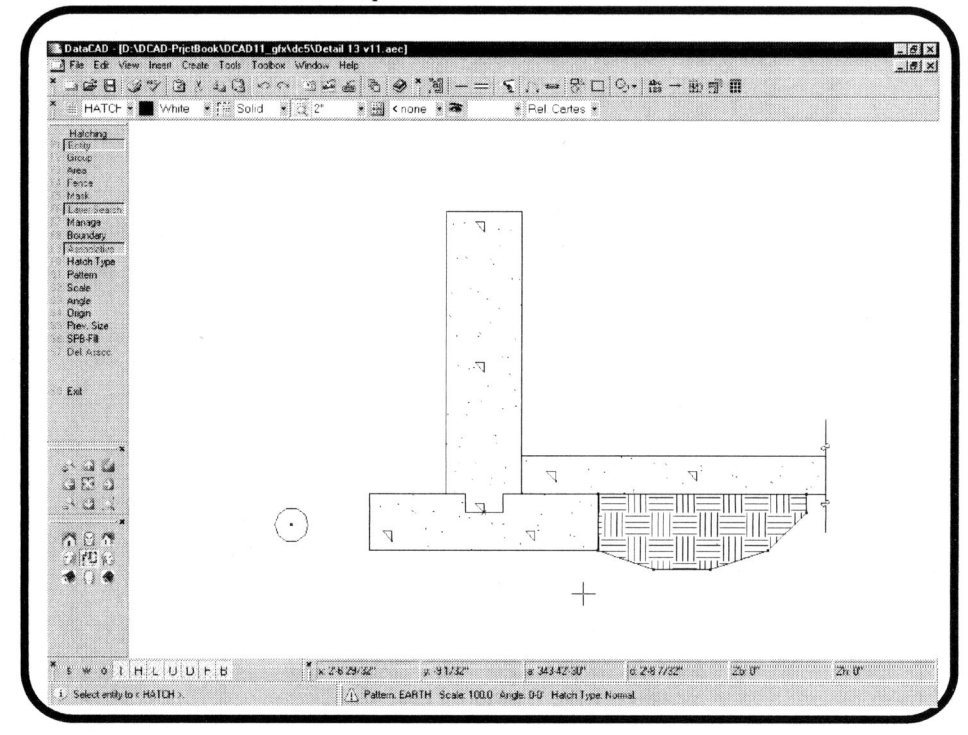

Figure 13-6, Adding earth fill.

•• **NOTE:** If you need to draw more than one instance of the same hatch you don't have to press the *right* mouse button to exit the **Boundary** menu every time. You can create two more hatched areas, and *then* press the *right* mouse button to exit the **Boundary** menu. All the hatches will then be drawn at once. Try it.

The boundary sketched for the earth fill is shown along with the hatch in Figure 13-6. If **Outline** would have been toggled OFF in the **Hatch Type** menu, the hatch would look the same, but no boundary would be shown.

Adding Hatching

Observe that the boundary that was drawn for the earth fill is actually a polyline. Since the **Associative** option was on, the hatch is actually associated with this polyline boundary. This means that if you use the **Stretch** function and move any of the points on the polyline, the hatch will be automatically adjusted with it.

44.__ Press **Alt+S** to jump to the **Stretch** menu. *Click* two places for the first and second points defining the distance to stretch, then capture one or more points on the earth fill polyline boundary using **Point, Area** or **Fence**, and the hatch will stretch to match the new boundaries.

Very cool! Next, some gravel fill will be added to the area around the drain pipe. First, a brief discussion of the **Hatch Type** option.

Non-Associative Hatching

Next we'll use non-Associative hatching. Unlike Associative hatching, non-Associative hatching is not defined by a polyline boundary, and cannot be edited after being drawn. The resulting hatch will be drawn of dozens or even hundreds of line segments. This makes editing and cleanup a real chore. It also makes your drawing file bigger, since there are more individual entities in the drawing. In most cases you will not want to use non-Associative hatching. The only options that non-Associative hatching has that Associative does not, are the **Normal, Outermost and Ignore**, described below.

There are three different hatch types available: **Normal, Outermost** and **Ignore**. The illustration to the right shows the differences between the different hatch types. When the **Normal** option is used, each time the system crosses a boundary, it will start or stop hatching. In a situation where there are islands inside of islands, this results in alternate hatched and un hatched areas.

When the **Outermost** option is used, the system will only hatch the outer boundary: any islands inside of islands will not be hatched. When the **Ignore** option is used, the system will hatch over any enclosed islands.

In the exercise presented here, the **Normal** hatch type option will work. However, both the hatch boundary and the polygon representing the drain pipe must be selected before beginning the hatch.

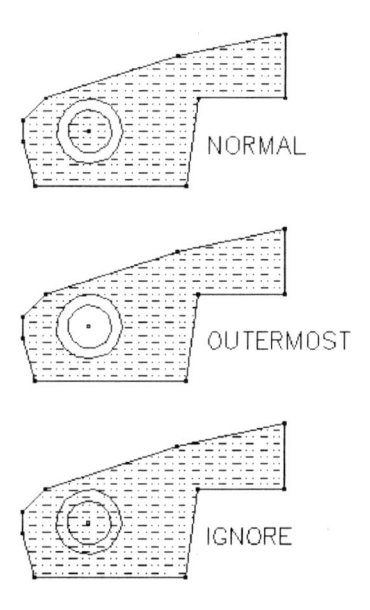

Chapter 13, Page 263

The DataCAD 11 Project Book

45.__ From the **Hatch** menu, select **Pattern;** then (use **Scroll Fwrd** as required) select the **mudst** pattern.

46.__ Set the hatch **Scale** to **40** and the **Angle** to **0**.

47.__ Toggle **Associated OFF**.

48.__ Select **Hatch Type** and toggle **Normal ON**.

49.__ From the **Hatch** menu, select **Boundary** and trace a hatch boundary similar to that shown in Figure 13-7. (Click the *right* mouse button to close the boundary.)

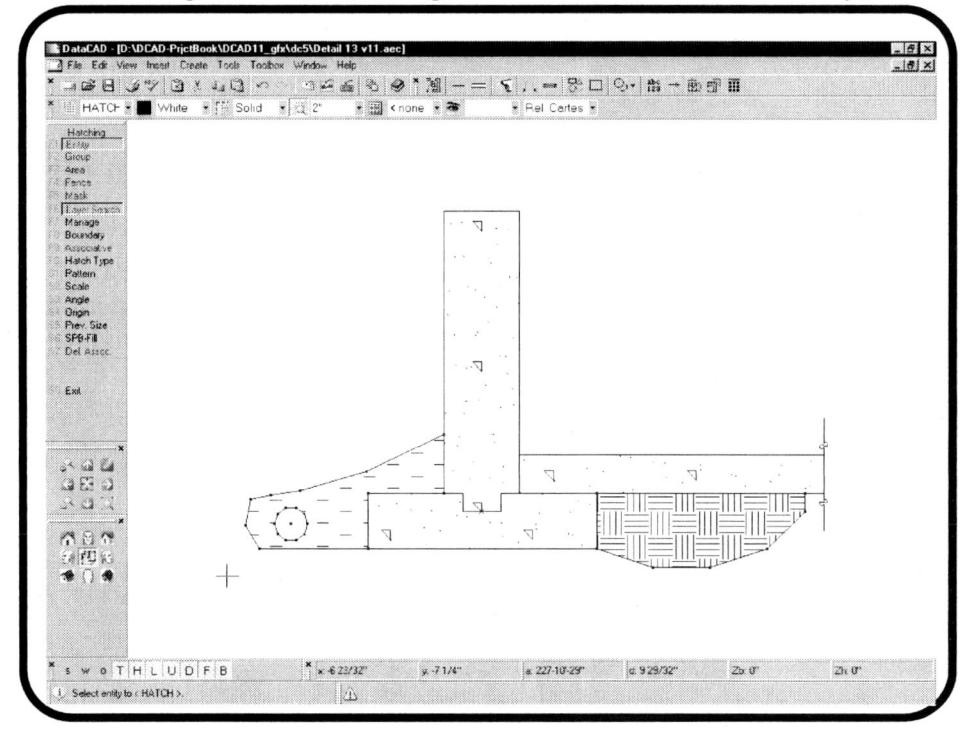

Figure 13-7, Creating a new hatch boundary.

50.__ Select **Cont. Search** and click on the drain pipe circle.

51.__ Click the *right* mouse button once to exit the **Boundary** menu and the hatch pattern is drawn.

Your display should now look similar to Figure 13-7. DataCAD should have left the hatching out of the drain pipe circle. Now that the hatching is complete, you should take some time to finish the detail by adding appropriate dimensions and notes.

Chapter 13, Page 264

Adding Hatching

52.__ Add dimensions and notes as shown in Figure 13-8. Be sure to place them on the **DIM** layer.

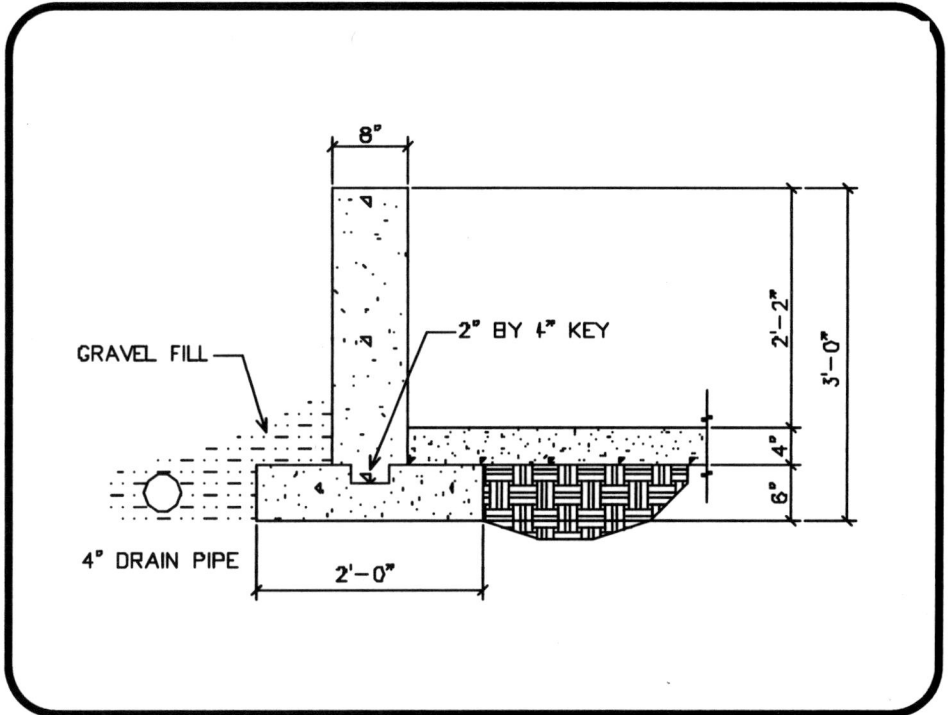

Figure 13-8, The completed foundation detail.

INFO STOP

If you haven't added notes with arrows to your drawings yet, here is how it is done. Use the **Text** menu to add the desired text. Then, from the **Text** menu, select **Arrows**. Start at the beginning of the leader line near the text and click points for the leader line. *Right click* to end the leader and place the arrowhead. The size of the arrowheads is drawn relative to the size of the text, so be sure your text size is appropriate to the scale of the drawing before drawing the arrows.

Chapter 13, Page 265

The DataCAD 11 Project Book

Review and Important Points

- When working on 2D drawings, set the **Z-base** and **Z-height** both to **0**.

- The **Layer Manager** dialog box [**Ctrl+L**] provides a quick and easy way to manage layers.

- When constructing a conventional break line, use the **Section** line style.

- Press the **q** or the **Q** key to scroll forward and backward through the different line types.

- If you want to use the same detail on different drawing projects you can cut and paste the detail between drawings, just as you would with any other Windows program: 1. In DataCAD, select **Edit, Clipboard Select** from the **Menu Bar**; 2. select all the entities that you want to copy by using **Entity, Group, Area, Fence** or **Mask**; 3. select **Copy** from the **Add To Set** menu; 4. open the drawing that you want to copy into; 5. select **Edit, Paste**; 6. select **Orig Layers, Active Layer** and/or **By Center**, as appropriate; 7. *click* the left mouse button once and the detail will be placed in the new drawing.

Adding Hatching

Practice Drawings

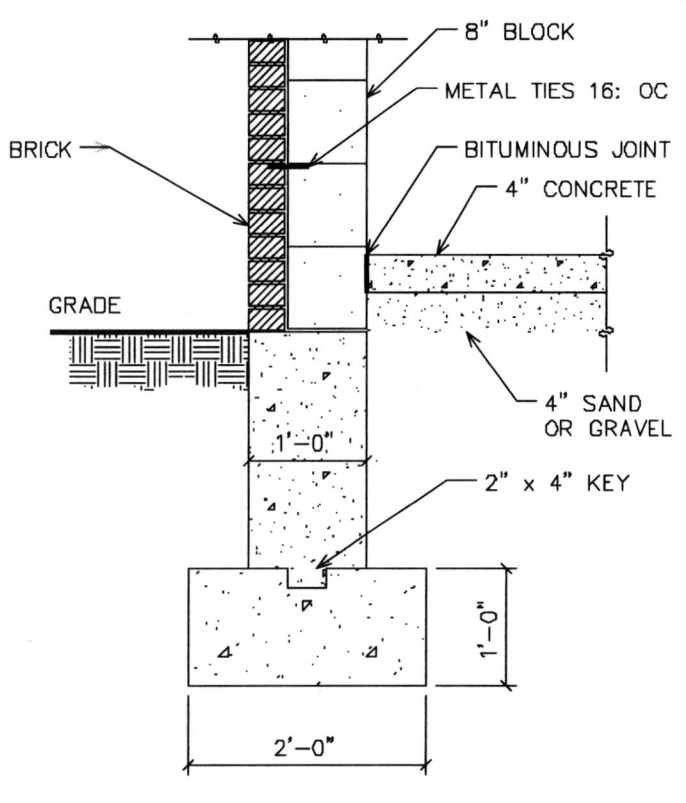

Assignment P13-A

- Create a foundation detail as shown above.
- Use the **cement** hatch pattern for sand.
- Use the **brick Line Type** for brick.
- Use **ansi31** hatch pattern for brick.
- Set **Line Weight** to **3** in the **Line Type** menu for metal ties and bituminous joint lines.

The DataCAD 11 Project Book

Notes

CHAPTER 14

Typical Wall Section

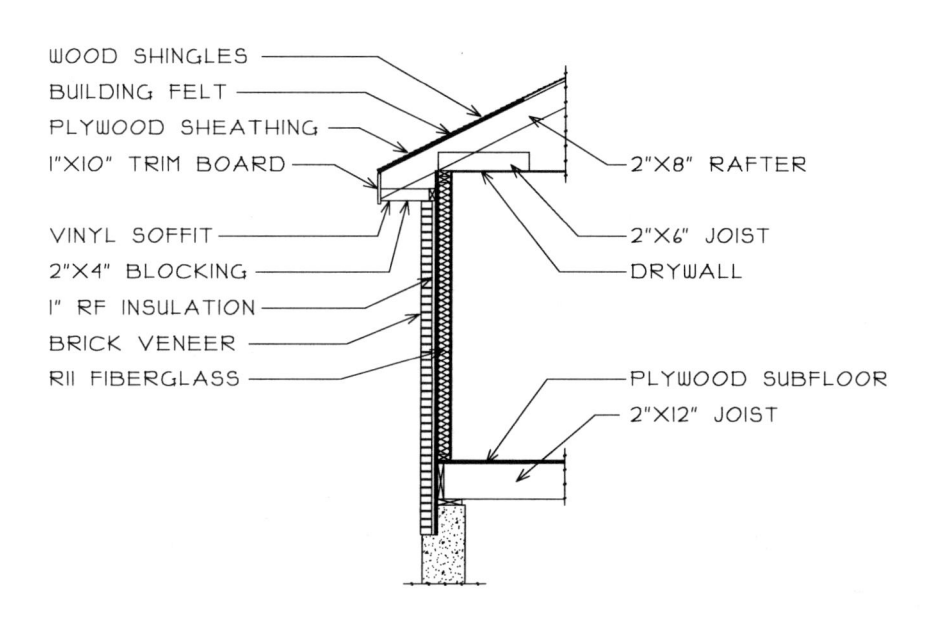

WOOD SHINGLES
BUILDING FELT
PLYWOOD SHEATHING
1"X10" TRIM BOARD
2"X8" RAFTER
VINYL SOFFIT
2"X6" JOIST
2"X4" BLOCKING
DRYWALL
1" RF INSULATION
BRICK VENEER
R11 FIBERGLASS
PLYWOOD SUBFLOOR
2"X12" JOIST

Project Goals

When you have completed this project, you will be able to:

1. Create a typical wall section.
2. Use different line types and spacing.
3. Trim and extend lines using editing commands.

The DataCAD 11 Project Book

Prerequisite: DataCAD 11 interface experience

Introduction

In Chapter 13, the topic of 2D detail drawing was introduced. A very common type of detail drawing is a typical wall section. In this project, you will create a typical wall section and, in the process, utilize some of the different line types and editing features in DataCAD that make drawing typical wall sections much easier.

The wall section to be constructed in this project is illustrated on the first page of the chapter. There are a lot of variations in wall design, and this example is not intended to be the last word on the subject. The wall section presented here is provided as a tool for introducing a number of DataCAD functions and features that have not been covered in earlier chapters.

1. __ Start DataCAD and create a new drawing file named **Detail14**.
2. __ Press **Ctrl+L** to open the **Layer Manager** dialog box [or click on the layer name in the STATUS AREA].
3. __ In the **Layer Manager** dialog box, select **New**. Then enter **1** in the pop up window and click **OK**.
4. __ Edit the layer names (by double clicking on the layer name column) so that they read as follows; layer 1 - **DETAIL**, layer 2 - **TEXT**.
5. __ Set the desired color for each layer by clicking on the color box in the **COLOR** column and then selecting a color from the **Color Palette** dialog box.
6. __ Make the **DETAIL** layer active and **ON**.
7. __ Select **Ok** to close the **Layer Manager** dialog box.

   ••• **WARNING**••• Have you saved your drawing lately? This warning will not be given again. Remember to save often. If you don't, you might have to start the project all over again if you make a mistake or have a power outage.

Since this drawing will be done completely in 2D, the **Z-base** and **Z-height** will both be set to 0.

8. __ Press the **z** [or the Z] key, and enter **0** for the **Z-base** and **Z-height**.
9. __ Press the **s** [not the S] key and set the snap to **0.2** for both X and Y.
10. __ Check the **w** in **SwOTHLUDFB** to be sure that walls are toggled **OFF**. (Walls can be toggled on and off from the **Architect** menu.)

Selecting a Line Type

Selecting a Line Type

11. ___ From the **Edit** menu, select **Line Type** [or position the cursor over the button to the left of the *line type* in the **STATUS AREA** and click the *left* mouse button].

12. ___ Select **Scroll Fwrd** to the next page of line types; then select **DShngl_L** as shown in Figure 14-1.

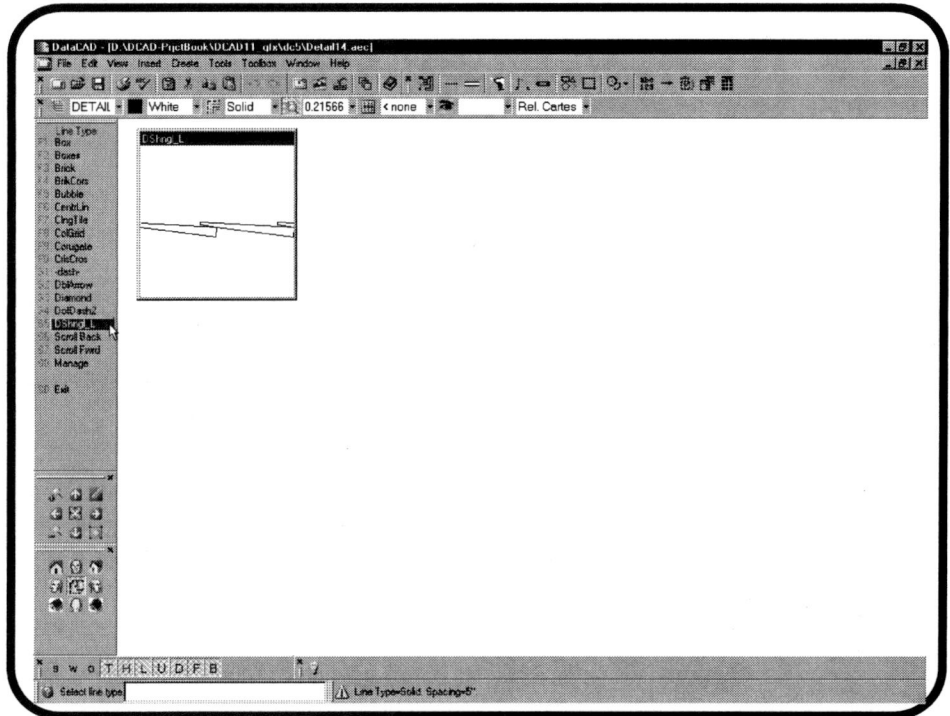

Figure 14-1, Selecting the shingle line style.

DataCAD allows the user to adjust the spacing to any desired value by changing the **Spacing** value in the **Line Type** menu.

The DataCAD 11 Project Book

> **INFO STOP**
> The **Spacing** value is determined by whomever created the particular line type you select. Therefore you may find that you need to use some trial and error to get the spacing and size of the line type that you desire. Many of the linetypes that represent materials (Insulation, Brick, Shingles, etc., were created around units of 1/32 inch. So a **Spacing** of **32** would give you a 1" width of height to your line type. Other linetype may have been created so that the **Spacing** value represents the size of the material. For instance, a **Spacing** of **8** for 8".

13.__ Click on **Spacing** and select a value of **0.4**. For this linetype that will give you an exposure of 4" for each shingle.

> **INFO STOP**
> Some line styles are evenly centered on the points selected for positioning the line, while other styles will be constructed to one side of the line. When you are using the various line styles you have to do an experiment to determine how the line details will be constructed.

The roof that will be created in this example will have a 6:12 pitch. This means that the roof drops 6 feet for every 12 feet of horizontal distance. (In math class, 6/12 is the same as 1/2 or 2/4.) Since a full 12 feet of roof is not wanted in the wall section, a piece of roof that is 4 feet wide will be drawn.

> •• **NOTE:** You may want to choose different colors for different parts of the wall section. For example, brown for shingles, light magenta for insulation, and so on. You can either select color from the **Line Type** menu, or press the **k** [or the **K**] key to cycle forward and backward through the color list.

Selecting a Line Type

INFO STOP The standard color list contains 15 colors. The **Custom** option found in the **Line Type, Color** menu, or the **Layers, Color** menu, allows for selecting from the full 256 color palette. In most cases you will want to stay within the standard 15 colors, as these are the only ones you can cycle through with the **K** or **k** keys.

14. __ From the **Line Type** menu select the desired color for shingles.
15. __ Right click to return to the **Edit** menu.
16. __ Anchor one end of a line. Use **Relative Cartesian** coordinate entry mode, press the **Space Bar**, and (watch the minus signs) enter values of **X=-4, Y=-2**.
17. __ Press the *right* mouse button to stop drawing lines.

After adjusting the display scale, your display should look similar to Figure 14-2

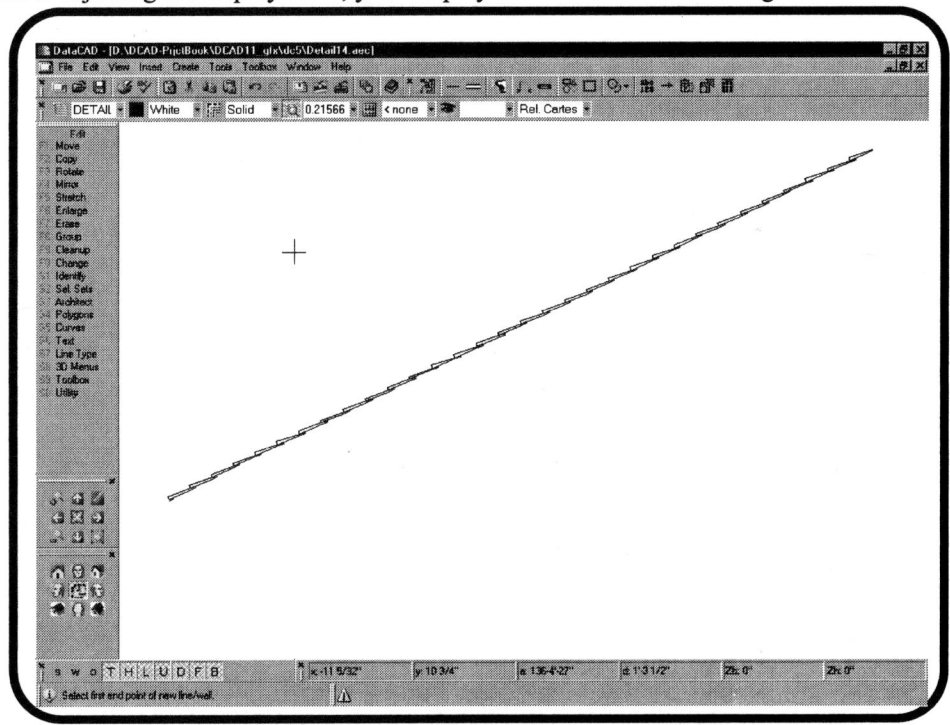

Figure 14-2, Creating a line of shingles.

Chapter 14, Page 273

The DataCAD 11 Project Book

Next, some plywood roof sheathing will be added.

18. __ From the **Edit** menu, select **Line Type** [or position the cursor over the button to the left of the *line type* in the **STATUS AREA** and click the *left* mouse button].

19. __ Scroll forward as required and select **Plywood2**.

20. __ Use the object snap to construct a line from the lower left to the upper right end of the shingle line.

21. __ Change the line type to **Solid** from within the **Line Type** menu.

Next, a 2 by 8 rafter will be added. Of course, only the 8 inch dimension of the rafter will be seen in the section view. Drawing the rafter in the section view requires drawing two lines that are 8 inches apart (a 2x8 is actually 7-1/4" wide, but let's keep it simple, here). A trick for this is to use the walls option.

22. __ Press the **a** [not the A] key to jump to the **Architect** menu.

23. __ From the **Architect** menu, toggle **Walls**, **2 Line Walls**, and **Outside ON**.

24. __ Select **Width** and set the width to **0.8**.(remember that you can actually omit the "0").

25. __ Right click out of the **Architect** menu, then press the **O** key to toggle **Ortho** mode **OFF**.

26. __ Press the unshifted ~ key and use the object snap function to set the relative 0,0 at the left end of the line of shingles.

27. __ Press the **Space bar** and enter values (watch the minus sign) **X=0, Y=-0.1**.

28. __ Press the **Space Bar**, and enter values of **X=4, Y=2**.

29. __ At the prompt *Select a point to define the Inside of the wall*, click below the shingle line. Then right click to stop drawing lines.

Chapter 14, Page 274

Selecting a Line Type

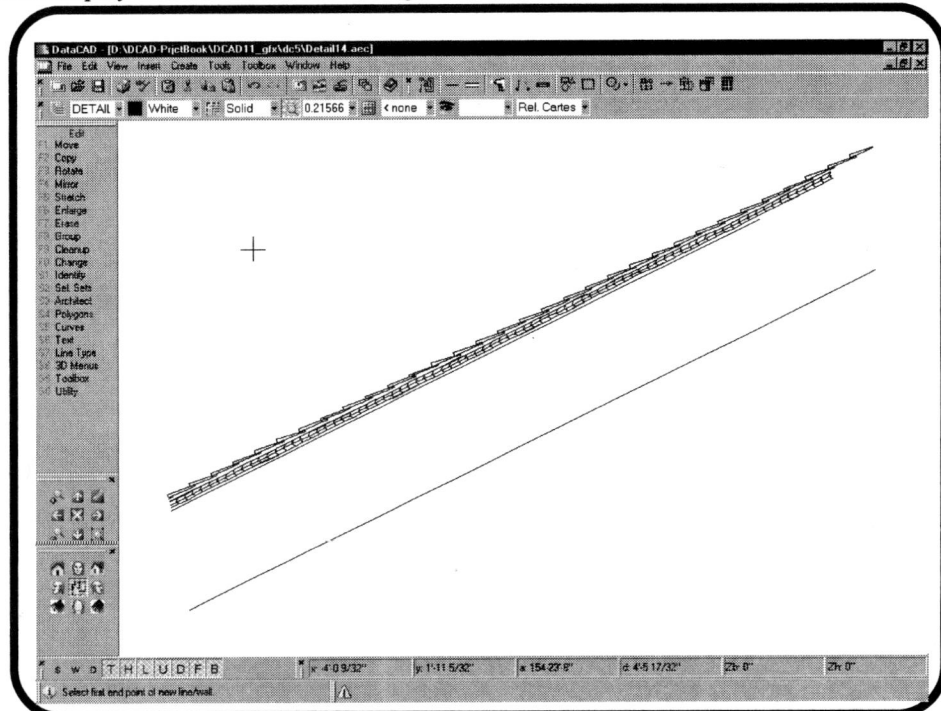

Your display should look similar to Figure 14-3.

Figure 14-3, Adding sheathing and a rafter.

Don't worry yet about the ends of the rafter lines. They will be edited after additional lines have been added. Next, a 1 by 10 trim board will be added.

30.___ Toggle **Walls OFF** by clicking on the "W" in **SWOTHLUDFB**.

31.___ Press the **s** [not the S] key and set the snap to **1** inch for X and Y (enter **0.1**).

32.___ Press **Alt+R** to jump to the **Polygons, Rectangle** menu [or select **Polygons, Rectangle** from the **Edit** menu].

Chapter 14, Page 275

The DataCAD 11 Project Book

33.___ Create a **1** inch by **10** inch rectangle at the left end of the roof as shown in Figure 14-4.

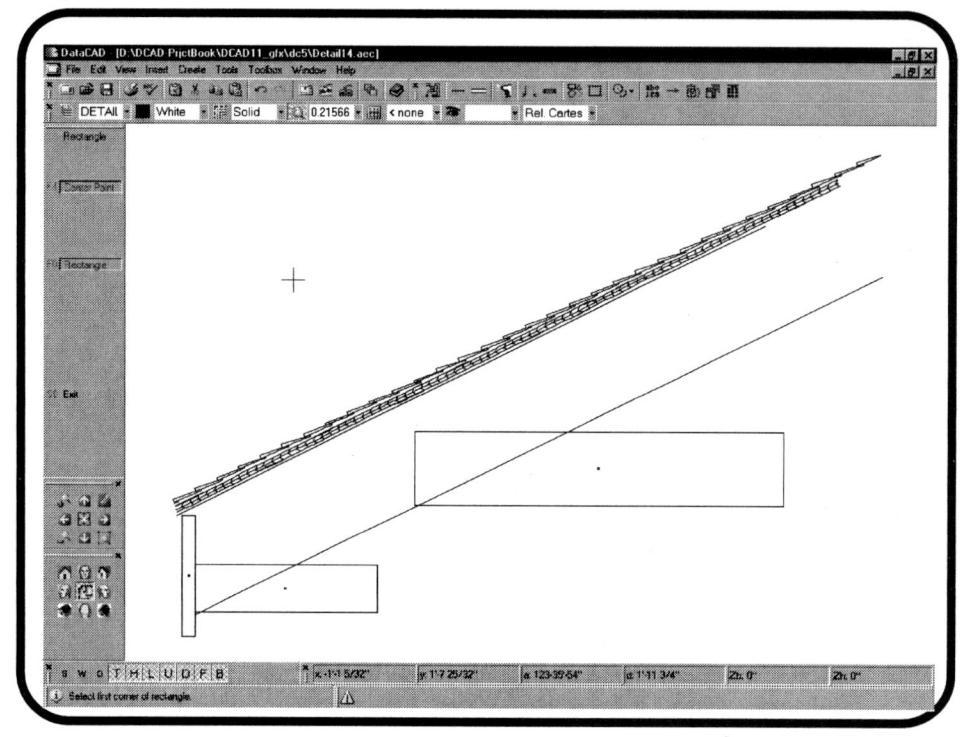

Figure 14-4, Adding a trim board.

Trimming Lines

Next, the bottom edge of the rafter will be cleaned up.

34.___ Press **Alt+B** to jump to the **1 Line Trim** menu [or select **Cleanup, 1 Line Trim**, from the **Edit** menu].

35.___ When the **1 Line Trim** menu is displayed, select **Entity**. Then click on the *right vertical* line on the 1 by 10 rectangle.

36.___ At the prompt, *Point to outside*, click somewhere to the *left* of the 1 by 10 rectangle.

37.___ From the **1 Line Trim** menu, select **Entity**; then click on the lower line of the rafter.

Adding Blocking

Adding Blocking

Next, a 2 by 4 blocking board for the soffit, and a 2 by 6 ceiling joist will be added (see Figure 14-5). The blocking board will be created 3 inches too short to allow for another 2 by 4 and some insulation to be added later.

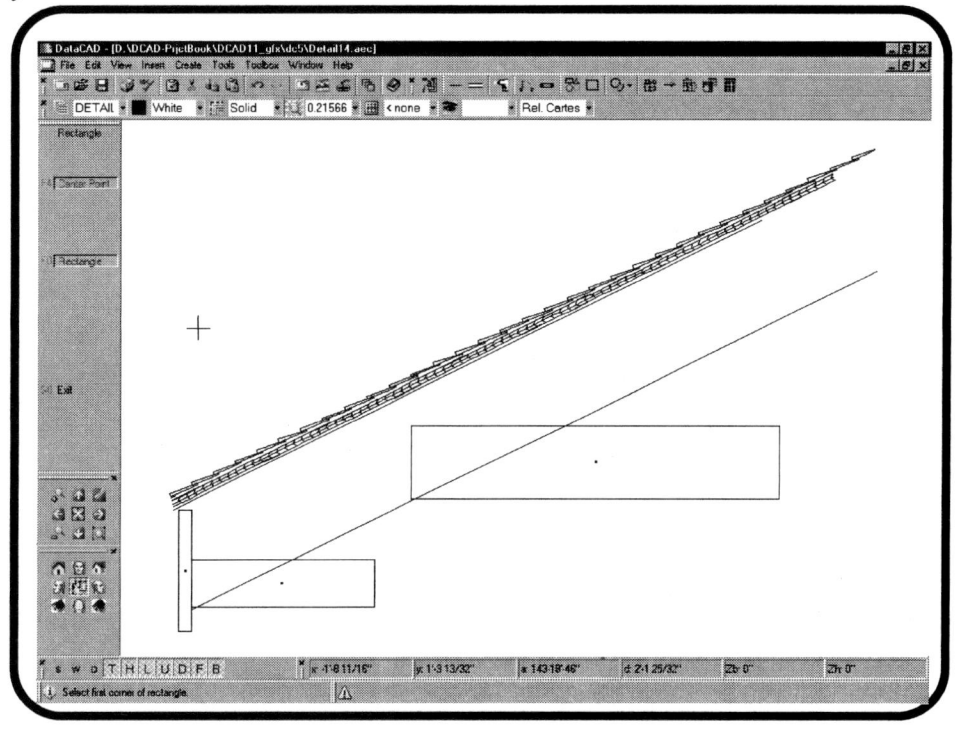

Figure 14-5, Adding a blocking board and ceiling joist.

38. ___ Use the **Polygons, Rectangle** option and construct a blocking board that is **4** inches high and extends **1** foot **3** inches to the right of the trim board (see Figure 14-4).

39. ___ Use the **Polygons, Rectangle** option and construct a **6** inch by **2** feet **6** inch rectangle that starts **3** inches to the right of the blocking board, and extends to the left, as shown in Figure 14-4.

Next, the double plate will be added. This consists of two 2 inch by 4 inch rectangles with lines drawn in an X pattern across each rectangle to indicate the end view of a framing member.

Chapter 14, Page 277

The DataCAD 11 Project Book

40. __ Use the **Polygons, Rectangle** option to construct two 2 inch by 4 inch rectangles under the left end of the ceiling joist (see Figure 14-6). Add crossing lines to each rectangle.

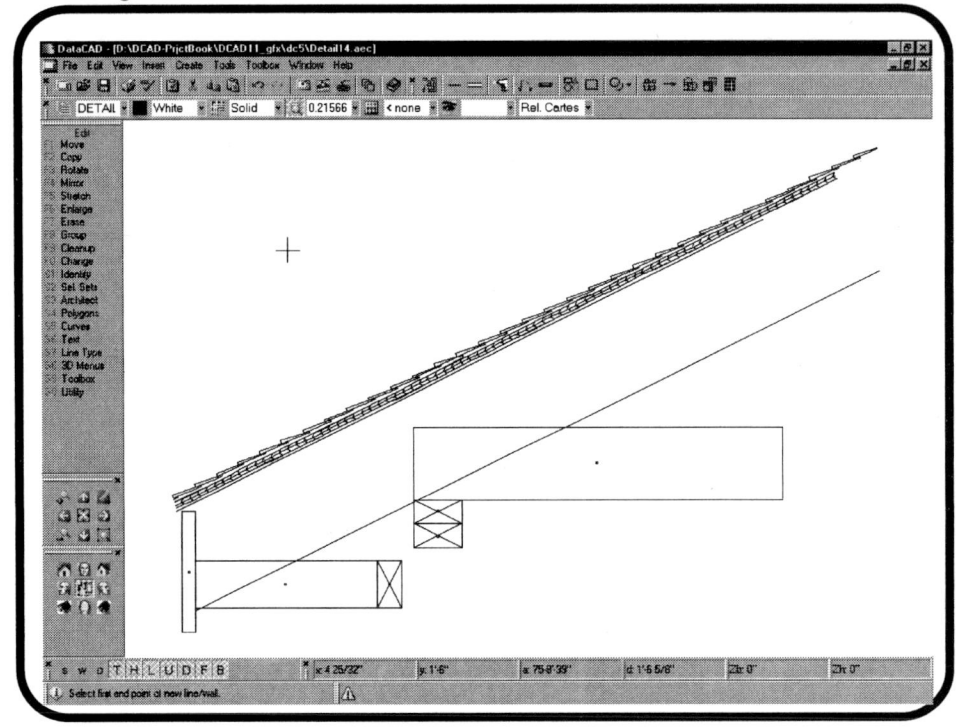

Figure 14-6, Adding a double top plate.

41. __ Use the **Polygons, Rectangle** option to construct a 2 inch by 4 inch rectangle at the end of the blocking board as shown in Figure 14-6. Add crossing lines to indicate the end view of a framing member.

Adding Insulation

Next, some vertical lines will be drawn for rigid insulation, fiberglass insulation, and a brick veneer. Don't worry about the length of these lines yet. They will be edited to length after the floor and foundation are constructed.

42. __ From the **Line Type** menu, select the **Inslu1** line type. Change the color as desired.

Adding Insulation

43.___ Since the **Insul1** line type is centered on the selected positions, construct a vertical line from the bottom center of the lower top plate rectangle downwards a couple of feet (see Figure 14-7).

44.___ From the **Line Type** menu, select the **RigdIns1** line type. Change the color as desired.

45.___ From the **Line Type** menu, select **Spacing** and enter a value of **0.1**.

46.___ Press the **s** [not the S] key, change the snap to 1/2 inch (**0.0.1/2**) and add a vertical rigid insulation line between the top plate and blocking board downwards a couple of feet.

47.___ From the **Line Type** menu, select the **Brick** line type. Change the color as desired.

> •• **NOTE:** The **Brick** line type constructs the bricks to one side (not the center) of the specified line.

48.___ From the **Line Type** menu, select **Spacing** and enter a value of **0.8** (since there are 3 bricks in every 8 inches, that's how the author of this line type created the **Spacing** value).

49.___ Change the snap to 1 inch (**0.1**) and add a vertical line of bricks from a point 1 inch to the left of the rigid insulation line, just under the blocking board downward.

Chapter 14, Page 279

The DataCAD 11 Project Book

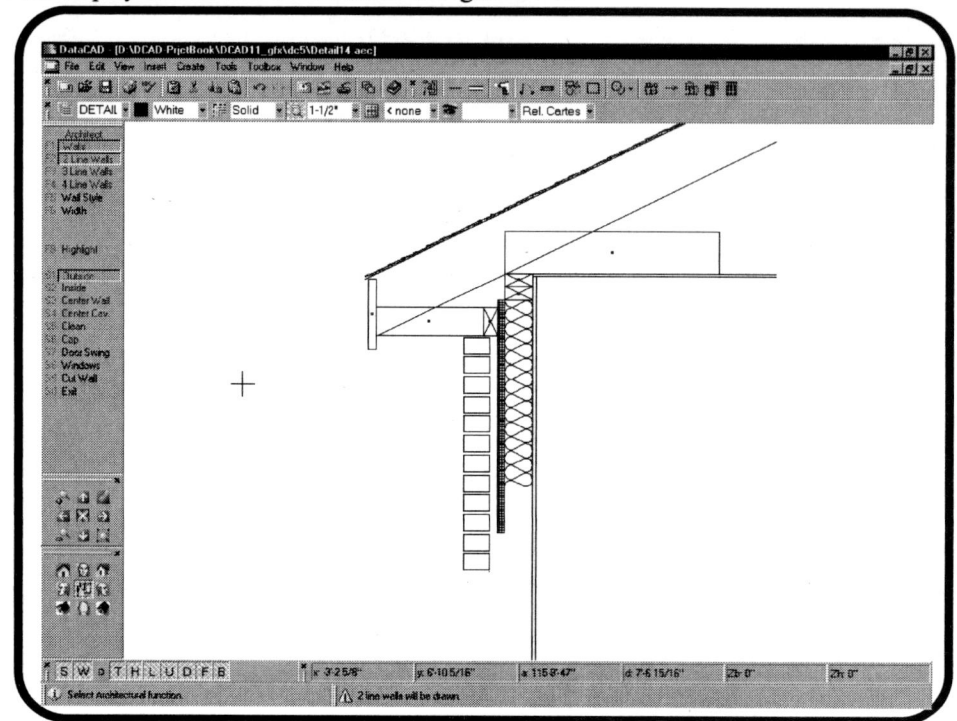

Your display should now look similar to Figure 14-7.

Figure 14-7, Adding insulation and bricks.

Next the wall thickness will be set to 1/2 inch and a double line will be added to show the drywall. (In this example, 1/2 inch drywall will be used for both the walls and ceiling, although different thicknesses might be used in most situations.)

50.___ Change the line type to **Solid**. Set the color as desired.

51.___ From the **Architect** menu, toggle **Walls, 2 Line Walls** and **Outside ON**.

52.___ Select **Width** and set the width to 1/2 inch (**0.0.1/2**).

53.___ Anchor one end of a line at the *upper right* corner of the top plate. Extend this line to the right of the ceiling joist. At the prompt *Select a point to define the Inside of the wall,* click below the line.

54.___ Use the **Object Snap** button to anchor the end of a line at the *lower left* end of the ceiling drywall. (Press the **x** [not the **X**] key to temporarily turn off the snap to make this easier.) Use **Relative Cartesian** coordinate entry to place the lower end of this line at (watch the minus sign) **X=0, Y=-8**.

Chapter 14, Page 280

Adding the Floor

55. ___ At the prompt *Select a point to define the Inside of the wall,* click to the right of the line.

56. ___ Press the **x** [not the **X**] key to turn the snap ON.

Adding the Floor

The drywall line has established the height of the wall. Next, the sub floor, bottom plate, and floor joist will be added.

57. ___ Scroll the display down until the bottom end of the drywall is in view.

58. ___ Toggle **Walls OFF**. Set the color as desired for the bottom plate.

59. ___ Use the **Polygons, Rectangle** option and construct a 2 inch by 4 inch rectangle with its lower right corner 1/2 inch below the lower left end of the left side of the drywall (see Figure 14-8).

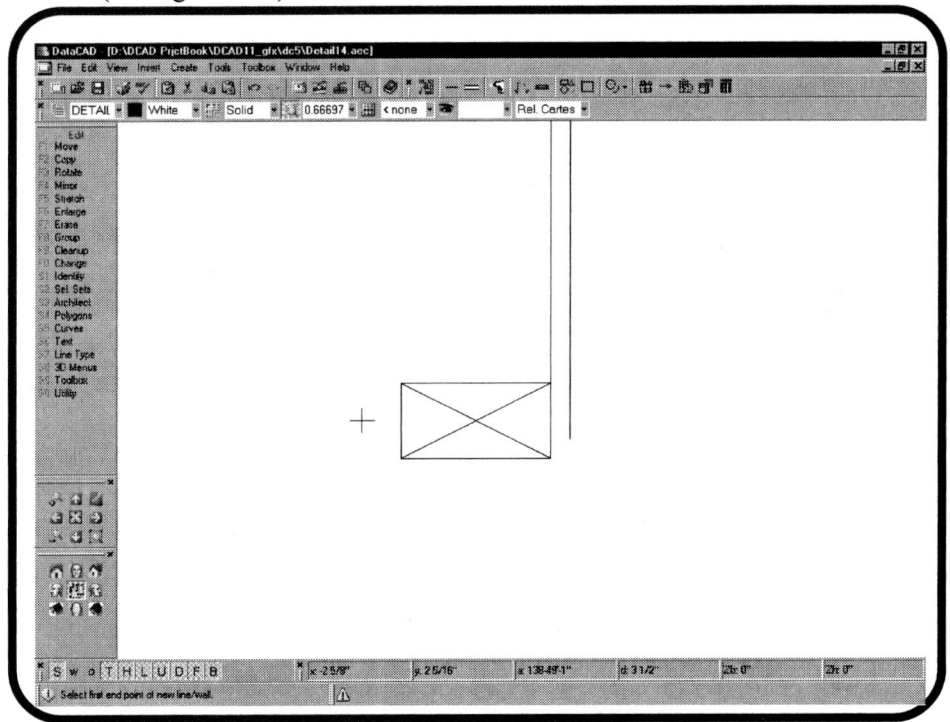

Figure 14-8, Adding a bottom plate.

60. ___ Construct crossing lines in the bottom plate rectangle.

The DataCAD 11 Project Book

61.___ From the **Line Type** menu, select the **Plywood2** line type. Change the color as desired.

62.___ From the **Line Type** menu, select **Spacing** and enter a value of **0.1**.

•• **NOTE:** To get the plywood thickness on the proper side of the line, the next line has to be drawn from left to right.

63.___ Construct a line along the bottom plate starting at a point at the lower left corner of the bottom plate and extending to the right (see Figure 14-9).

64.___ Change the line type to **Solid**.

65.___ From the **Architect** menu, toggle **Walls, 2 Line Walls** and **Outside ON**.

66.___ Select **Width** and set the width to **1** foot.

67.___ Construct a floor joist just under the plywood sub floor. (Don't worry if the top of the joist does not line up with the plywood. We are not being perfect here, and there would ordinarily be additional flooring added.)

68.___ Toggle walls OFF.

69.___ Use the **Polygons, Rectangle** option and construct a **2** by **12** end block, and a **2** by **8** sill plate as shown in Figure 14-9.

Adding the Floor

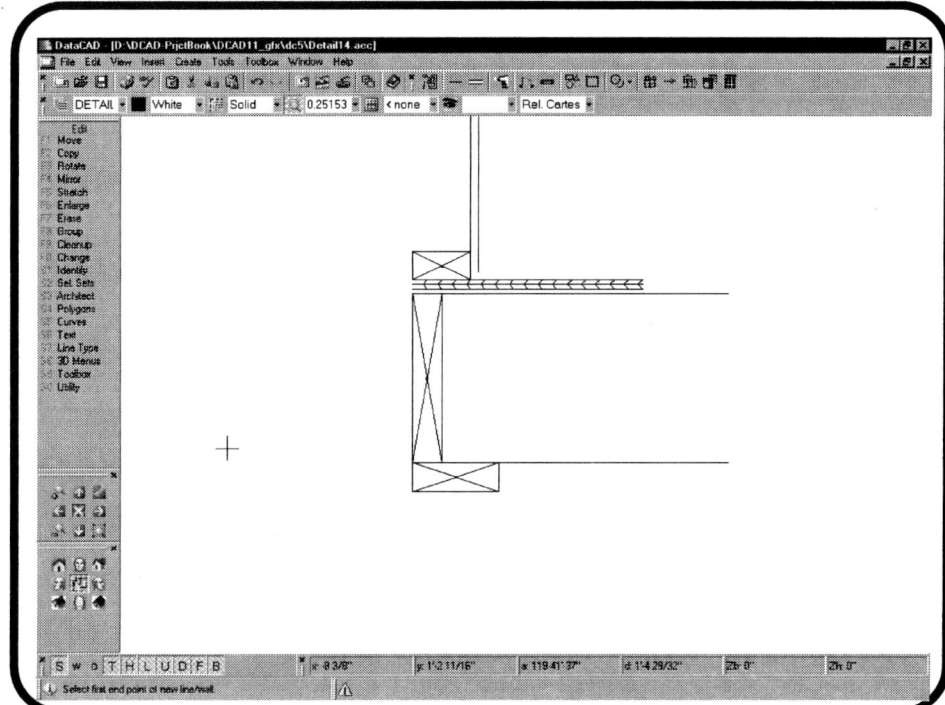

70.___ Add crossing lines to indicate the end view of the framing members.

Figure 14-9, Adding floor framing.

Next, a foundation wall will be added. Then the insulation and brick lines can be extended to meet it.

The DataCAD 11 Project Book

71.__ Change the drawing color and add a portion of the foundation wall as shown in Figure 14-10. Make the foundation wall **1'-2"** inches thick and create a brick shelf in the upper left corner of the foundation wall that is **5** inches wide and **10** inches deep.

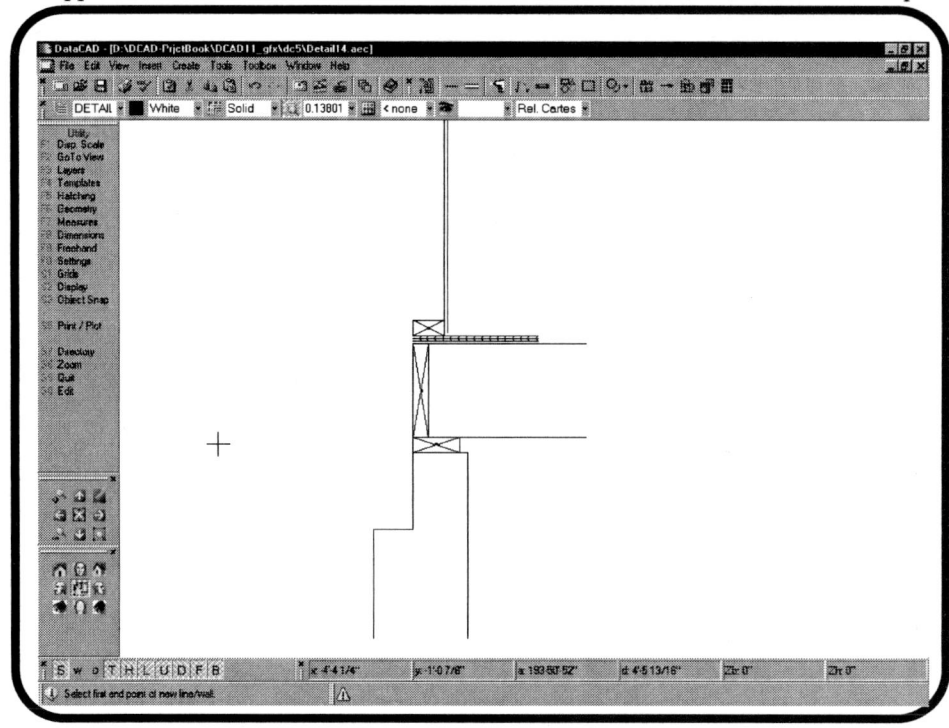

Figure 14-10, Adding the foundation.

72.__ Use the **1 Line Trim** function to extend the brick and insulation lines downward.

Adding Text

Your display should look similar to Figure 14-11 after extending the brick and insulation lines.

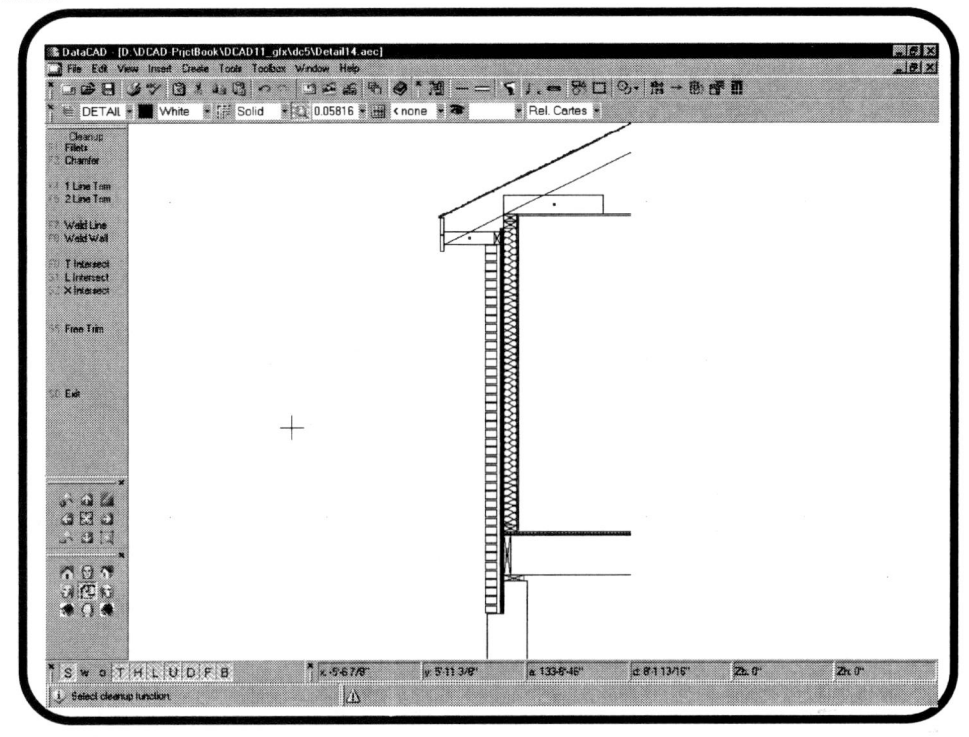

Figure 14-11, Editing lines.

A few conventional break lines will finish the job.

73.___ From the **Line Type** menu, select the **NewSectn** line type. Change the color as desired.

74.___ From the **Line Type** menu, select **Spacing** and enter a value of **0.5**.

75.___ Add break lines below the foundation wall and to the right of the roof and floor. Edit the lines as required using the **1 Line Trim** function.

76.___ Change the line type to **Solid**.

77.___ Use the **Hatch** function to add a **concrete** hatch pattern to the foundation wall. (A hatch **Scale** of **20** will work out well for this example.)

Adding Text

78.___ Change to the **TEXT** layer; then change the color as desired for text.

The DataCAD 11 Project Book

79.__ Add text and arrows to identify the different wall section features, as shown in Figure 14-12.

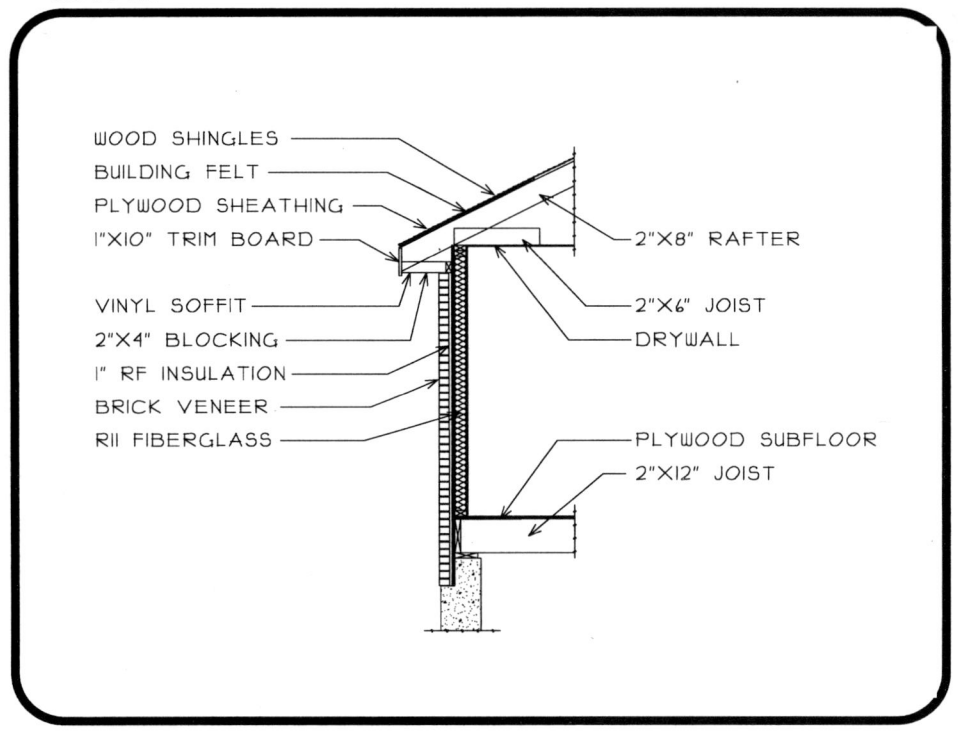

Figure 14-12, Completing the wall section.

By taking advantage of the different line types, typical wall sections are much easier to construct with DataCAD than with traditional manual drawing methods. Unlike real construction, where the floor has to be made before walls can be erected, a CAD wall section can be created in whatever order you wish. Sometimes it is easier to work from the top down than from the bottom up.

Adding Text

Review and Important Points.

- Different line types reduce the amount of time required to complete a drawing.

- Some of the line types are drawn completely to one side of the specified line position (asymmetrical), while other types are centered on the specified line (symmetrical). You will have to do a quick experiment on each line type to determine how it is arranged.

- For asymmetrical line types, the direction in which the line is drawn will determine which side of the line the enhanced line pattern is placed.

- The **Spacing** value is determined by whomever created the particular line type you select. Therefore you may find that you need to use some trial and error to get the spacing and size of the line type that you desire.

- To edit a line using the **1 Line Trim** function, first define a trimming line (either by selecting an existing entity, or by selecting two points). Next, point to the outside of the trimming line (this is the side of the trimmed line that will be removed), and then select the line to be trimmed.

- The **1 Line Trim** function can also be used to extend a line.

The DataCAD 11 Project Book

Practice Drawings
P14-A

- Consult an architecture text book, and create an alternate typical wall section.

The DataCAD 11 Project Book

Notes

The DataCAD 11 Project Book

CHAPTER 15

My Old House

Project Goals

When you have completed this project, you will be able to:

1. Create a 3D model of a two story house.
2. Create slabs and create voids in slabs.
3. Use the **3D Stairs** macro.

The DataCAD 11 Project Book

Prerequisite: Chapters 10, 12

Introduction

One of the common questions asked by users of earlier versions of this book is how to deal with multiple story houses. To answer this question, I have decided to add this chapter that uses a real house as the basis for creating a two story house model with DataCAD.

A number of years ago, I acquired an old abandoned two story house that sits on 75 acres of unglaciated Appalachian Plateau in south eastern Ohio. The best I can determine by examining old county records, is that the house was built around 1836. The house is used by my family as a weekend retreat. I have spent quite a bit of time and effort modernizing this house, but there is still a lot of inside work to be done. I have added a new bathroom and a first floor bedroom, remodeled the kitchen, added an attached garage, done a lot of rewiring, and installed a natural gas furnace. The house at this point is livable all year, but I may not live long enough to complete all of the work I envision.

For purposes of this chapter, only the original core of the house will be modeled. In 1836, they did not have indoor plumbing, and for heat there was a small fireplace in every room. It may surprise you to learn that back in 1836 they were building houses with basements in the midwest and northern USA. As a matter of fact, I have visited houses built by the famous William Johnson near Amsterdam, New York, around 1750, that included basements. These basements were probably used as winter kitchens or cooking areas. Anyway, the purpose of this chapter is to discuss techniques for modeling multistory houses, so not much attention will be paid to interior details.

Defining Layers

1. Start DataCAD and create a new drawing file named **House15**.
2. Press **Ctrl+L** [or click on the **layer** button to the left of the layer name in the **STATUS AREA**] to open the **Layer Manager** dialog box.
3. In the **Layer Manager** dialog box, select **New**. Then enter **13** in the pop up window and click **OK**.

Creating the First Floor Slab

4. __ Edit the layer names (by double clicking on the layer name column) so that they read as follows: **Slab_1, Walls_1, Wind_1, Doors_1, Roof_1, Slab_2, Walls_2, Wind_2, Doors_2, Roof_2, Slab_B, Walls_B, Stairs_1, Stairs_2.**
5. __ Set the desired color for each layer by clicking on the **COLOR** column and then selecting a color from the **Color Palette** dialog box.

Creating the First Floor Slab

6. __ Highlight the **Slab_1** layer, and click the **Active Only** button.
7. __ Select **Ok** to close the **Layer Manager** dialog box.

> ••• **WARNING**••• Have you saved your drawing lately? This warning will not be given again. Remember to save often. If you don't, you might have to start the project all over again if you make a mistake or have a power outage.

This old house has very high ceilings.

8. __ Press the **z** [or the **Z**] key, and enter **0** for the **Z-base** and enter **9.6** for the **Z-height.**
9. __ Press the **s** [not the **S**] key and set the snap to **1** for both X and Y.
10. __ Press the **J** [not the **j**] key to jump to the **3D Entity** menu.
11. __ From the **3D Entity** menu, select **Slabs, Horizontal, Base+Thk.**
12. __ Select **Thickness** and (watch the minus sign) enter **-1.0.**

The reason for making the thickness negative is so the first floor walls can be built at a **Z-base** of **0**. For the purposes of this project, a solid slab will be used for the floor. If you are more ambitious, you can replace this with a floor created with the **FrameIt** macro.

> **INFO STOP** When you are creating a slab, remember that the system will automatically close the boundary of the slab. Always right-click to finish a slab without selecting the starting point.

The DataCAD 11 Project Book

13.__ Click near the lower left corner of the drawing window to start the slab. Then use **Relative Cartesian** entry mode to complete the slab with the dimensions shown in Figure 15-1.

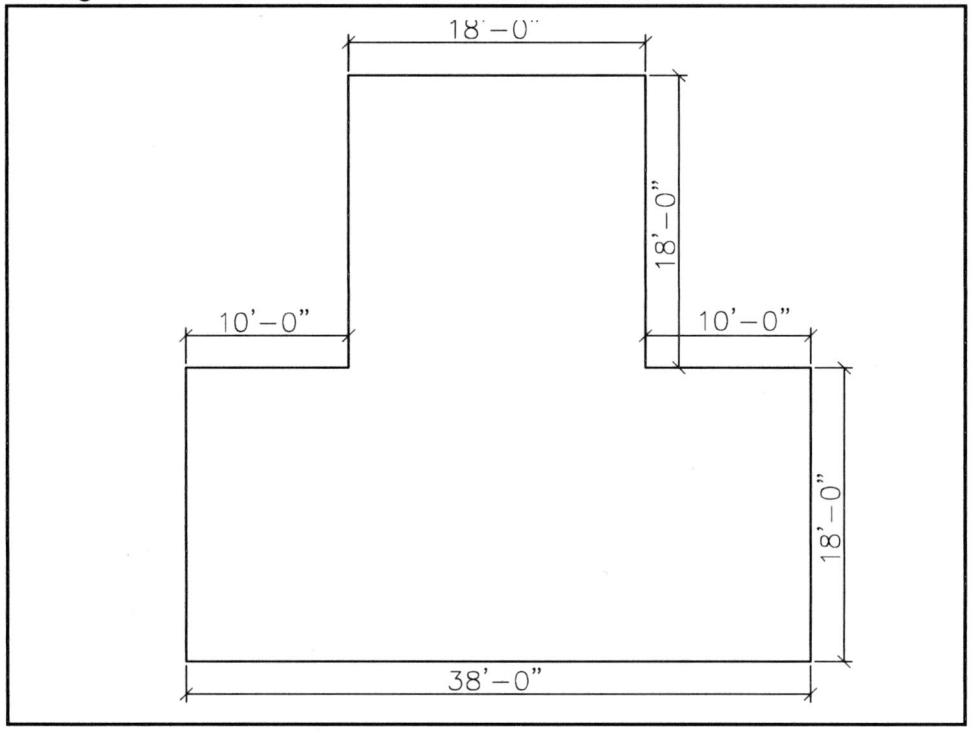

Figure 15-1, Creating the first floor slab.

Now that the slab outline has been created, a second slab will be created that will be used to create a void (a hole in a slab).

INFO STOP

In order for this slab/void method to work both slabs need to be the same thickness, and at the same Z-base and Z-height. If you are not creating both slabs at the same time, the quick way to make sure your second slab matches the first, is to select Identify, click on the first slab, then select the Set All option from the side menu. Now your Z-base and Z-height will match.

Creating the First Floor Slab

This void will be used for adding the cellar stairs later.

14.___ Press the unshifted ~ key and locate a new 0,0 at the *lower left* corner of the slab.

15.___ From the **Slabs** menu, select **Rectangle**, and use **Relative Cartesian** coordinates to locate the first corner of the rectangle at **X=17.4, Y= 18.0**.

16.___ For the second corner of the rectangle, press the **Space Bar** and (watch the minus sign) enter **X=3.4** and **Y=-8.0**.

17.___ From the **Slabs** menu, select **Voids**. At the prompt *Select master polygon or slab to process voids*, click on a line in the large slab.

18.___ From the **Voids** menu, select **Group**. At the prompt *Select Group to add void*, click on the small rectangular slab.

A shaded view of the slab with the rectangular void is shown in Figure 15-2.

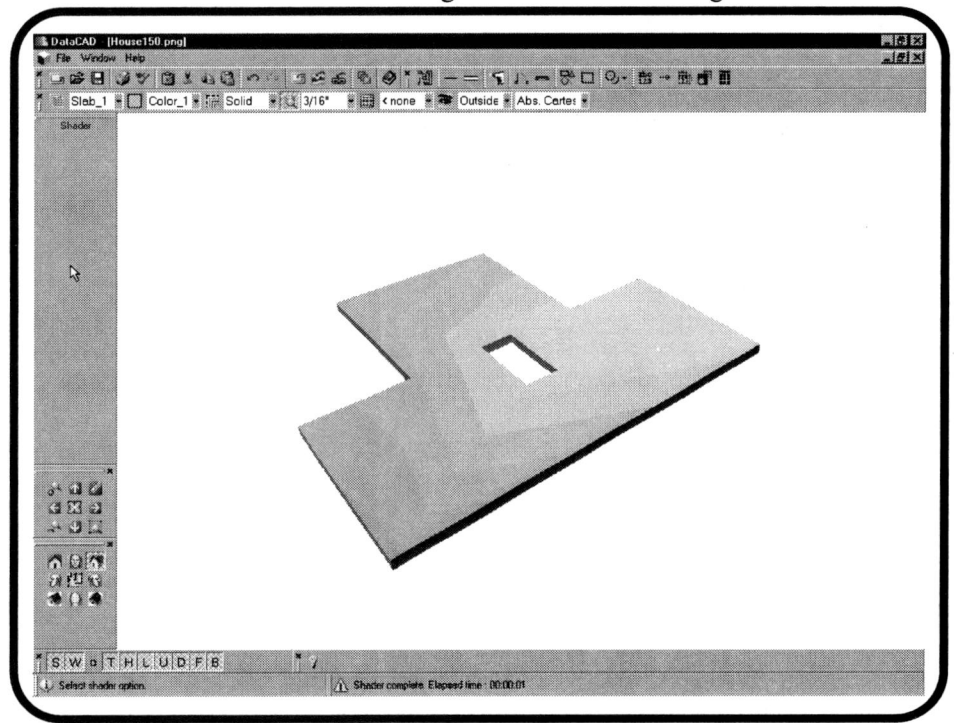

Figure 15-2, Slab with void.

Chapter 15, Page 295

The DataCAD 11 Project Book

> **INFO STOP** You might have noticed that several different slabs can be added to create multiple voids in a master slab.

Creating First Floor Walls

Next, the first floor walls will be added.

19.__ Use the **Layer Manager** to turn the **Slab_1** and **Walls_1** layers **ON**, and make **Walls_1** *active*.

20.__ Choose a color for **Walls_1**.

21.__ Press the **a** [not the **A**] key to jump to the Architct menu.

22.__ Toggle **Walls, 2 Line Walls, Outside**, and **Clean ON**.

The original house was built with masonry walls three bricks thick on the first floor. The insides of the brick walls were covered with plaster that has horse hair mixed in for strength.

23.__ Select **Width** and set the wall width to **1.0**.

24.__ With the **Z-base** set to **0**, and the **Z-height** set to **9.6**, create a 38' by 18' rectangle aligned with the lower part of the slab. Click inside the rectangle for the inside of the wall.

25.__ Create three additional walls aligned with the *upper* extension of the slab.

Creating First Floor Walls

26. __ Add a **6** inch wide wall that extends across the rectangular portion of the house and is **12'7"** from the lower left inside corner of the house (see Figure 15-3).

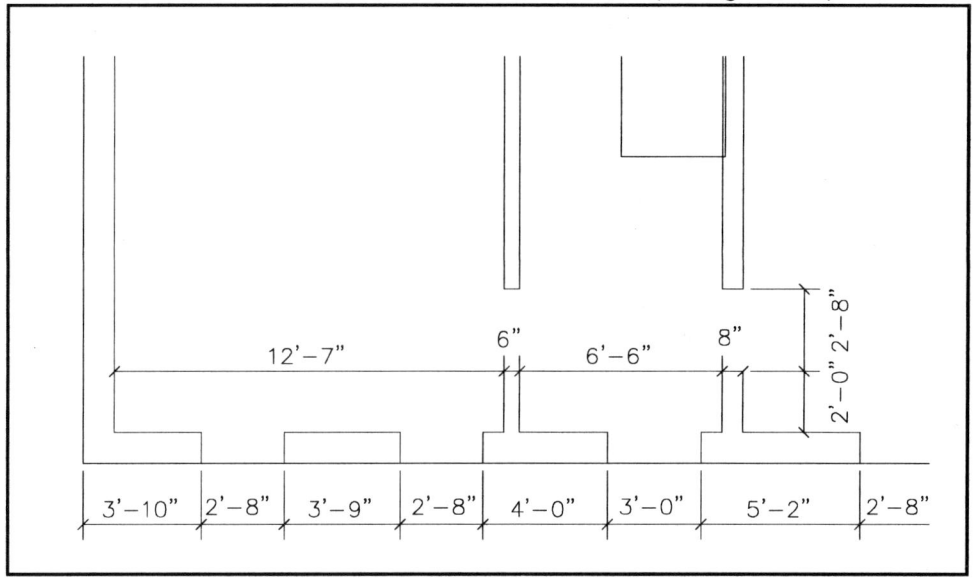

Figure 15-3, Adding interior walls.

27. __ Add an **8** inch wide wall **6'6"** to the right of the interior wall, as shown in Figure 15-3.

28. __ From the **Architect** menu, use **Cut Wall** and add *two* openings 2' from the inside front wall and 2'8" wide to the 6" and 8" walls, as shown in Figure 15-3.

Next, the openings for the doors and windows will be made. Since the **AEC_MODL** tool will be used to create nice 3D doors and windows, the **Cut Wall** option will be used here to create the door and window openings.

29. __ From the **Architect** menu, select **Cut Wall** and add openings to the front wall of the house using the dimensions shown below. (*Hint*: Start on the left side and use **Relative Cartesian** entry mode.)

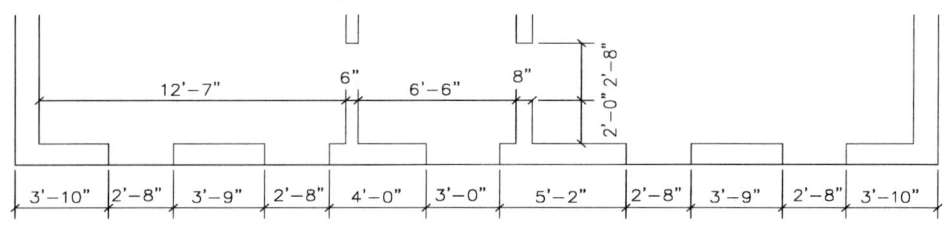

Chapter 15, Page 297

The DataCAD 11 Project Book

30.___ From the **Architect** menu, select **Cut Wall** and add openings to the *middle* wall of the house using the dimensions shown below.

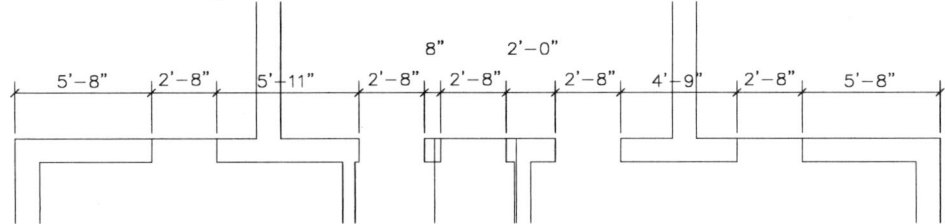

31.___ From the **Architect** menu, select **Cut Wall** and add openings to the *left* wall of the house using the dimensions shown. Add a 3' door opening to the right wall that is aligned with the 3' door opening in the left wall.

••• **WARNING** •••
Have you saved your drawing lately?

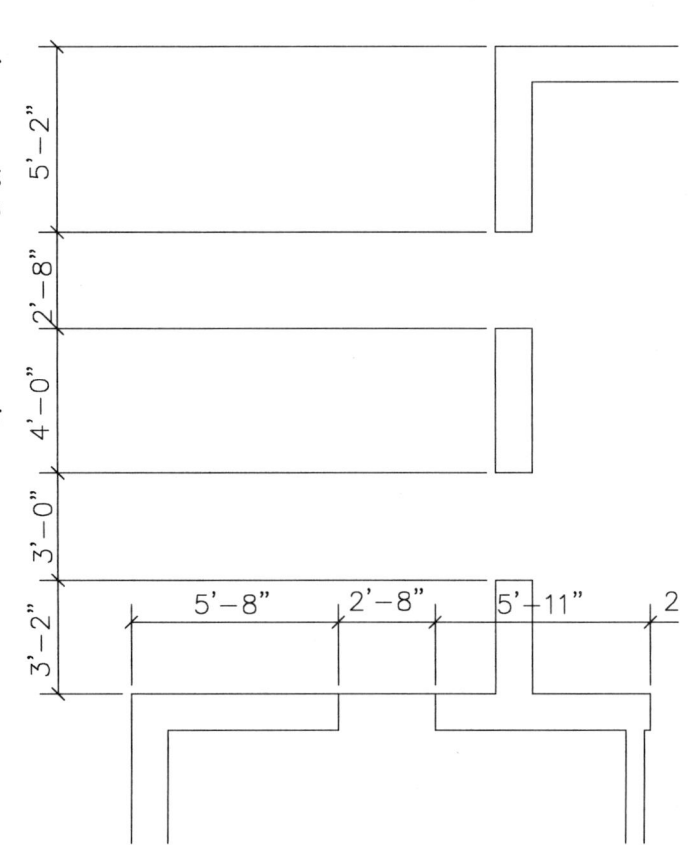

Chapter 15, Page 298

Creating First Floor Walls

This takes care of the first floor openings. An pictorial view should look similar to Figure 15.4.

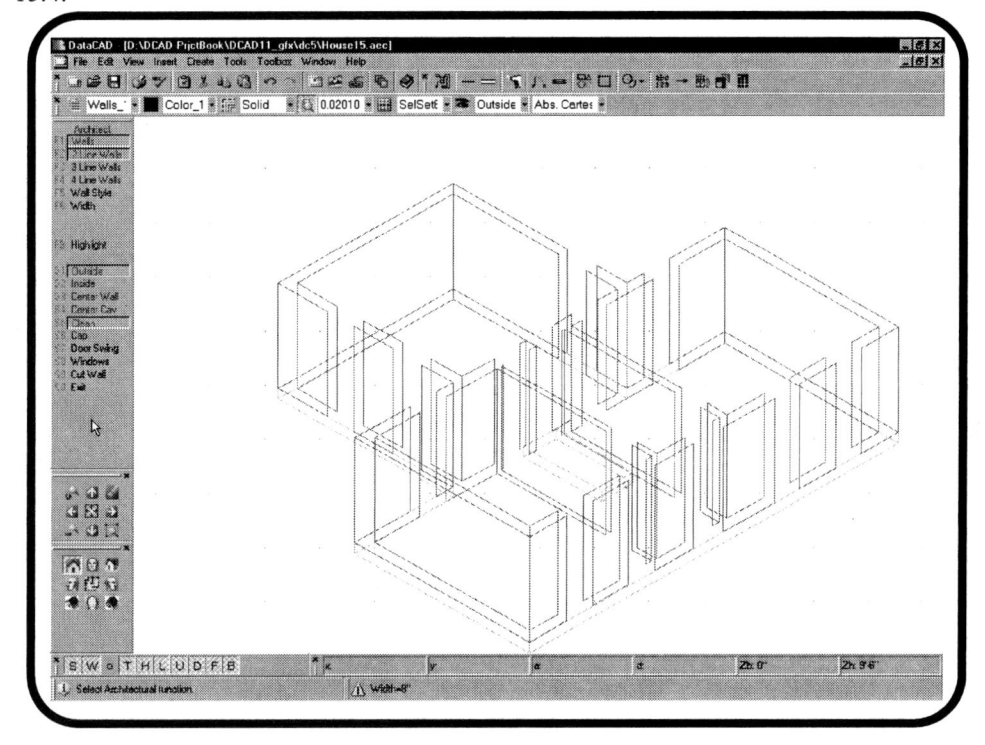

Figure 15-4, Creating wall openings.

Next, the spaces above the doors and windows and below the windows will be filled with short wall sections. Remember that you are working from "as built" dimensions of an 1836 farmhouse. Dimensions might not correspond to today's standards.

Since the new short wall sections to be added will tend to disappear if they are the same color as the walls, you might want to use a different drawing color. You can always change this color after the model is complete.

32.___ Change to a new drawing color.

33.___ Toggle **Walls ON,** and set the width to **1.0**.

The new wall sections will be constructed in the plan view; however, a pictorial view is provided to identify the window openings.

Chapter 15, Page 299

The DataCAD 11 Project Book

34.___ Set the **Z-base** to **0** and the **Z-height** to **2.0**. Add short walls to the 7 windows shown in Figure 15-5.

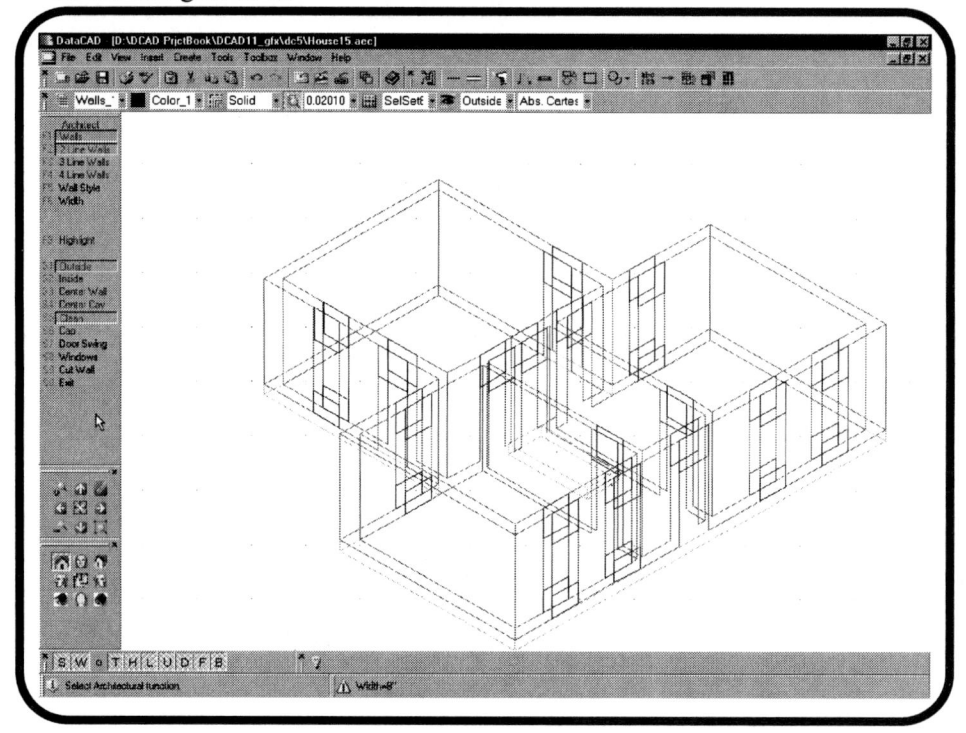

Figure 15-5, Filling spaces below windows, and above doors and windows.

35.___ Set the **Z-base** to **6.8** and the **Z-height** to **9.6**. Add short walls above all window and door openings.(Don't forget to change the wall width and add fillers above the two interior doors.)

36.___ Check a shaded pictorial view to make sure that all wall openings have been properly filled.

Creating the Second Floor Slab

Next, a slab will be added for the second floor. Recall that the first floor walls are 9' 6" high.

37.___ Use the **Layer Manager** to set the **Slab_2** layer active. Choose a new color for the second floor slab.

38.___ Press the **J** [not the **j**] key to jump to the **3D Entity** menu.

Creating the Second Floor Slab

39. ___ From the **3D Entity** menu, select **Slab, Horizontal, Base+Thk**.
40. ___ Select **Thickness** and enter **1.0**.
41. ___ Press the **z** [or the **Z**] key and set the **Z-base** to **9.6** and the Z-height to **19**.
42. ___ Trace the perimeter of the house to create a second floor slab. (Remember to right click to close the slab.)

Figure 15-6 shows a pictorial of the house with a second floor slab.

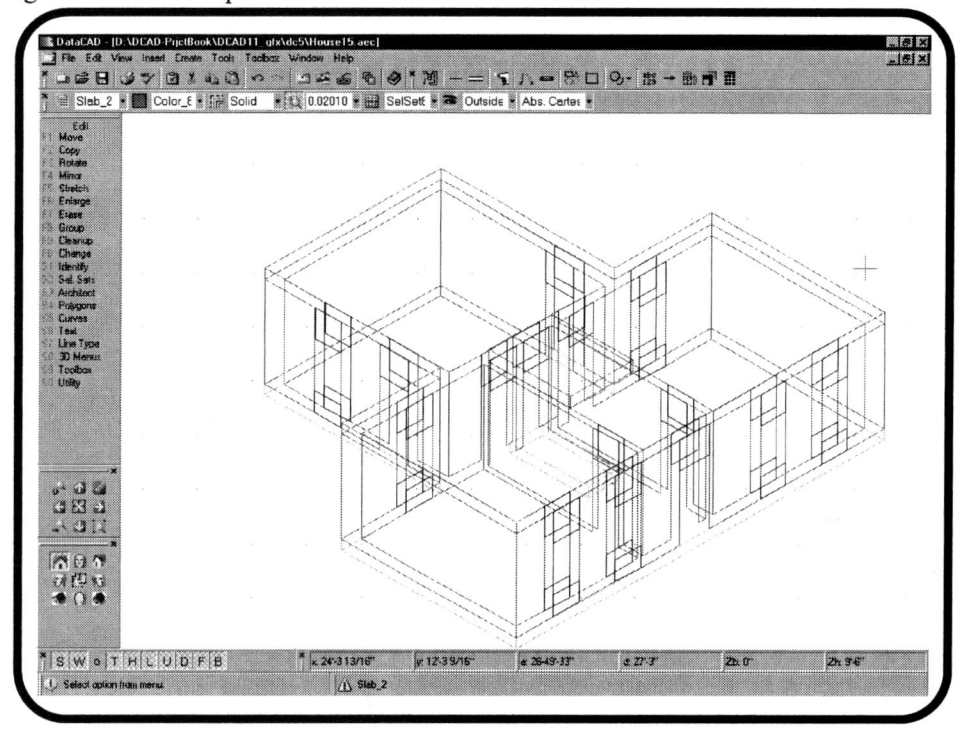

Figure 15-6, Adding a second floor slab.

A void has to be created in this slab for the stairs. (A void is a hole in a slab.) This void will allow for adding the stairs later in this exercise.

43. ___ Press the unshifted ~ key and locate a new 0,0 at the *upper left inside* corner of the hall in the center of the house.
44. ___ From the **Slabs** menu, select **Rectangle**, and use **Relative Cartesian** coordinates to locate the first corner of the rectangle at **X=0, Y= 0**.

The DataCAD 11 Project Book

45.__ For the second corner of the rectangle. press the **Space Bar** and (watch the minus sign) enter **X=6.6** and **Y=-8.4**.

46.__ Use the **Layer Manager** dialog box to turn all layers off except **Slab_2**.

47.__ From the **Slabs** menu, select **Voids**. At the prompt *Select master polygon or slab to process voids*, click on a line in the large slab.

48.__ From the **Voids** menu, select **Group**. At the prompt *Select Group to add void*, click on the small rectangular slab.

A pictorial view of the walls and two slabs is shown in Figure 15-7.

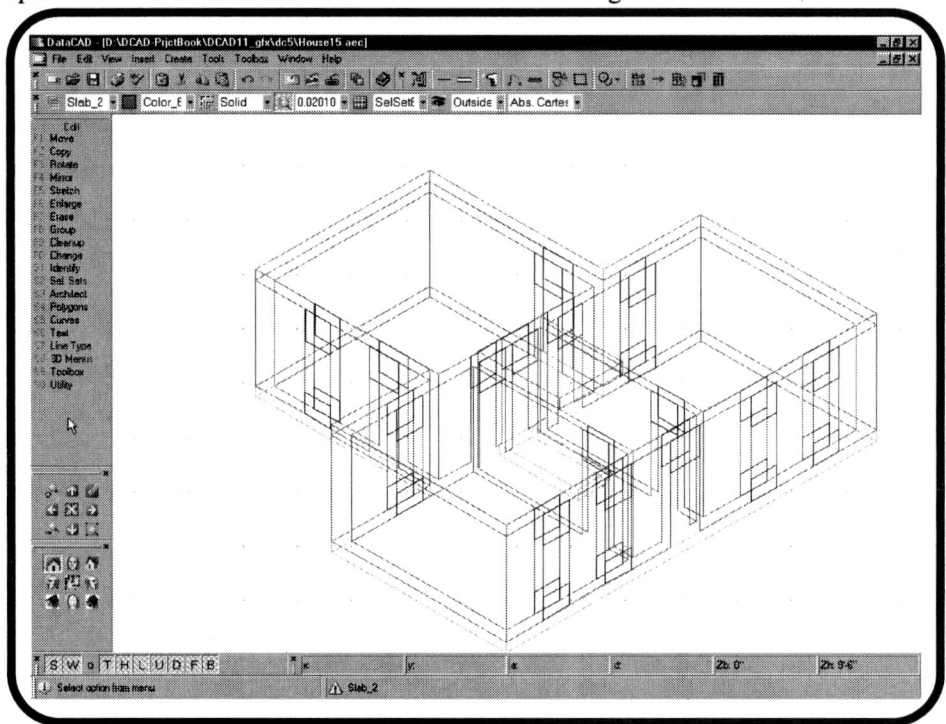

Figure 15-7, Construction progress.

Adding Second Floor Walls

Next, the second floor walls will be built. These walls are two bricks thick (plus the plaster). Also, the second floor rooms are only over the rectangular portion of the house.

49.__ Use the **Layer Manager** to turn on the **Slab_2** and **Walls_2** layers. Make **Walls_2** active.

Chapter 15, Page 302

Adding Second Floor Walls

50.__ Choose a new color for **Walls_2**.

51.__ Press the **a** [not the **A**] key to jump to the **Architect** menu. Set the wall **Width** to 8" (**0.8**).

52.__ Press the **z** [or the **Z**] key and set the **Z-base** to **10.6** (which is equal to the first floor height of 9.6 plus the slab thickness) and the **Z-height** to **19**.

53.__ Create a **38** foot by **18** foot rectangle aligned with **Slab_2**.

Next the second floor window openings will be added. All but one of these is aligned with the first floor windows.

54.__ Use the **Layer Manager** to turn on only the **Walls_1** and **Walls_2** layers. Make **Walls_2** active.

55.__ From the **Architect** menu, use **Cut Wall** (make sure that **Layer Search** is toggled **OFF**), and create **6** window openings aligned with the first floor windows.

There is one more window opening to create. This opening is 2 feet 8 inches wide and is in the center of the house. The **Object Snap** function will be used to create half of the window, and then it will be stretched to form the other half.

56.__ From the **Architect** menu, use **Cut Wall**. Move the cursor close to the center of the space between the two front windows and click the **Object Snap** button.

57.__ Press the **Space Bar** and enter values of **X=1.4, Y=0**.

58.__ Press the **Alt+S** to jump to the **Stretch** menu. Click at any point; then press the **Space Bar** and (watch the minus sign) enter values of **X=-1.4, Y=0**.

59.__ Select **Area** from the **Stretch** menu.

60.__ Toggle **Layer Search OFF**.

61.__ Capture the *left* edge of the *center* window opening in an area box.

The DataCAD 11 Project Book

A shaded view of the second floor walls is shown in Figure 15-8. (Note the areas above and below the windows need to be filled.)

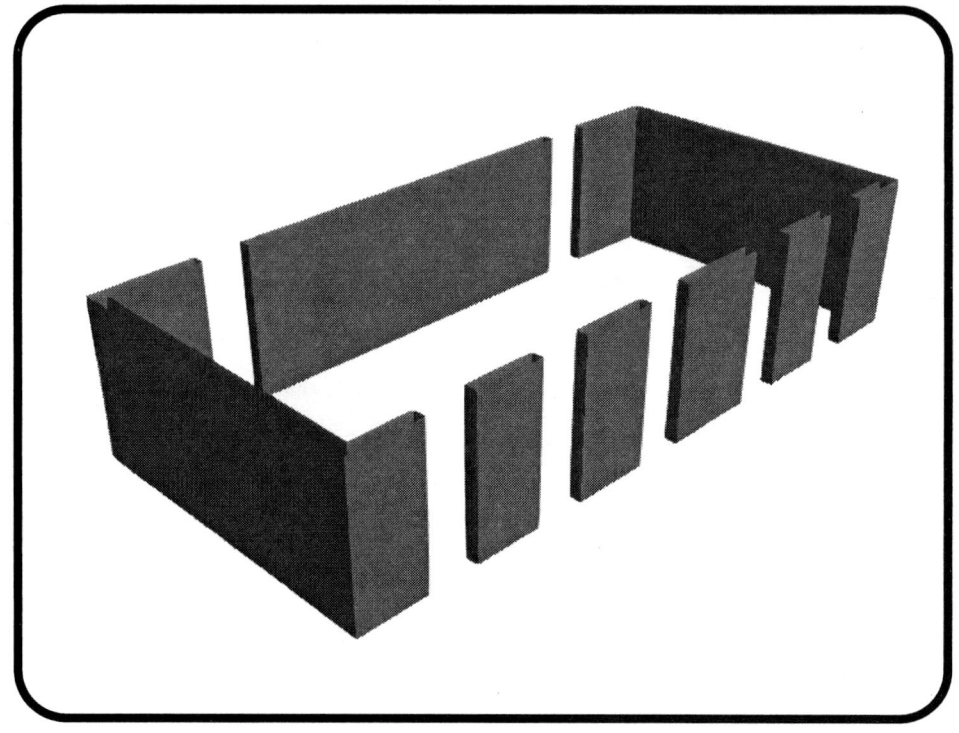

Figure 15-8, Second floor walls.

Next, short wall sections will be added to fill the space above and below the windows.

62.__ Change the drawing color. Press the **z** [or the Z] key and set the **Z-base** to **10.6** and the **Z-height** to **12.6**. Add wall sections to the window openings.

The top of the second floor windows is 6' 8" above the second floor slab. Since the top of the second floor slab is 10'6" inches, some arithmetic shows this to be at a Z elevation of 17'2".

63.__ Press the **z** [or the Z] key and set the **Z-base** to **17.2** and the **Z-height** to **19**. Add wall sections to the window openings.

64.__ Check a pictorial view to make sure all openings have been properly filled.

Adding Windows

Adding Windows

Next, some 3D windows will be added.

65. __ Use the **Layer Manager** to turn on only the **Walls_2** and **Wind_2** layers. Make **Wind_2** active.

66. __ Press the **z** [or the **Z**] key and set the **Z-base** to **10.6** and the **Z-height** to **19**.

67. __ Press the **M** [not the **m**] key to jump to the **Enter macro filename** dialog box. Select **AEC_MODL** and then **Open** the macro.

68. __ From the **AEC_Modl** menu, select **Windows.**

69. __ From the **WindPlan** menu, select **UnitType** and then select **DblHung**.

70. __ From the **WindPlan** menu, toggle **CutWall OFF**.

INFO STOP

The **CutWall** option only works if the walls have been created as *Slabs*. Since in this example, standard 2 line walls were created, the option must be turned off. In your future 3D modeling projects, you might decide to create walls as slabs.

71. __ From the **WindPlan** menu, select **Sill Hgt** and enter **2.0** as the new sill height.

72. __ From the **WindPlan** menu, select **Head Hgt** and enter **6.8**.

•• **NOTE:** Use the **Object Snap** button to make it easier to select points on the wall openings for the inside and outside window positions. Watch the prompts and be sure to get the inside of the windows on the proper side of the wall.

73. __ Add windows to the **7** second floor window openings. Check a pictorial view to make sure they are correctly positioned.

Next, some 3D windows will be added to the first floor.

74. __ Use the **Layer Manager** to turn on only the **Walls_1** and **Wind_1** layers. Make **Wind_1** active.

75. __ Press the **z** [or the **Z**] key and set the **Z-base** to **0** and the **Z-height** to **9.6**.

76. __ Add windows to the **7** first floor window openings. Check a pictorial view to make sure they are correctly positioned.

The DataCAD 11 Project Book

Adding Doors

Next, the doors will be added to the first floor plan.

77.__ Use the **Layer Manager** to turn on only the **Walls_1** and **Doors_1** layers. Make **Doors_1** active.

78.__ Press the **M** [not the **m**] key to jump to the **Enter macro filename** dialog box. Select **AEC_MODL** and then **Open** the macro.

79.__ From the **AEC_Modl** menu, select **Doors**.

80.__ From the **DoorPlan** menu, select **UnitType**.

81.__ From the **DoorUnit** menu, select **Single**. Then select **% Open** and enter **0**.

82.__ Right click back to the **DoorPlan** menu. Select **Door, Toggle DoDoor, HingRht, and Flush ON**.

83.__ Right click back to the **DoorPlan** menu and add doors to the front and left and right sides of the house.

Adding the Roof

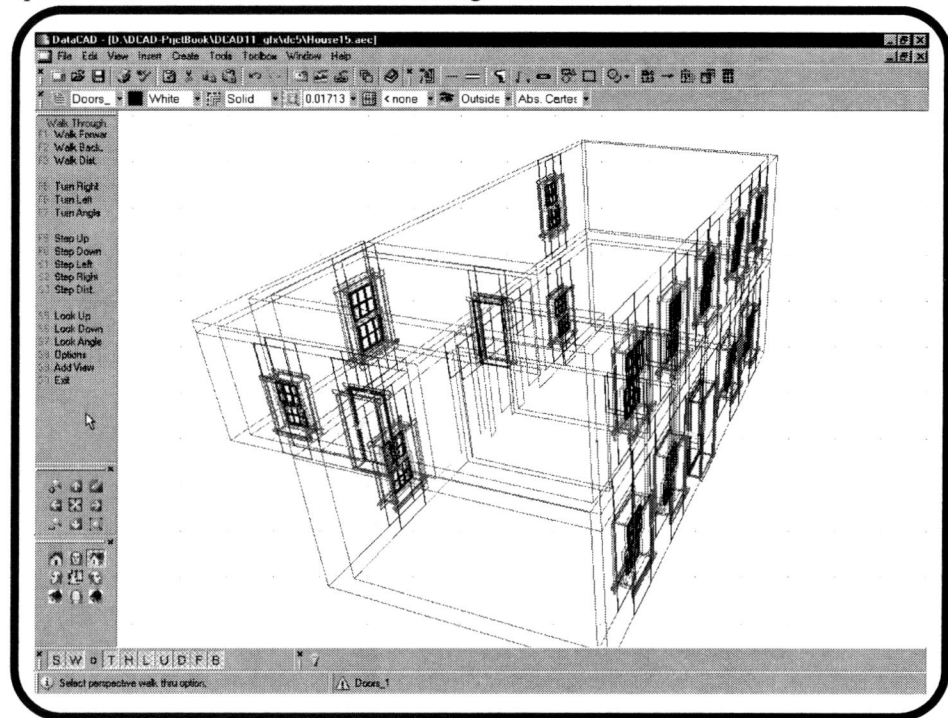

A pictorial view of the house is shown in Figure 15-9.

Figure 15-9, Doors and windows.

Adding the Roof

The last things to add are a couple of roofs. These will be created using the **RoofIt** macro.

84.__ Use the **Layer Manager** to turn on only the **Walls_2** and **Roof_2** layers. Make **Roof_2** active.

85.__ Press the **M** [not the **m**] key to jump to the **Enter macro filename:** dialog box. Select **ROOFIT,** and then **Open** the macro.

Note the current roof settings on the information line. These will be adjusted for the second floor roof.

86.__ From the **RoofIt** menu, make sure that **F1** shows **RectRoof** (not PolyRoof).

87.__ In the **RoofIt** menu, select **Settings**.

88.__ Set **PlatHgt** to **19**.

The DataCAD 11 Project Book

89.___ Set the **Pitch** to **6:12**.

90.___ Set **RoofThck** to **1.0**.

91.___ Set **Overhng** to **1.0**.

92.___ Set **WallThck** to **0.8**.

93.___ Toggle soffits (**F0**) to **SoffHorz** (not SoffIncl) and (**S1**) fascia to **FaciaPlb** (not FaciaSqr).

> •• **NOTE:** If the **EndWall** button is already ON in the next step, toggle it OFF, then ON again. Otherwise, you will not be able to control the **Color** of the end wall.

94.___ Toggle **EndWall ON**, and set the end wall color to **Match** the wall color.

95.___ Right click to return to the **RoofIt** menu. Toggle **Gable ON** and at the prompt *Pick the 1st point of roof boundary*, select the *upper left* corner of the wall.

96.___ At the prompt *Enter second point of the roof length*, select the *upper right* corner of the wall.

97.___ At the prompt *Enter second point of roof width*, select the *lower right* corner of the wall.

98.___ Check your roof construction in a pictorial view.

INFO STOP
The entire roof is a **Group**. If you want to erase your roof and replace it. Press the **<** key [or press the **e** key, pick the **Group** option] and select a line in the **Group**.

Next, a roof will be added over the one story section of the house.

99.___ Use the **Layer Manager** to turn on only the **Slab_1** and **Roof_1** layers. Make **Roof_1** active.

100.___ Press the **M** [not the **m**] key, or select **Edit, Toolbox** to jump to the **Enter macro filename** dialog box. Select **ROOFIT,** and then **Open** the macro.

101.___ From the **RoofIt** menu, select **Settings**, and set **PlatHgt** to **10.6**.

102.___ Right click to return to the **RoofIt** menu. At the prompt *Pick the 1st point of the roof boundary*, select the *lower left* corner of the upper extension of the slab.

103.___ At the prompt *Enter second point of the roof length*, select the *upper right* corner of the slab.

Adding the Roof

104.___ At the prompt *Enter second point of roof width*, select the *lower right* corner of the upper extension of the slab.

105.___ Check your roof construction in a pictorial view.

There is one problem with the first floor roof. It sticks into the second floor wall. This will be fixed using the **Stretch** function.

106.___ Use the **Layer Manager** to turn on only the **Roof_1** layer.

107.___ Press the **Alt+S** to jump to the **Stretch** menu. Click at some convenient point; then press the **Space Bar** and enter values of **X=0, Y=1**.

108.___ Select **Area** from the **Stretch** menu. Capture the *bottom* edge of the *roof* in an area box.

109.___ Turn all layers on and admire your work.

You can use the **Shader** function to create different lighting for rendered images. You can also view the house using the **Object Viewer**.

INFO STOP
The colors used by the Object Viewer program do not have to be the same colors as those used by DataCAD. It is possible to set the Object Viewer color for any DataCAD color to any other color in the palette.

Here is a trick for adjusting colors used by the **Object Viewer** program.

110.___ Adjust the display so that you can select one of the lines on the second floor slab. Then hold the **Ctrl** key down and *right* click on a line in the slab. Select **Rendering Settings** from the pop up menu.

This opens the **DcadWin.dmf** (DataCAD Materials File) dialog box.

111.___ Make sure **Color** is selected in the **Entity Color** area of the dialog.

112.___ Click on the **Rendered Color** color box, and then pick the desired color from the **Color Palette**.

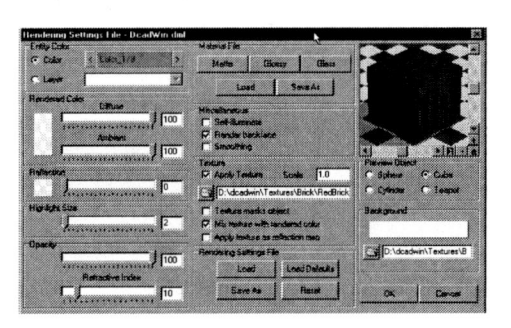

Chapter 15, Page 309

The DataCAD 11 Project Book

Observe that the **DcadWin.dmf** dialog box also allows for adjusting the **Opacity** of a rendered color. This makes it possible to create a phantom see-through rendering of a model.

> **INFO STOP**
> New in DataCAD 11 is the ability to render all materials of a particular color, or all materials by their layers, but not both. By selecting **Color** in the **Entity Color** area of the dialog, all **Red** entities, for instance, will be rendered alike, no matter which layers they are on. If you pick **Layer** in the **Entity Color** area, then all entities on **Walls_1**, for instance, will be rendered alike, regardless of their colors.

Adding Stairs

There is no end of things that could be done to dress up the model of *My Old House*. However, there is limited space in this book, so the last topic covered will be adding stairs.

1. Use the **Layer Manager** to turn on only the **Slab_2** and **Stairs_2** layers. Make **Stairs_2** active.
2. Press the **z** [or the **Z**] key and set **Z-base** to **0** and **Z-height** to **9.6**.
3. Press the **M** [not the **m**] key to open the **Enter macro name** dialog box. **Open** the **3DSTAIRS** macro.
4. From the **3DStairs** menu, toggle **DoubBck ON**.

Be patient; there are a lot of parameters in the **3DStairs** macro to set. Note that although **Color** is not mentioned in the following steps, you can set the color separately for most of the stair components.

5. From the **3DStairs** menu, select **Settings**.
6. From the **Settings** menu, select **BaseElev** and set the base elevation to **0**.
7. From the **Settings** menu, select **TrdLngth** and set the tread length to **2.9**.
8. From the **Settings** menu, select **Trd/Run1** and set the number of treads to **10**.
9. From the **Settings** menu, select **Trd/Run2** and set the number of treads to **5**.
10. From the **Settings** menu, toggle **Trd/Clkwise OFF**.

Adding Stairs

11. __ From the **Settings** menu, select **Chase** and set the chase width to **0.6**. (The stairs will go up to a landing, double back, and then continue up. The chase width is the space between the two sets of passing stairs.)
12. __ From the **Settings** menu, select **Landing** and toggle **DoLndTp** and **DoLndBt OFF**.
13. __ From the **Settings** menu, select **Riser**. Toggle **DoRiser ON** and set the riser height to **0.7.5/16**. Select **Thicknss** and set the thickness to **1"**.
14. __ From the **Settings** menu, select **Stringer**. Toggle **DoLeft** and **DoRght ON**.
15. __ From the **Settings** menu, select **Handrail**. Toggle **DoLeft ON** and **DoRght OFF**. Select either a cylindrical or rectangular handrail.
16. __ From the **Settings** menu, select **Baluster**. Toggle **DoBalus** and **OnTread ON**.
17. __ From the **Settings** menu, select **NewelPst**. Toggle **DoNlPst ON**.
18. __ Right click to return to the **3DStairs** menu. Select **Begin** and position the stairs outline so that the upper edge of the outline is lined up with the upper line in the rectangular void in the slab. Click the *left* mouse button once to locate the stairs, and click the *left* mouse button once more to set the orientation of the stairs.

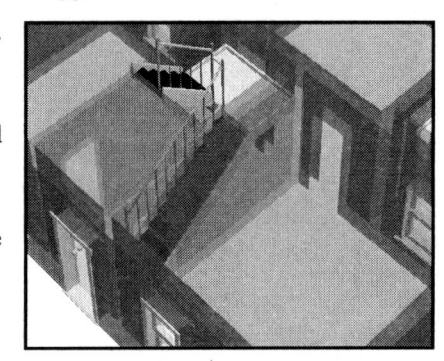

> **INFO STOP**
> It is possible to dynamically rotate the orientation of **3DStairs** by moving the mouse after setting the stair location.

19. __ From the **Menu Bar**, select **View, Object Viewer** and dynamically rotate your stairs and slab model to confirm that the location and orientation are correct.

The original old house has a set of stairs to the basement.

20. __ Use the **Layer Manager** to turn on only the **Slab_1** and **Stairs_1** layers. Make **Stairs_1** active.
21. __ Press the **M** [not the **m**] key to open the **Enter macro name** dialog box. **Open** the **3DSTAIRS** macro.
22. __ From the **3DStairs** menu, toggle **Single ON**.

Chapter 15, Page 311

The DataCAD 11 Project Book

23. ___ From the **3DStairs** menu, select **Settings**.
24. ___ From the **Settings** menu, select **BaseElev** and set the base elevation to **-8**.
25. ___ From the **Settings** menu, select **Handrail**. Toggle both **DoLeft** and **DoRght** *OFF*.
26. ___ Right click to return to the **3DStairs** menu. Select **Begin** and position the stairs outline so that the upper edge of the outline is lined up with the upper line in the rectangular void in the slab. Click the *left* mouse button once to locate the stairs, and click the *left* mouse button once more to set the orientation of the stairs.

A transparent view of the house with stairs added is shown in Figure 15-10.

Figure 15-10, A phantom view of My Old House.

Adding Stairs

INFO STOP

In Figure 15-10, the colors of the walls, the fillers above and below windows, and the second floor slab, were all set to render the same color and with the same opacity in the o2c viewer. This is done by holding the **Ctrl** key down and *right clicking* on an entity in the DataCAD model. This opens a dialog box where the o2c **Rendered Color** and **Opacity** can be set. It is easier to select entities if they are isolated by temporarily turning off all layers except the layer that contains the entity whose color is to be adjusted. (For detailed information about setting o2c colors, see **Appendix C**.)

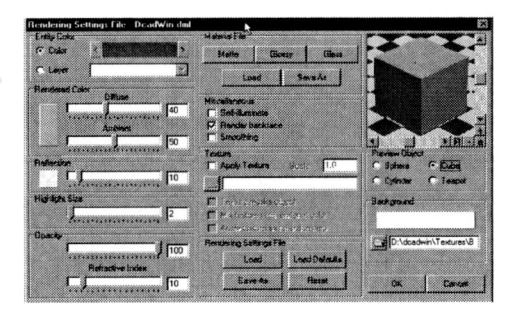

This is as far as I will go in this project. I leave it to you to add whatever additional embellishments you would like to make to the old house. You can add basement walls, front and side porches, finish the interior rooms, modernize the place by adding indoor plumbing and a garage, or whatever.

If you have faithfully completed the steps in all 15 projects in this book, you should now be able to do some serious architectural designs of your own.

Have fun!

Chapter 15, Page 313

The DataCAD 11 Project Book

Notes

CHAPTER 16

Creating and Modifying Symbols

Project Goals

When you have completed this project, you will be able to:

1. Create a 2D symbol.
2. Create a 3D symbol.
3. **Explode** a symbol (it won't be a symbol any more).
4. Modify symbols using **Redefine** and **Replace**.

The DataCAD 11 Project Book

Prerequisite: Chapters 2 and 6

Introduction

In Chapters 2 and 6 you placed 3D symbols in Ferd's Playhouse. Those symbols were created by sombody else. In this chapter you will learn how to make your own 2D and 3D symbols. And once placed in your drawing we will see how to replace one or several symbols with another.

In DataCAD 11 a new feature was added to allow saved symbols to "remember" which entities were on which drawing layers when created. If you later "**Explode**" the symbol back into its original entities, you have the option to put them all back on their original layers, or to put them all on the active layer.

Create a 2D Symbol

We need to draw something to make into a symbol, so let's just draw a simple living room couch. It's not imortant what it looks like, so let's just pick some arbitrary dimensions like 6' long and 3' deep. Although this is not going to be a 3D symbol it's still good practice to set the Z-base and Z-height to something real-world.

1.__ Start DataCAD and create a new drawing file named **Symbol16**.
2.__ To set the Z-base and Z-height, press the **z** [or **Z**] key.
3.__ You are prompted to **Enter new Z-base**. Type **0** then press **Enter**. You are then prompted to **Enter new Z-height**. Type **3.0** (or just **3**) and press **Enter**.

4.__ Draw a rectangle 6' wide by 3' deep, then add some lines to represent the arms, back and cushions. It should look something like Figure 16-1.

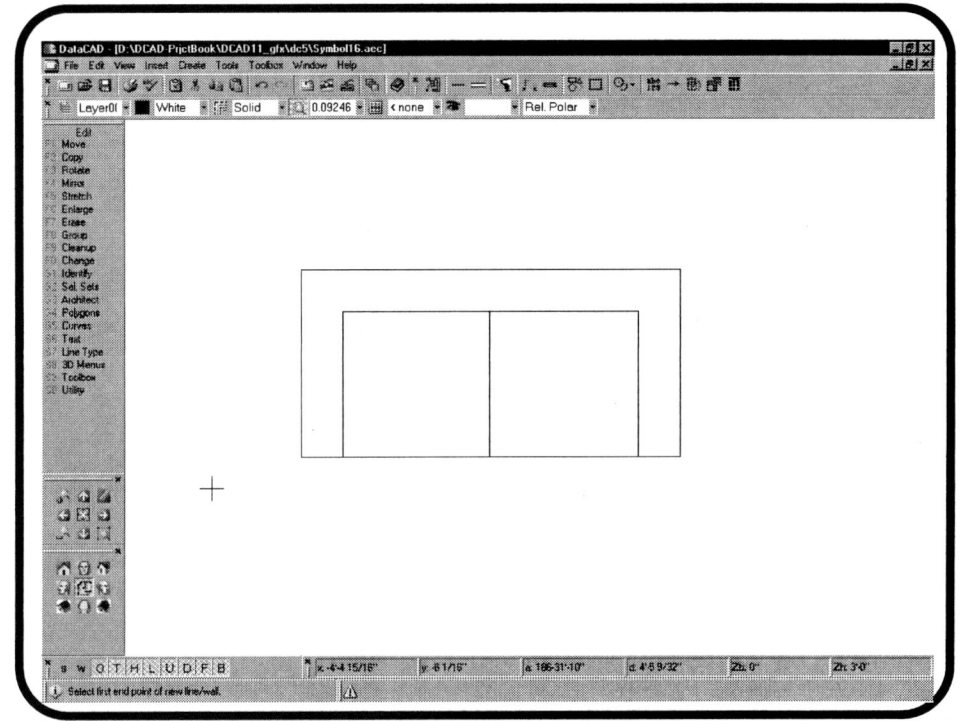

Figure 16-1, A simple couch.

5.__ Open the **Symbol Browser** by pressing **Ctrl+T**. Place your cursor over the yellow folder icon at the upper left corner of the **Symbol Browser** and click the left mouse button.

Chapter 16, Page 317

The DataCAD 11 Project Book

6.__ In the resulting drop-down menu pick the Folder option. A Browse dialog will appear, as in Figure 16-2.

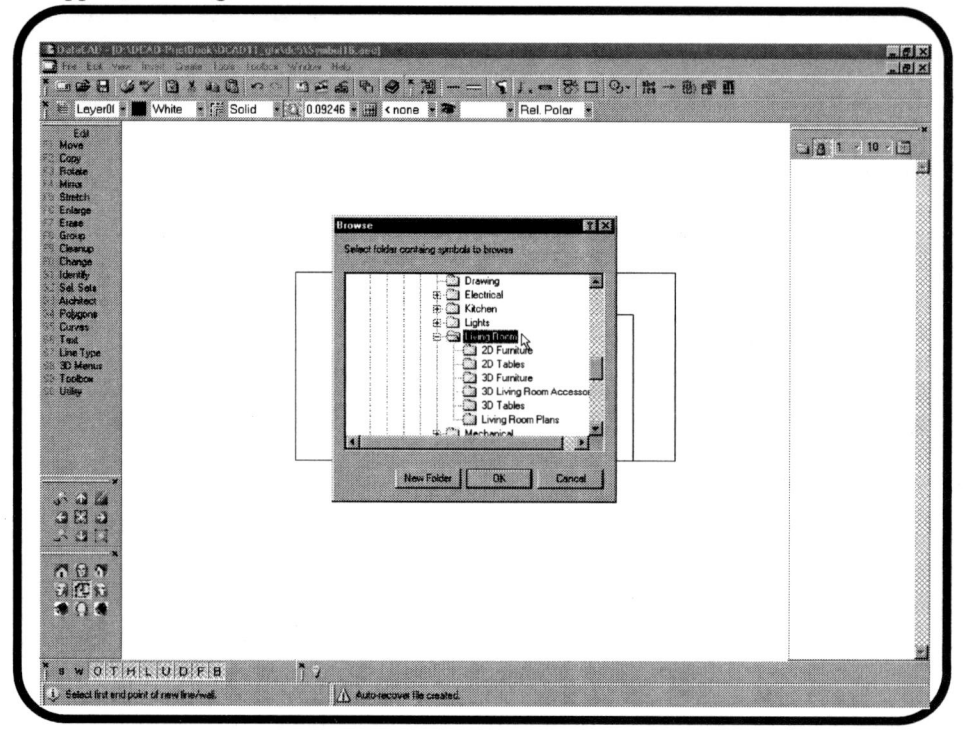

Figure 16-2, The Browse dialog of the Symbol Browser.

We need to select a folder to put the new symbol in. DataCAD stores its symbols in sub-folders in the Symbols folder, though you can actually store them anywhere on your computer that you want, or even onto a floppy disk or rewritable CD. We'll stick to the standard Symbols folder.

7.__ Under the **Symbols** folder find the sub-folder called **Living Room**.

Under the **Living Room** folder are several more sub-folders. If you don't see them click on the small white square with the + in it, to the left of the yellow **Living Room** folder. We could put our new symbol in the existing **2D Furniture** folder, but that folder is already full of other symbols and it will be hard to see our new symbol, so we're going to create our own, new folder, though we still want to make the new folder a sub-folder under **Living Room**.

Chapter 16, Page 318

8. __ Highlight the **Living Room** folder by placing your cursor on either the name or the yellow folder to the left of it, then clicking on it with the left mouse button.
9. __ At the bottom of the **Browse** dialog click on the **New Folder** button. The **New Folder** dialog appears with the words **New Folder** highlighted in it.
10. __ Type in the the new folder name, **Chapter 16** (see Figure 16-3), and press the **OK** button. Our new folder appears as a sub-folder of the **Living Room** folder, and it should already be highlighted.

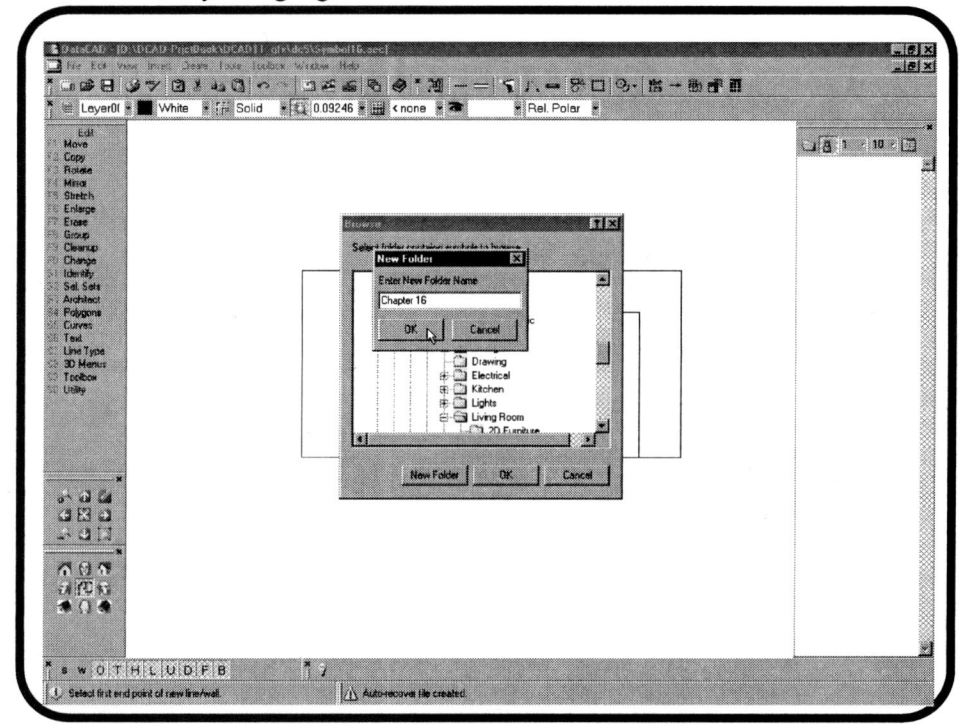

Figure 16-3, Adding a new folder.

11. __ Press the **OK** button to close the **Browse** dialog.

By highlighting the **Chapter 16** folder before pressing the **OK** button, we've told DataCAD to show that folder in the **Symbol Browser**. You will notice that the **Symbol Browser** is currently empty, so let's fill it with our couch symbol.

12. __ Click on the **Options** button in the upper right corner of the **Symbol Browser** (the one that looks like a green folder).

The DataCAD 11 Project Book

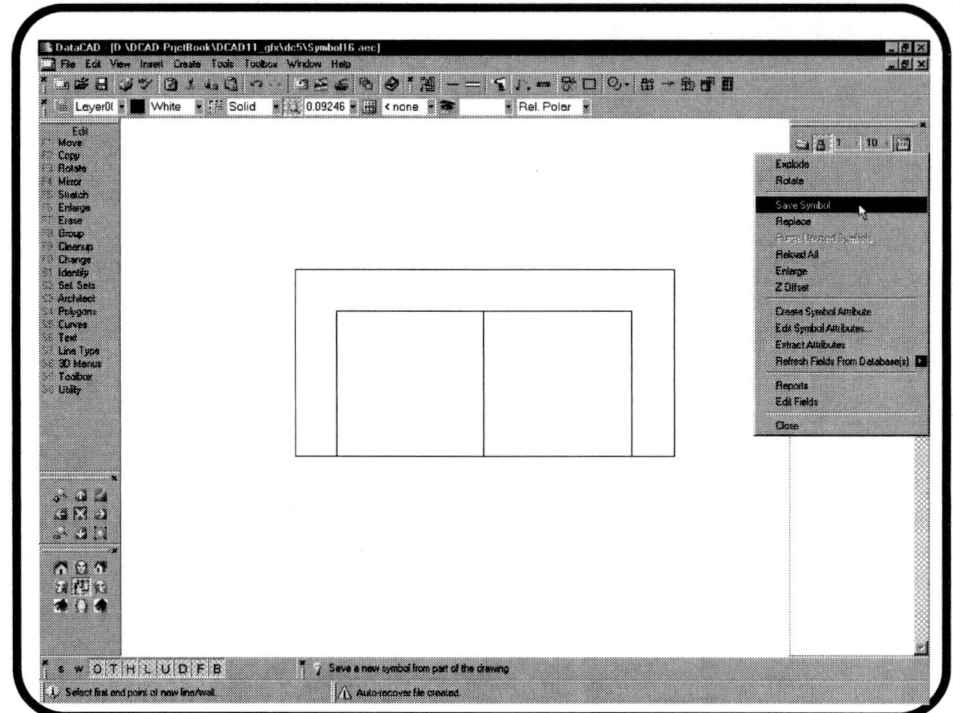

13. __ In the sub-menu that appears below, select the **Save Symbol** option (Figure 16-4).

Figure 16-4, Picking Save Symbol.

A new dialog box appears with the title **Enter name of symbol file to be saved:**. Notice that the **Save in:** folder is the new **Chapter 16** folder that we created and highlighted earlier.

14. __ In the **File name:** box type the name of our new couch symbol, **2D Couch**, and press **Save**. A new **Save Symbol** menu appears on the left side of your screen.

15. __ To pick all the lines of the couch select the **Area** option from the **Save Symbol** menu.

16. __ Capture the entire couch in an **Area** box. All the lines of the couch will appear gray and dashed, showing that they have been selected.

It is important to notice that at the bottom of your screen DataCAD is prompting you to *Select an insertion point for symbol*. When saving symbols it is very important to read all the prompts at the bottom of your screen. It will make the process run a lot smoother.

As you have seen in Chapters 2 and 6, every symbol has an insertion point, so that's what we need to do here, though it's not important exactly where that point is for our couch.

17.__ Snap to the upper right corner of the couch by placing your cursor close to that point, then clicking your mouse's middle button, or pressing the **n** [or **N**] key.

18.__ A new dialog box appears, prompting you for an **Item name** (Figure 16-5).

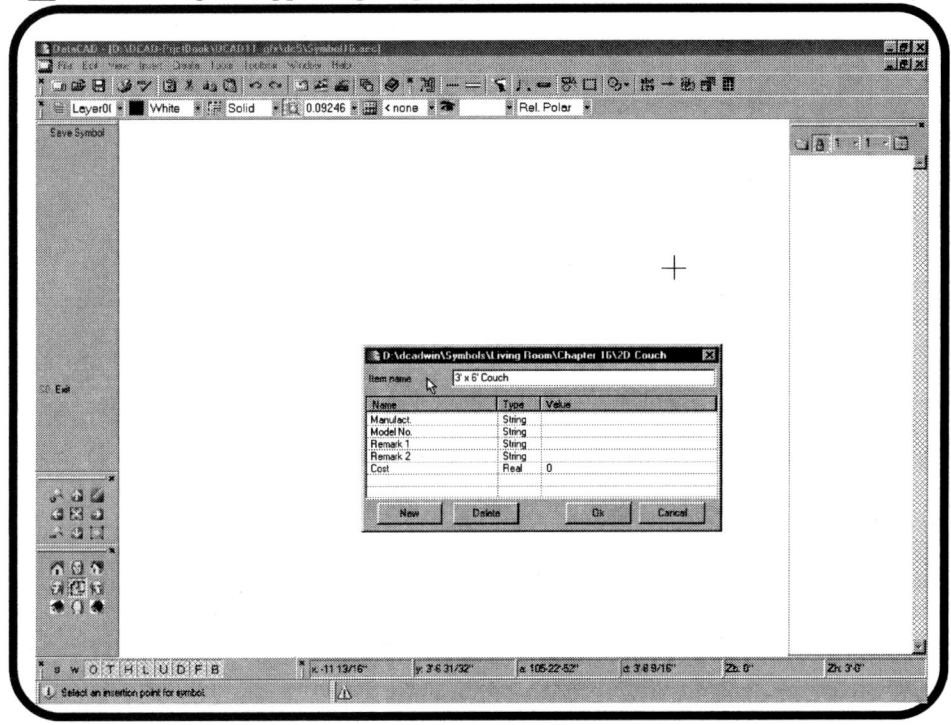

Figure 16-5, Adding a symbol description.

Chapter 16, Page 321

The DataCAD 11 Project Book

> **INFO STOP**
> The name you type into this dialog box will be the name displayed by DataCAD when your cursor is over the symbol in the **Symbol Browser**. This name can be completely different from the name you used earlier. In fact, you could type a long description like: **3' x 6' couch with arms and 2 seat cushions**. The name that you typed in Step 14 is the name of the symbol on your computer, but the name you type here is a descriptive name that is unique to the symbol in DataCAD.

19.__ In the **Item Name** box type, **3' x 6' Couch**, then press **OK** to close the dialog box. The symbol appears in the **Symbol Browser**, ready for you to use.

> **INFO STOP**
> Notice also that the original couch that you drew has disappeard from the Drawing Window. DataCAD assumes that because you are making a symbol of something you don't need the original anymore. However, if you do need it then there is a trick that you can use to get it back. Immediately after creating the symbol, press **Ctrl+Z** (the Windows command for Undo), and the couch will reappear without removing the symbol from the **Symbol Browser**.

20.__ Now that you have your very own symbol, pick it from the **Symbol Browser** and place a few copies of it in your drawing window.

Create a 3D Symbol

Creating a 3D symbol is identical to creating a 2D symbol except that you need to be more careful about the Z-location of the insertion point. Like the 2D example we need to create a 3D symbol to work with.

1.__ In the **Symbol16** file from the previous example, use the **Layer Manager** to add one new layer, and call it **3D Symbol**. While you're at it, rename the original **Layer001** (from the previous example) to **2D Symbol**.

2.__ Press the **Tab** key until the new **3D Symbol** layer is the active layer.

Create a 3D Symbol

We're going to use one of the 2D couch symbols on your screen as a guide for drawing a simple 3D couch with 3D slabs.

3. __ To make it easier to see the new 3D couch press the **k** key to change the active color to **Red**.

4. __ Press the **j** key [not the **J** key] to get to the **3D Menus** menu.

5. __ Select **Settings, Z-Base**, type in **0**, and press **Enter**.

> •• **NOTE:** It is important to draw your 3D object at the correct Z-elevation, as it makes placement in later drawings easier and more accurate.

6. __ Right-click twice to get to the **3D Entity** menu.

7. __ We're going to build our couch out of Slabs, so select **Slabs** from the menu.

8. __ We can build the slabs that make up the couch from either Horizontal or Vertical slabs. Select **Horizontal** from the **Slabs** menu.

9. __ Select **Base+Thk.**. This will create the slabs at the Z-base that we defined (**0**), and allow us to select a thickness, or z-height.

10. __ Click on **Thickness** and type **3.0** [or just **3**], then press **Enter**. DataCAD prompts you to *Select the 1st point of the <SLAB>*.

The DataCAD 11 Project Book

11.__ Refering to Figure 16-6, we're going to draw over one of the 2D couches on your screen. Use your middle mouse button or the **N** key to snap to the upper left corner of one of the 2D couches (point 1 in Figure 16-6).

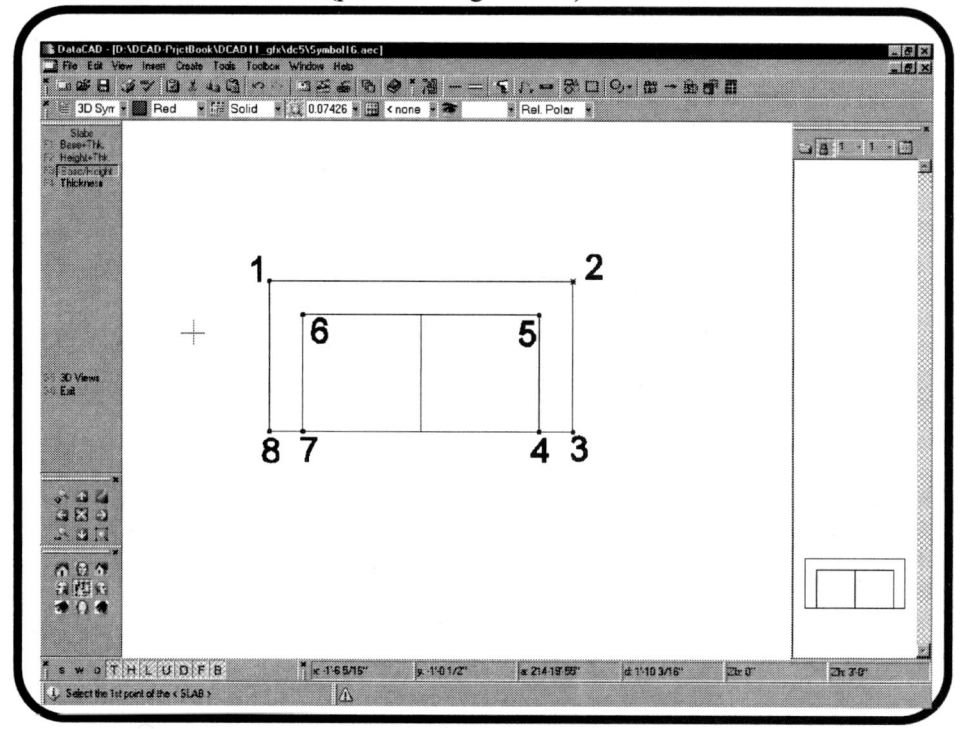

Figure 16-6, Creating the arms and back of the couch.

12.__ Snap to the next corner of the couch (point 2 in Figure 16-6).

13.__ Continue snapping to the rest of the points in order, until you get to point 8.

14.__ After snapping to point 8, right-click your mouse and the slab will be automatically closed from point 8 to point 1.

We'll look at the results in 3D by first temporarily turning off the **2D Symbol** layer. This will make the 3D symbol easier to see.

15.__ Press the **L** or **l** key, **On/Off, 2D Symbol**.

Create a 3D Symbol

16.__ Click on the **Isometric** icon button in the **PROJECTION PAD**, then click on the **Zoom extents** button in the **NAVIGATION PAD**. Your display should look something like Figure 16-7.

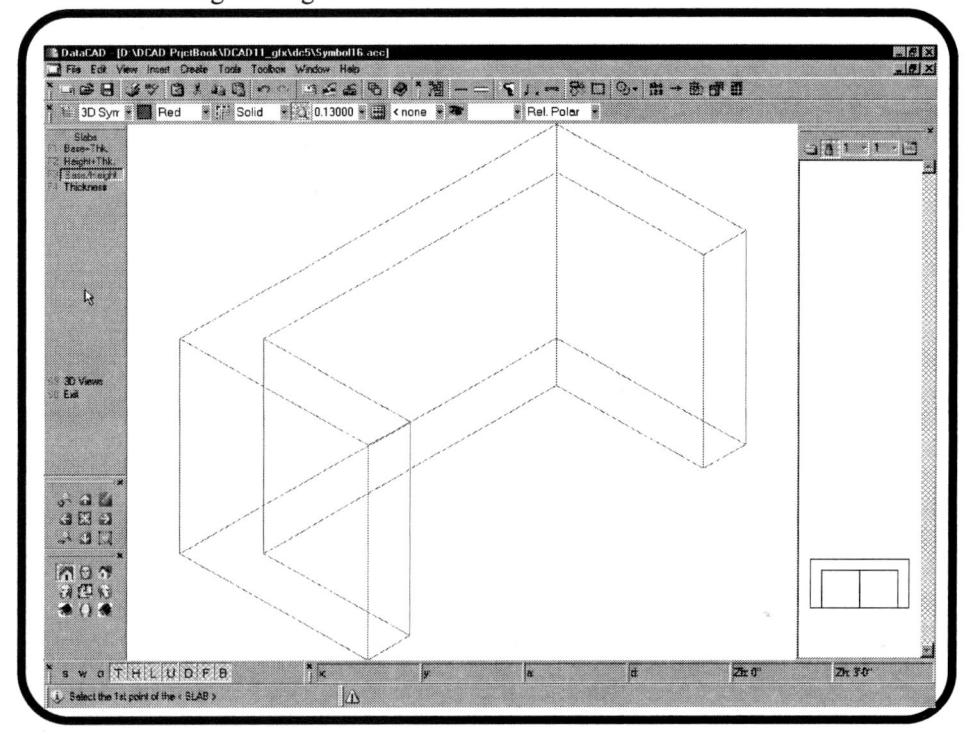

Figure 16-7, Isometric view of the couch in progress.

Let's complete the couch by drawing the seat area.

17.__ Click on the **Plan view** button in the center of the **PROJECTION PAD**.
18.__ Turn the **2D Symbol** layer on again (**l, On/Off, 2D Symbol**).
19.__ Press the **J** key [not the **j** key] to get to the **3D Entity** menu.
20.__ Select **Slabs, Horizontal** from the menu.
21.__ Select **Base+Thk.** again.
22.__ Click on **Thickness** and type **1.6** (for 1'-6"), then press **Enter**. DataCAD prompts you to *Select the 1st point of the <SLAB>*.

Chapter 16, Page 325

The DataCAD 11 Project Book

We will now be drawing the seat area of the couch with the same Z-base of 0, and a height (thickness) of 1'-6".

23._ Refering to Figure 16-6, use your middle mouse button or the **N** key to snap to point 4.

24._ Snap to the next corner, point 5, in Figure 16-6.

25._ Continue snapping to points 6 and 7.

26._ After snapping to point 7, right-click your mouse and the slab will be automatically closed from point 7 to point 4.

We'll look at the results again in 3D by first temporarily turning off the **2D Symbol** layer, making the 3D symbol easier to see.

27._ Press the **L** or **l** key, **On/Off, 2D Symbol**.

28._ Click on the **Isometric** icon button in the center of the **PROJECTION PAD**, then click on the **Zoom extents** button in the **NAVIGATION PAD**.

Create a 3D Symbol

29.__ Right-click until you see the **3D Menus** menu, and select **Shader, Shade**. Your display should look something like Figure 16-8

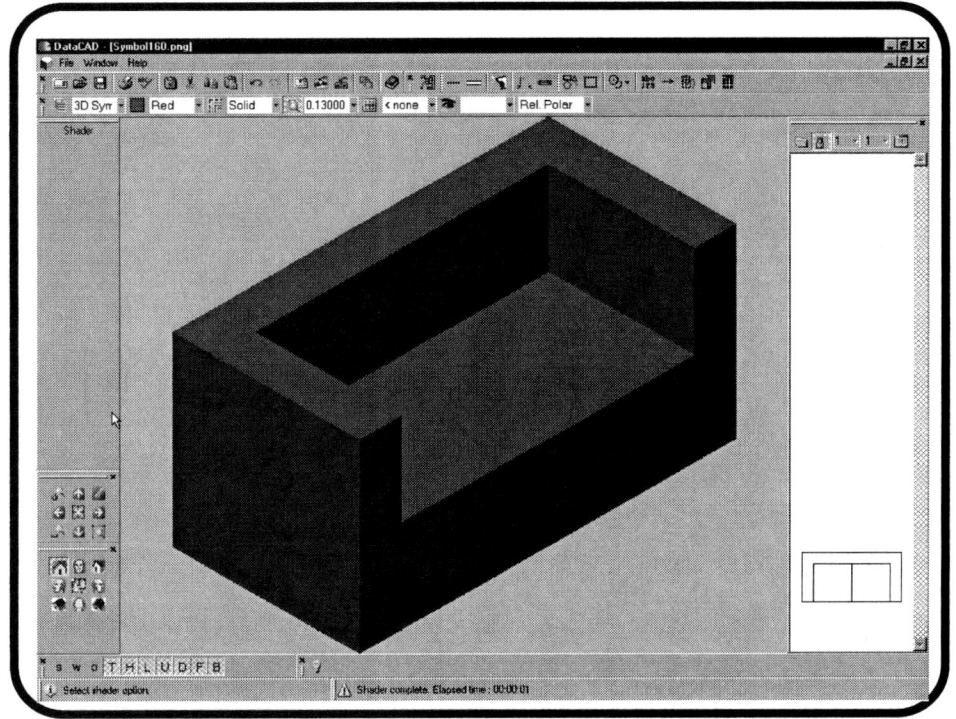

Figure 16-8, Isometric, shaded view of the completed couch.

30.__ Right-click your mouse until you get back to the **3D Menus** menu, then press **PgUp** a couple of times to give some room around the couch.

Now we can finally make a symbol of this couch.

31.__ If the **Symbol Browser** is not already open, press **Ctrl+T** to open it.

21.__ Click on the **Options** button in the upper right corner of the **Symbol Browser** (the one that looks like a green folder).

32.__ In the sub-menu that appears below, select the **Save Symbol** option.

A new dialog box appear with the title **Enter name of symbol file to be saved:**. Notice that the **Save in:** folder is the new **Chapter 16** folder that we created and highlighted earlier.

The DataCAD 11 Project Book

22. __ In the **File name:** box type the name of our new couch symbol, **3D Couch**, and press **Save**. A new **Save Symbol** menu appears on the left side of your screen.

23. __ To pick all the lines of the couch select the **Area** option from the **Save Symbol** menu.

24. __ Capture the entire couch in an **Area** box. All the lines of the couch will appear gray and dashed, showing that they have been selected.

INFO STOP

Notice also that the couch in your Drawing Window returns to a plan view. DataCAD always reverts to plan view when saving symbols. There is nothing you can do about that.

It is important to notice that at the bottom of your screen DataCAD is promting you to *Select an insertion point for symbol.* When saving symbols it is very important to read all the prompts at the bottom of your screen. It will make the process run a lot smoother.

As you have seen in Chapters 2 and 6, every symbol has an insertion point, so that's what we need to do here.

25. __ Snap to the upper right corner of the couch by placing your cursor close to that point, then clicking your mouse's middle button, or pressing the **n** [or **N**] key.

33. __ A new dialog box appears, prompting you for an **Item name** (Figure 16-5).

34. __ In the **Item Name** box type, **3D 3' x 6' Couch**, then press **OK** to close the dialog box. The symbol appears in the **Symbol Browser**, ready for you to use.

35. __ Now that you have this symbol saved, pick it from the **Symbol Browser** and place a few copies of it in your drawing window

Symbols always appear in the **Symbol Browser** in plan view, even when those symbols are 3D. But if you want to see a symbol in 3D you before placing it in a drawing, you can do the following.

1. __ Place the cursor over the symbol in the **Symbol Browser** and right-click your mouse. A new new dialog will appear.

Explode A Symbol

2.__ Select the **o2c Viewer** option and the o2c Object Viewer will appear on your screen, showing the symbol in 3D (see Figure 16-9). Try it.

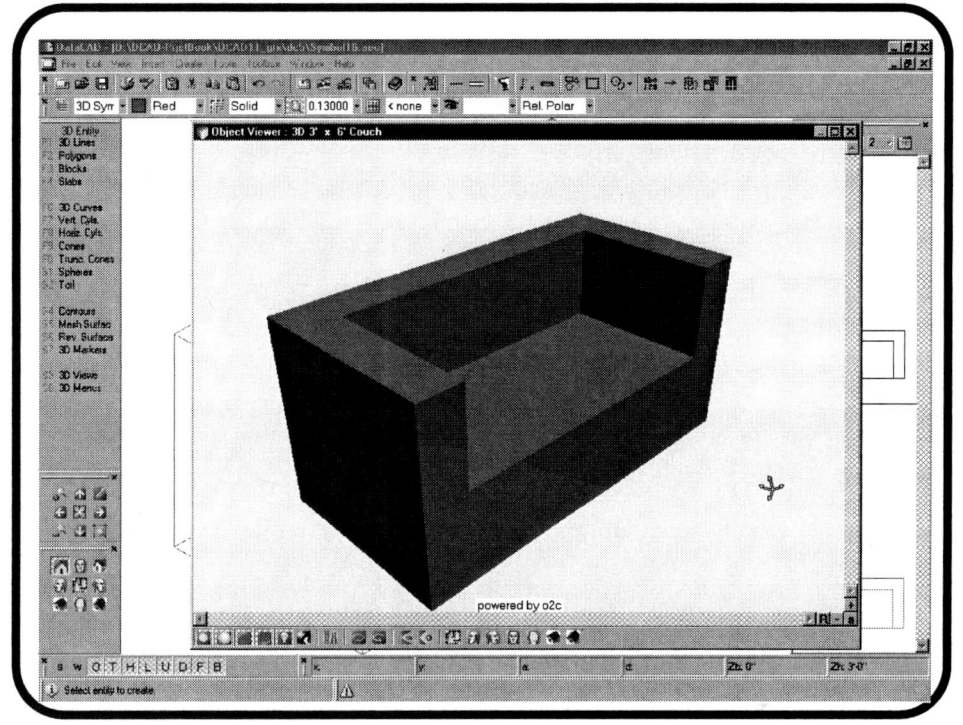

Figure 16-9, Object Viewer depiction of the couch symbol.

See Appendix C if you need more information on using the Object Viewer.

Explode A Symbol

To "explode" a symbol means to change it back into its original 2D or 3D parts, so that it is not a symbol anymore. If you need to make any changes to a symbol; its size, colors, insertion point, etc., you must first explode it back into its original entities. After making those changes you would then redefine the symbol. There will be times when you will want to put a symbol in your drawing, explode it, modify it in place, and not replace the original symbol with the changes.

There are two ways to explode a symbol: during the insertion of the symbol, or after placing the symbol.

The DataCAD 11 Project Book

Placing Symbols Exploded

1. __ Using the **Symbol16** drawing file from the previous example, open the **Symbol Browser** by pressing **Ctrl+T**.
2. __ Open the **Symbols, Living Room, Chapter 16** folder.
3. __ In the **Symbol Browser** place your cursor over the **3D Couch** symbol and click your left mouse button to select it.
4. __ The **Ins Symbol** menu appears on the left side of your screen. Pick the **Explode** option from that menu.

> **INFO STOP**
> When **Explode** is selected another option appears underneath, called **Current Layer**. In DataCAD 11 a new feature was added to allow a saved symbol to "remember" the drawing layers that all the symbol entities were on when created. So when you "**Explode**" the symbol back to its original entities, you have the option to put them all back on their original layers, or to put them all on the active layer. With **Current Layer** turned OFF DataCAD will put all the entities back on their original layers. If those layers don't already exist in the drawing file, then DataCAD will create them.

5. __ Make sure **Current Layer** is ON in the **Ins Symbol** menu.
6. __ Place the new couch below the original couch symbol in the Drawing Window, then right-click once to stop adding symbols.
7. __ Click on the **Identify** button in the **NAVIGATION PAD**, then use your left mouse button to click on any part of the 3D couch that you just placed. In the **Identify** menu at the left side of your screen you should see that DataCAD is telling you that the selected object is a **SLAB**.

> **INFO STOP**
> The word next to the **F1** in the **Identify** menu always tells you the type of entity that is being identified: a **LINE**, **SLAB**, **SYMBOL**, etc.

8. __ If you use your left mouse button to click on the first **3D Couch** symbol (the one that is still a symbol) in the Drawing Window. The **Identify** menu tells us this object is a **SYMBOL**.

Redefining Symbols

Exploding Symbols After Placement

To see how to explode a symbol after it is already placed in the Drawing Window do the following.

1. __ In the **Symbol Browser** place your cursor over the **3D Couch** symbol and click your left mouse button to select it.
2. __ The **Ins Symbol** menu appears on the left side of your screen. Make sure the **Explode** option is turned OFF.
3. __ Place the couch in the Drawing Window by clicking the left mouse button, then right-click your mouse to stop placing symbols.
4. __ Press the **j** key [not the **J** key] to go to the **3D Menus** menu.
5. __ Pick **Explode, Entity, Sym Explode, CurrentLayer** from the **3D Menus** menu.
6. __ Put your cursor on the copy of the couch and click your left mouse button. The symbol is exploded into its original 3D slabs.
7. __ To prove it click on the **Identify** button in the **NAVIGATION PAD**, then use your left mouse button to click on any part of the 3D couch that you just exploded. In the **Identify** menu at the left side of your screen you should see that DataCAD is telling you that the selected object is a **SLAB**, rather than a **SYMBOL**.

Redefining Symbols

Using the previous example, now that you have two exploded symbols in the Drawing Window, let's pick one to change its color and then use it to replace, or "Redefine", the 3D Couch symbol in the Symbol Browser.

1. __ In the **Symbol16** file, go to the 2D **Edit** menu and select **Change, Color, Cyan**.
2. __ From the **Change** menu select **Area**, then draw an area selection box around one of the already exploded couch symbols. All the 3D slabs that make up the couch will change to **Cyan**.
3. __ If the **Symbol Browser** is not already open, open it by pressing **Ctrl+T**, and browse to the **Symbols, Living Room, Chapter 16** folder.
4. __ In the **Symbol Browser** place your cursor over the **3D Couch** symbol and click your RIGHT mouse button.

The DataCAD 11 Project Book

5.__ A new dialog box appears at the point that you click your mouse. Pick the **Redefine** option from that menu (Figure 16-10).

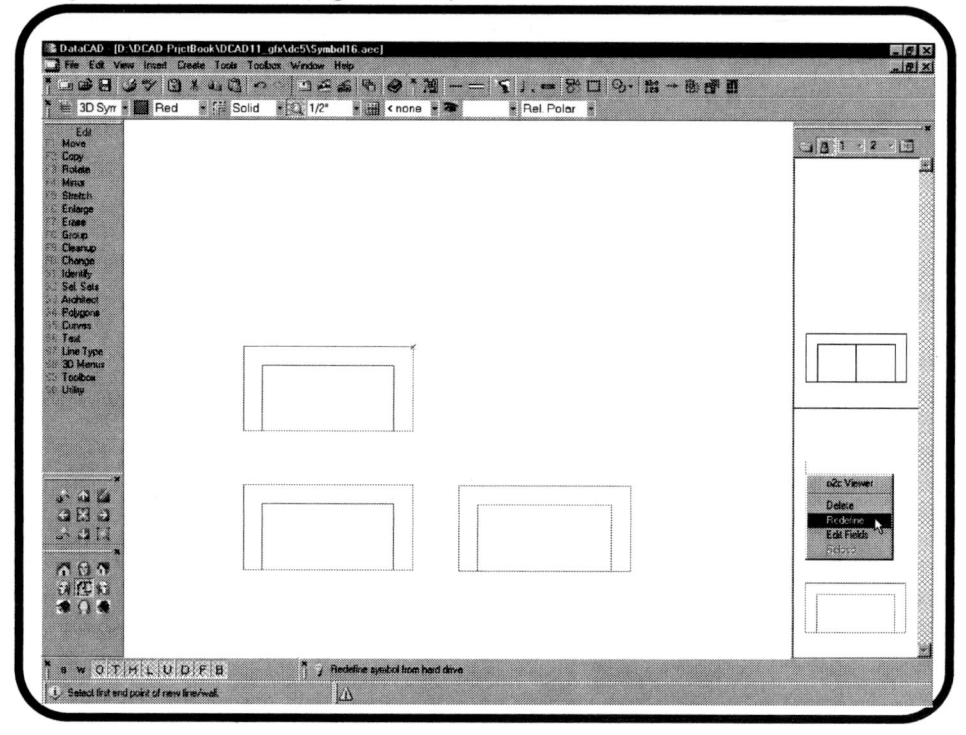

Figure 16-10, Selecting the symbol Redefine option.

6.__ A new **Redefine** menu appears at the left side of your screen, and at the bottom of your screen DataCAD prompts you to *Select first corner of area to <CREATE SYMBOL>*. Select **Area**, then draw an area selection box around the Cyan colored exploded couch symbol. All the 3D slabs that make up the couch will change to dashed gray lines to show they are selected.

A new prompt appears at the bottom of your screen: *You may select "Continue" to create symbol OR...*". The **OR** simply means you can keep picking more objects. For symbols that are more complex than this you might need to use **Entity**, **Area** or any of the other selection methods to keep picking more entities. In this case we're done picking entities, so we're going to pick the **Continue** option in the **Redefine** menu.

7.__ Pick **Continue** from the **Redefine** menu. Another prompt tells you to *Select an insertion point for symbol* (just like when you were defining the symbol the first time).

Replacing Symbols

Once again it is important to pick exactly the same insertion point as the original symbol, otherwise any of these symbols which have already been placed will shift, or move, to match the new insertion point.

8. ___ Snap to the upper right corner of the couch by placing your cursor close to that point, then clicking your mouse's middle button, or pressing the **n** [or **N**] key.

9. ___ The symbol will disappear from the Drawing Window and appear, with its new Cyan color, in the **Symbol Browser**.

Also notice that the 3D couch symbol that was already in your Drawing Window has changed to Cyan, too. This is how **Redefine** works in DataCAD. Every instance of that symbol will change in the current drawing. However, if you open any other drawing with that symbol in it, DataCAD will NOT have changed those symbols. DataCAD does not presume that you would want to make the change to every drawing you've ever used that symbol in.

But if you open another drawing and you do want to change all the symbols to match the newly defined symbol, right-click the symbol in the **Symbol Browser**, then select the **Reload** option.

Replacing Symbols

Symbols are a great time saver, but you might wonder if it isn't just as easy to copy a drawing of a bed, table, or lamp from one drawing to another, without ever using symbols. Well, one of the great benefits of symbols is that after you place multiple instances of the same symbol all over your drawing, you can change them all to another symbol with just a few keystrokes. Here's an example.

You have 8 rooms in your hotel from Chapter 7. Let's say you have it filled with several symbols, including a chair, bed, dresser, side table, lamp, tub, toilet and sink. Now let's say your client tells you she wants a smaller bed, more ornate side table, and a different style of sink and toilet. You could erase every one of those symbols (4 symbols per room x 8 rooms = 32 symbols to have to redo). Or you could use DataCAD's symbol **Replace** option to replace them.

Let's see how this would work using the 3D couch that we created. But first we need to make another 3D symbol to demonstrate with.

1. ___ Open the **Symbol16** drawing file we created earlier.
2. ___ Make the **3D Symbol** layer the only layer that is on.
3. ___ Press the **C** or **c** key to go to the **Copy** menu.

Chapter 16, Page 333

The DataCAD 11 Project Book

4. ___ Select **Drag, Entity**.

5. ___ Click on the 3D couch in the Drawing Window to pick it, then click anywhere near the couch.

6. ___ Drag the copy of the couch up above the original couch and click the left mouse button once to place it in the Drawing Window.

7. ___ Right-click three times to stop copying symbols and to exit the **Copy** menu. Your screen will look something like Figure 16-11.

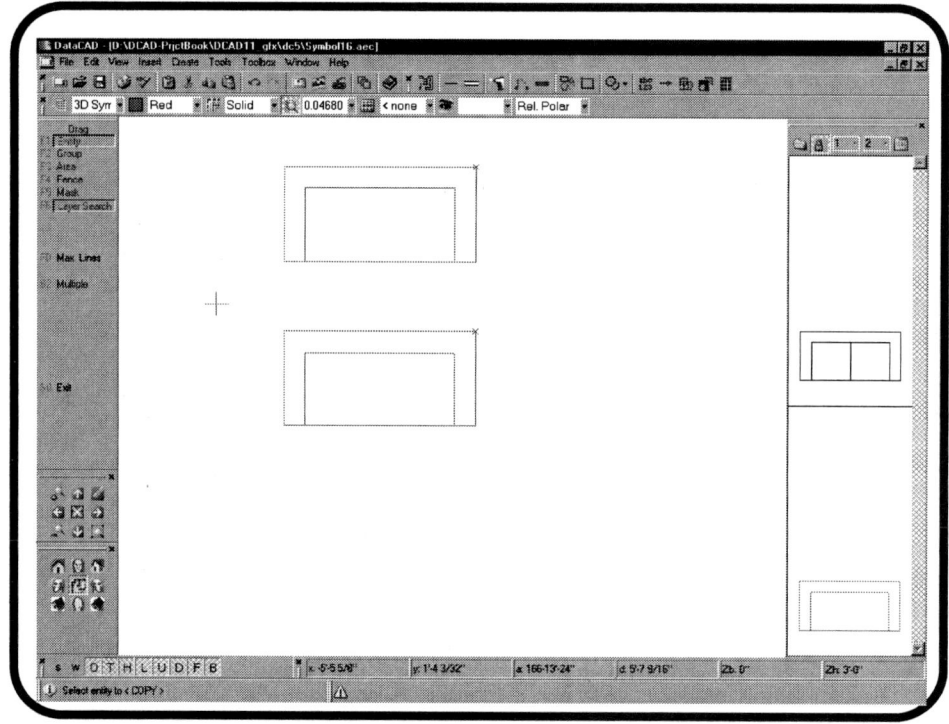

Figure 16-11, Two couch symbols.

As we learned in the previous section of this chapter, in order to make changes to a symbol it must first be exploded, so let's do that to the copy of our couch.

8. ___ Press the **j** [not the **J**] key to go to the **3D Menus** menu.

9. ___ Pick **Explode, Entity, Sym Explode, CurrentLayer** from the **3D Menus** menu.

10. ___ Put your cursor on the copy of the couch and click your left mouse button. The symbol is exploded into its original 3D slabs.

Chapter 16, Page 334

Replacing Symbols

Now we can make the exploded couch into a chair, instead.

11.__ Press **Alt+S** to open the **Stretch** menu.

12.__ Place your cursor near the upper right corner of the couch and click the left mouse button. DataCAD prompts you to *Select second point of the distance to stretch*.

13.__ Drag the cursor to the left about 2'-6" and click the left mouse button (the actual distance is not important, as long as we can see a clear difference between the size of the couch and the size of the chair).

14.__ Make sure **Area** is selected in the **Stretch** menu. DataCAD prompts you to *Select the first corner of box to <STRETCH>*.

15.__ Draw an area box around the entire right side of the couch. The 3D couch is now a 3D chair (Figure 16-12).

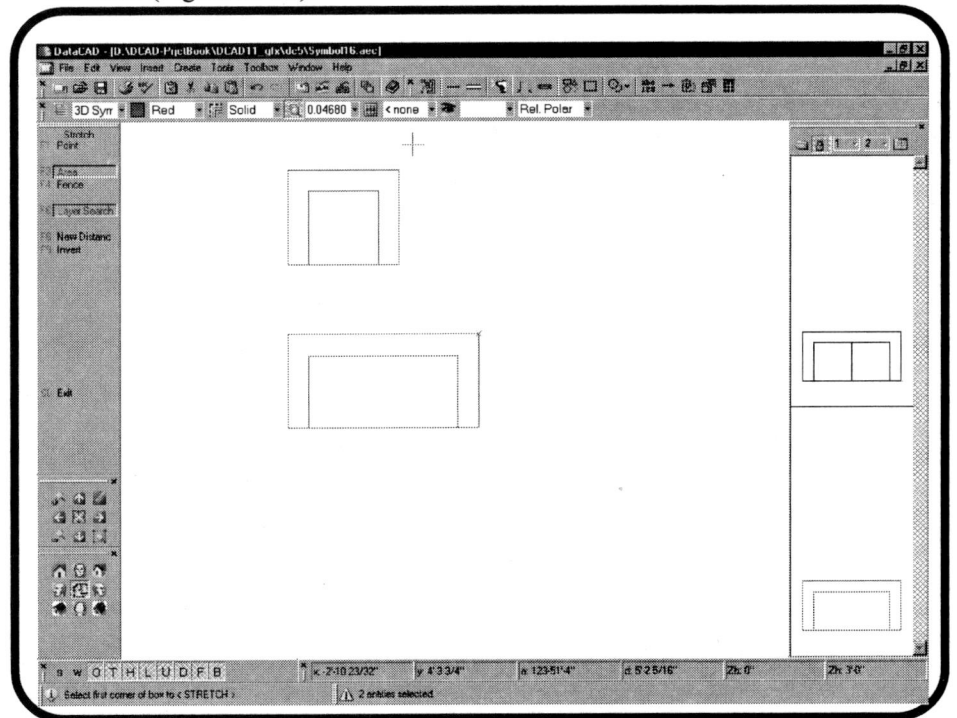

Figure 16-12, Our new 3D chair, above the couch.

16.__ If you don't already have the **Symbol Browser** open with the **Living Room, Chapter 16** folder we created earlier in this chapter, press **Ctrl+T** and browse to the **Living Room, Chapter 16** folder.

Chapter 16, Page 335

The DataCAD 11 Project Book

17. ___ Click on the **Options** button in the upper right corner of the **Symbol Browser**.

18. ___ In the sub-menu that appears below, select the **Save Symbol** option.

19. ___ Type a file name of **3D Chair**, then press **Save**.

20. ___ From the **Save Symbol** menu pick **Area**, and draw an area box around the newly created chair.

21. ___ DataCAD prompts you to *Select an insertion point for symbol*.

   •• **NOTE:** This is where the insertion point for a symbol becomes critical. Since every symbol is located in a drawing based on its insertion point, you want to make sure that similar symbols have identical insertion points, or the new symbol will be shifted to a different location when you use the **Replace** command.

22. ___ Snap to the upper right corner of the chair by placing your cursor close to that point, then clicking your mouse's middle button, or pressing the **n** [or **N**] key.

Replacing Symbols

23.___ In the **Item Name** dialog box type **3D Chair**, then press **OK** to close the dialog box. The symbol appears in the **Symbol Browser**, ready for you to use (Figure 16-13).

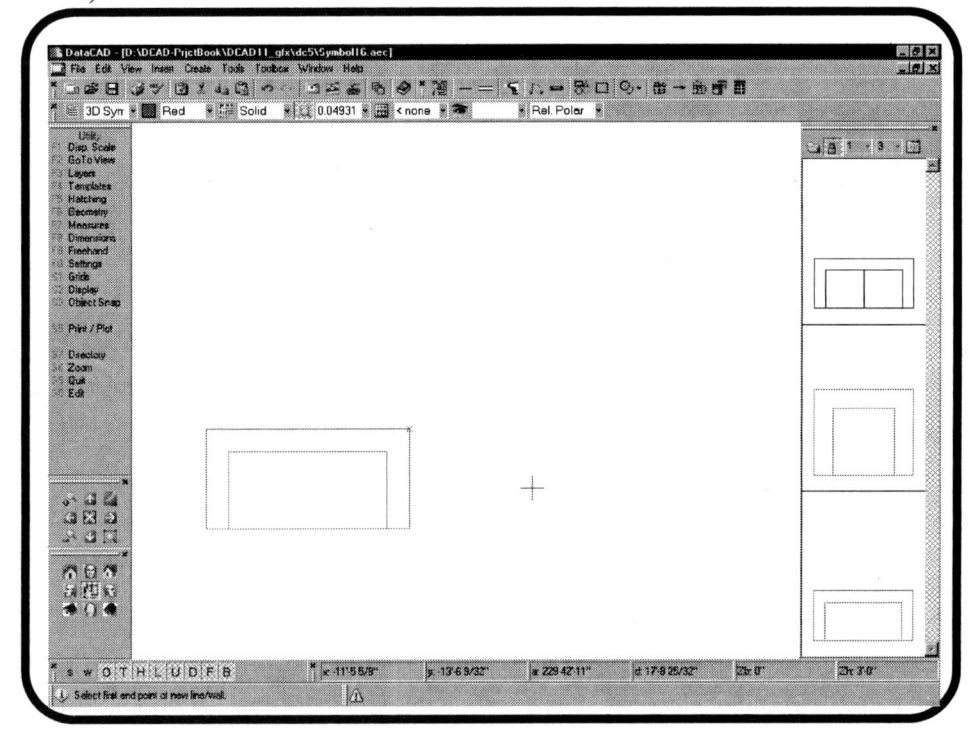

Figure 16-13, The Symbol Browser with the 3D chair.

The DataCAD 11 Project Book

24.___ Instead of placing copies of your new 3D chair in the Drawing Window, place a half dozen or so of the 3D couches, at different angles, instead (Figure 16-14).

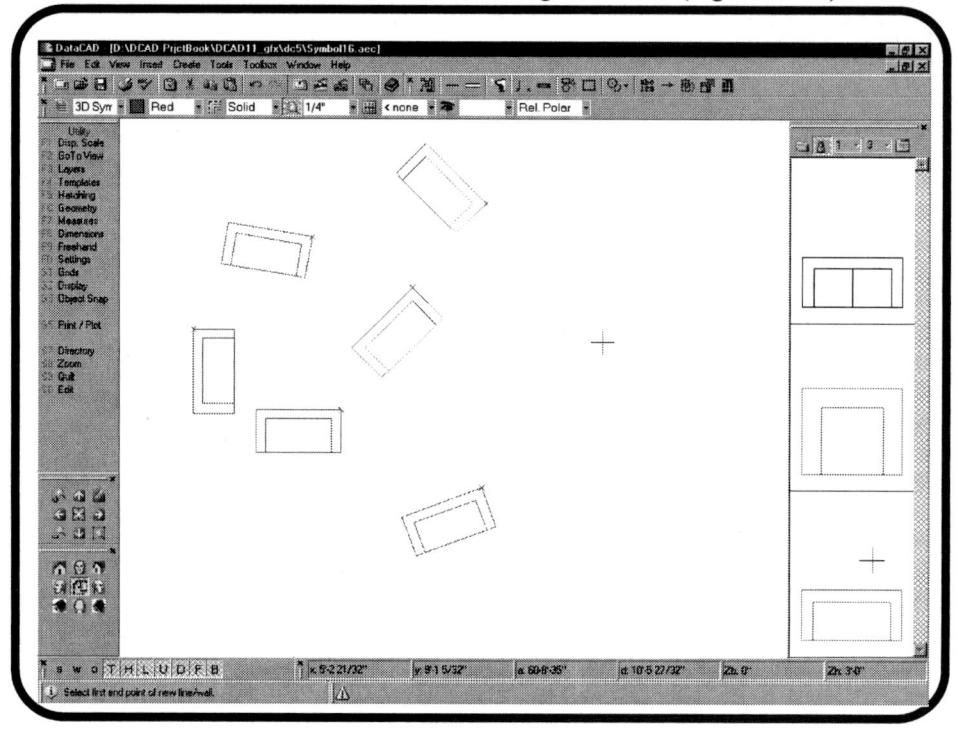

Figure 16-14, Lots of couches.

Now we can finally get to the **Replace** command. We will use it to replace all the 3D couches in the Drawing Window with the 3D chairs, instead.

25.___ Click on the **Options** button in the upper right corner of the **Symbol Browser**.

26.___ In the sub-menu that appears below, select the **Replace** option.

A new **Replace** menu appears on the left side of your screen, and at the bottom of your screen DataCAD prompts you to *Select symbol to replace*. Here again, reading the DataCAD prompts will help you through the process.

27.___ Select any one of the couch symbols in the drawing window by puting the cursor on it and clicking the left mouse button. DataCAD prompts you to *Select symbol to replace with*.

Replacing Symbols

DataCAD is now asking you to pick a symbol in the **Symbol Browser**. This symbol will be the one to replace all the 3D couches, selected in Step 27.

28.__ Place your cursor over the **3D Chair** symbol in the **Symbol Browser** and click on it with your left mouse button.

Now we're going to see two different ways to replace symbols.

29.__ In the **Replace** menu on the left side of your screen select the **Entity** option. DataCAD prompts you to *Select entity to <REPLACE>*.

30.__ Use your left mouse button to click on just one of the couches in the Drawing Window. The 3D couch is replaced with the 3D chair!

31.__ Now pick the **All** option from the **Replace** menu. When you do, all the 3D couches (because that was the one selected in Step 27) are replaced with the 3D chair.

Using this last method of replacement in the hypothetical hotel example at the beginning of this section would allow you to quickly replace all the bed symbols with a different bed symbol, all the sinks with another sink symbol, and so on. Now think of the time savings if you had a large office building with 400 round ceiling light fixture symbols that you wanted to change to 400 square ceiling light fixture symbols.

There are many more aspects to symbols, and many more tricks that they can do. These were just the basics. You can read more about symbols in the DataCAD online Help file (from the **Menu Bar**, pick **Help, Reference Manual**, or just press **Ctrl+F1**).

The DataCAD 11 Project Book

Review and Important Points.

- It is important to read the prompts DataCAD gives you at the bottom of your screen. It's easy to get lost in the process if you don't.

- You can create your own symbol folders with the **New Folder** button in the **Symbol Browser**'s **Browse** dialog.

- When naming symbols, remember that there is the name that the computer sees, and then there is the name, or description, that you will see when your cursor is over the symbol in the **Symbol Browser**.

- Symbol names should be descriptive enough to ensure that you always pick the correct symbol.

- You can see a symbol displayed in 3D in the **Symbol Browser** by using the **o2c Viewer** option.

- Symbols can be 2D or 3D.

- Symbols must be **Exploded** before they can be edited. They can be exploded before or after placing them in the drawing.

- Symbols can be **Redefined** after they have been created. **Redefined** symbols will automatically be updated in the current drawing.

- The **Replace** option lets you replace one or hundreds of symbols at a time.

- Symbols can be **Rotated** and **Mirrored**, just like other DataCAD entities.

Replacing Symbols

Practice Drawings
P16-A

- Draw a living room like the one below, in 2D.
- Create your own 2D furniture symbols to fill this living room.
- Make the entire Living Room a symbol. Yes, you can have symbols inside of symbols (it's called "nesting").
- Place several copies of the Living Room in your drawing.
- Pick one of the Living Room symbols to Explode. After doing so, what happens to the furniture symbols?

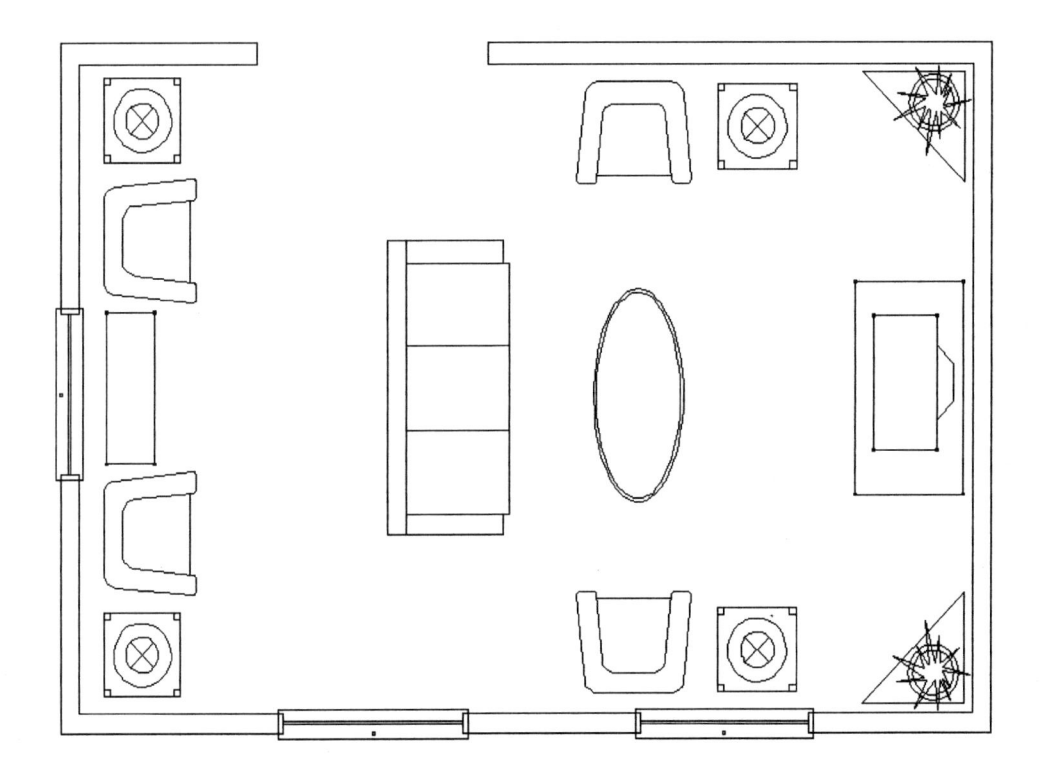

The DataCAD 11 Project Book

P16-B

- Make a copy of the plan from P16A.

- Create your own side chair, side table, and plant designs, and use the **Replace** command to update the previous symbols with the new ones.

The DataCAD 11 Project Book

Notes

The DataCAD 11 Project Book

APPENDIX A

The DataCAD 11 Project Book Quick Keys

Key	Action
A	Append Selection Set (toggle)
a	**Architect**
B or b	**Tangents**
C or c	**Copy**
D or d	**Dimension**
E or e	**Erase**
F	Save current drawing file
f	Line **Spacing**
G or g	**Grids**
H or h	**Hatch**
I	Identify
i	Digitizer (toggle)
J	**3D Menus**
j	**3D Edit**
K	Color selection scroll (backward)
k	Color selection scroll (forward)
L or l	**Layers**
M	**Macros**
m	**Move**
N or n	Near point snap (same as middle mouse button)
O or o	Ortho mode (toggle)
P or p	Previous view
Q	Line type selection scroll (backward)
q	Line type selection scroll (forward)
R or r	**Rotate**
S	**Edit Sets**
s	Set snap grid size
T or t	**Template**
U or u	Regenerate display list
V	**3D Views**
v	**GoToView**
W	Decrease line weight by 1
w	Increase line weight by 1
X	**Object Snap**
x	Snap on/off (toggle)
Y	**Hide** (3D)
y	**3D Views**
Z or z	Enter Z base value

Special Keys

Key	Action
/	**Zoom To Area**
=	Walls on/off (toggle), set width
\	Clean-up T intersection (toggle)
|	Cap wall ends (toggle)
-	Display line overshoot (toggle)
+	Big cursor (toggle)
?	Select point for coordinate identification
Space Bar	Enter coordinate entry mode
PgDn	Select next smaller scale
PgUp	Select next larger scale
Ins	Select coordinate system (scroll)
Esc	Redraw display
End	Interrupt action (redraw, plot, hide)
Del	Interrupt redraw
Home	Center display at cursor location
Ctrl+Home	Select new display center with cursor
TAB	Layer selection scroll (forward)
Shift+TAB	Layer selection scroll (backward)
Shift+(num)	Display saved view (num)
,	Erase last entity created (the comma key)
.	Restore last entity erased (the period key)
<	Erase last group created
>	Restore last group erased
`	Reset coordinate 0,0 (the lower-case ~ key)
~	Select snapping point
'	Layer search on/off (toggle) (the lower case " key)
"	Enter symbol name to load
:	**Utility**
;	**Edit**
Ctrl+C	**Edit**
Ctrl+e or E	Copy to clipboard
Ctrl+l or L	Open **Layer Manager** dialog box
Ctrl+n or N	Create new drawing file
Ctrl+o or O	Open drawing file
Ctrl+p or P	**Plotter**
Ctrl+r or R	**Reference File Manager**
Ctrl+s or S	Save file
Ctrl+w or W	Toggle multiple view windows
Ctrl+ -	Zoom ectents
Ctrl+arrow key	Previous View
Ctrl+Tab	Switch to other open drawing file
Ctrl+F1	On-line help
Arrow Keys	Pan display
F11	Left elevation
F12	Right elevation

The Alt Keys

Key	Action
Alt+A	**Curves**
Alt+B	**1 Line Trim**
Alt+C	**Change**
Alt+D	**Divide**
Alt+E	**Enlarge**
Alt+F	**Fillet**
Alt+G	**Offset**
Alt+H	**Add To**
Alt+I	Adjust settings to entity under cursor
Alt+J	**2 Line Trim**
Alt+K	**Link Entities**
Alt+L	**L Intersect**
Alt+M	**Mirror**
Alt+N	**New Drawing**
Alt+O	**1 Line Trim**
Alt+P	**Plotter**
Alt+Q	**Quit**
Alt+R	**Polygons, Rectangle**
Alt+S	**Stretch**
Alt+T	**Text**
Alt+U	**T Intersect**
Alt+V	**Change**
Alt+W	**Weld Line**
Alt+X	**Measures**
Alt+Y	**Layers, Name**
Alt+Z	**Change**
Alt+(num)	Change active selection set
Alt+-	Switch to previous view
Alt_Tab	Switch to other Window Application

BOLD TYPE Indicates menu name

The DataCAD 11 Project Book

Notes

APPENDIX B

Using Selection Sets and Masking

Project Goals

When you have completed this project, you will be able to:
1. Add entities to a selection set.
2. Make a selection set active.
3. Use a selection set and the **Mask** option to change colors.

Prerequisite: DataCAD 11 interface experience

Introduction

A very powerful pair of tools in DataCAD are the **Selection Set** and **Mask** functions. In some ways these two functions are similar to an old fashioned erasing shield. By covering some of the lines on a drawing with an erasing shield, you could rub hard and erase just the lines that appeared in the various holes in the shield. A selection set can be used in a similar way to selectively erase certain entities from a drawing. However, it is a lot more powerful and complex than the old erasing shield. Working through an example will give you a peek at how selection sets and masking work. Let's start by creating a 3D window that has many different entities and colors.

1.__ Start the **DataCAD** software and create a new drawing file named **AppxB**.
2.__ Press the **M** [not the **m**] key to jump to the **Toolbox (Enter macro filename:** dialog box).
3.__ Open the **AEC_MODL.DCX** file.
4.__ From the **AEC_Modl** menu, select **Windows**.
5.__ From the **WindPlan** menu, select **UnitType**. Then toggle **DblHung ON**, and press the *right* mouse button to return to the **WindPlan** menu.
6.__ Click the *left* mouse button somewhere near the center of the display to position the first inside corner of window.
7.__ Press the **Insert** key until **Relative Cartesian** coordinate mode is active; then press the **Space Bar** and enter **X=3, Y=0**.

The DataCAD 11 Project Book

8.__ At the prompt, *Enter a point on outside surface of wall*, press the **Space Bar** and enter **X=0, Y=0.6**.

9.__ Press the **y** [not the **Y**] key to jump to the **3D Views** menu, and select **Isometric**.

Your display should look similar to Figure B-1.

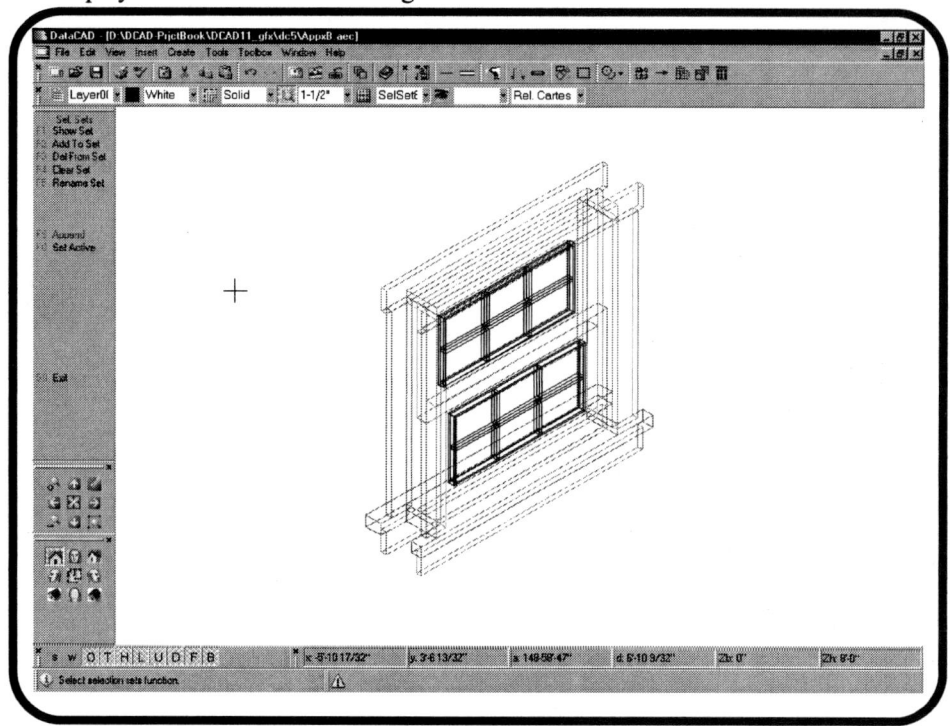

Figure B-1, The window model.

The **AEC_MODEL** macro has created an assembly of 3D objects (mostly slabs) of various colors. It has also created the entire assembly as one group. The challenge is to change the color of various individual components. One way to do this is to use a **Selection Set** and **Mask** to select all the parts of the window to be changed.

The Selection Sets Menu

10.__ Press the **S** [not the **s**] key to go to the **Sel. Sets** menu (or pick **Edit, Sel. Sets**).

The **Sel. Sets** menu is shown in Figure B-1. The first thing that must be done is to create a *Selection Set* that contains all of the entities of the inside trim. The inside trim color will be changed later to an off white color. (The inside trim default color is **LtRed**.)

The Selection Sets Menu

To determine the properties of an entity, select the **Identify** button on the **NAVIGATION PAD** (the one that looks like a finger) and then click on the entity. Use this if you need to verify the color of the inside trim.

11.___ From the **Sel. Sets** menu, select **Add To Set**. Then press **F1** or click on the **SelSet1** menu button.

12.___ From the **Add To Set** menu, select **Mask**. Then select **Color**.

13.___ Toggle only **Lt Red** on, press the *right* mouse button *twice* to get back to the **Add to Set** menu, select **Area**, and capture the entire window in an area box.

Your display should look similar to Figure B-2.

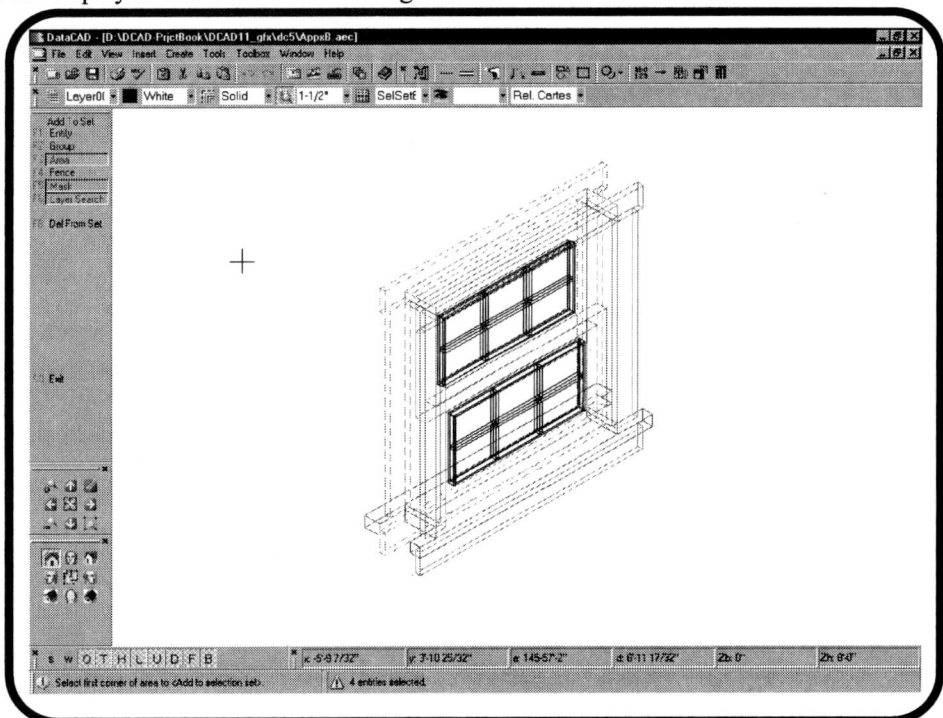

Figure B-2, Selecting the inside trim.

Appendix B, Page 349

The DataCAD 11 Project Book

If everything is going OK, 4 entities have been added to **SelSet1**, and they will appear as dashed lines on the display. Next, **SelSet2** will be created for the outside trim. (The outside trim default color is **Red**.)

14.__ Press the *right* mouse button to return to the **Edit Sets** menu.

15.__ From the **Sel. Sets** menu, select **Add To Set**. Then press **F2** or click on **SelSet2**.

16.__ From the **Add To Set** menu, select **Mask**. Then select **Color**. Note that if **Mask** and **Color** are already toggled on, you will have to toggle them off then back on to get to the next menu.

17.__ Toggle **Lt Red** off (it should still be on from the previous selection you made), then toggle only **Red** on.

18.__ Press the *right* mouse button *twice*, select **Area**, and capture the entire window in an area box.

> **INFO STOP** You can toggle one, several or all of the colors on or off to create a **Mask**.

> **INFO STOP** Note that the **Mask** menu allows for masking using **Entity, Color, Line-Type**, or **Weight**. You could create a **Selection Set** that includes all of the stuff that is both green and dashed. You can use **AddTo** to expand the items in a **Selection Set**. The **Edit Sets** menu also includes options to delete from, or clear a **Selection Set**. You can also change the name of a **Selection Set**.

The 4 outside trim pieces should be shown dashed. We could continue to define additional **Selection Sets**. However, this will be enough to demonstrate the concept. Next the **Selection Sets** will be used together with the **Change** menu to change some colors.

19.__ Right click once to get back to the **Sel. Sets** menu. From there select **Set Active**. Then select **SelSet1**.

Observe that the **STATUS AREA** shows the name of the currently active **Selection Set**. And in the **Sel. Sets** menu there is now a dollar sign (**$**) in front of **SelSet1**, signifying that it is the active selection set.

The Selection Sets Menu

20. __ Press **Alt+c** to jump to the **Change** menu (or pick **Edit, Change**).
21. __ From the **Change** menu, select **Color**. Then select **Custom** and pick the a light gray color like **Color_28** from the **Color Palette**. Select **OK**.
22. __ From the **Change** menu select **Mask, SelSet1**.

   •• **NOTE:** If **SelSet2** were the currently active selection set, then the menu option would have said **SelSet2**, instead.

The color of all of the inside trim pieces have been changed. Let's change the color for the outside trim.

23. __ Press the **S** [not the **s**] key to jump to the **Sel.Sets** menu. Select **Set Active**, and then select **SelSet2**. The **STATUS AREA** shows the name of the currently active **Selection Set**. And in the **Sel. Sets** menu there is now a dollar sign (**$**) in front of **SelSet2**,
24. __ Press **Alt+c** to jump to the **Change** menu.
25. __ From the **Change** menu, select **Color**. Then select **Custom,** and pick a nice green color like **Color_56** from the **Color Palette**. Select **OK**.
26. __ From the **Change** menu, select **Mask, SelSet2**.

The color of all of the outside trim pieces have been changed. If you like, you can use **Shader** or the **Object Viewer** to admire your new window trim colors.

INFO STOP
By now you have probably noticed that there is a "persistence" to Selection Sets. In other words, the entities you select will stay in that set until you use the **Clear Set** option in the **Sel. Sets** menu to clear them out of the Selection Set. This allows you to apply multiple changes to the entities in a Selection Set.

There is one more **Selection Set** function that can be very handy when you are creating a new design. The following example has nothing to do with the window currently in the drawing, so just ignore it for now. Or, if you are a neat freak, you can start a new drawing.

27. __ From the **NAVIGATION PAD**, select the **Ortho** button to display a plan view and adjust the display.
28. __ Press the **Page Up** key a few times to give yourself some drawing room.

The DataCAD 11 Project Book

29. ___ Press the **S** [not the **s**] key to jump to the **Sel. Sets** menu. Select **Set Active** and then select **SelSet3**.
30. ___ In the **Sel. Sets** menu, toggle the **Append** option **ON**.
31. ___ Press the *right* mouse button a few times until the *Select first endpoint of new line/wall* prompt appears, and draw a few lines on the display.

> **INFO STOP**
> Every entity that you add while the **Append** option is active will be added to the current **Selection Set**. Pressing the **A** [not the **a**] key will toggle the **Append** option on and off.

32. ___ Press the **S** [not the **s**] key to jump to the **Sel. Sets** menu. Select **Set Active**, and then select **SelSet4**. Toggle the **Append** option **OFF**.
33. ___ Press the *right* mouse button a few times until the *Select first endpoint of new line/wall* prompt appears, and draw a few lines on the display.
34. ___ Press the **A** [not the **a**] key to toggle the **Append** option **ON**, and draw a few more lines.

Next, the **Selection Sets** will be used with a couple of different functions.

35. ___ Press the **S** [not the **s**] key to jump to the **Sel. Sets** menu. Select **Set Active**, and then select **SelSet3**.
36. ___ Press the **m** [not the **M**] key to jump to the **Move** menu. Click at *two* places on the display to set the first and second points of the distance to move, and then select **Mask, SelSet3** from the **Move** menu.
37. ___ Press the **Zoom Extents** button in the center of the **NAVIGATION PAD** to regenerate the display.

Observe how all of the entities in **SelSet3** have been moved. Here is one more example before quitting this exercise.

38. ___ Press the **S** [not the **s**] key to jump to the **Sel. Sets** menu. Select **Set Active**, and then select **SelSet4**.
39. ___ Press the **e** [or the **E**] key to jump to the **Erase** menu. Then select **Mask, SelSet4**.

Appendix B, Page 352

The Mask Function

Everything in **SelSet4** has been erased. You should now appreciate how powerful the **Selection Set** function really is. Changing colors, erasing and moving entities are only some of the things you can do with the entities once they are in a selection set. And, if you can remember a few quick key presses to navigate around the menus, it can speed up your work considerably.

The Mask Function

In DataCAD 11 the **Mask** function was altered to allow you to use it without having to create a selection set. Selection sets are still very useful for grouping entities, but for simple masking functions let's use our window example to see how how much quicker it is now.

40.__ Press **Alt+c** to jump to the **Change** menu.

41.__ From the **Change** menu, select **Color**. Then select **Brown**. This is the color that you are going to change some entities TO.

42.__ From the **Change** menu, select **Mask, Color**. Now pick **White**. This is the current color of the window muntins that you want to change. Right-click twice to go back to the **Change** menu.

•• **NOTE:** Remember that if you have a white screen in your DataCAD Drawing Window then all "White" entities will actually appear as Black.

43.__ Select **Area**, and capture the entire window in an area box.

Unlike with Selection Sets, instead of the selected White window muntins being highlighted, they are immediately changed to **Brown**.

It was noted earlier that the **Mask** menu allows for masking using **Entity, Color, LineType,** or **Weight**. In fact, if you select **Entity** you will see that there are actually many TYPES of entities that can be selected. There are 12 types there on the first menu, and if you click on the **Scroll Fwrd** button you will see 14 more! Try experimenting with all of them.

The DataCAD 11 Project Book

APPENDIX C

Using Object Viewer

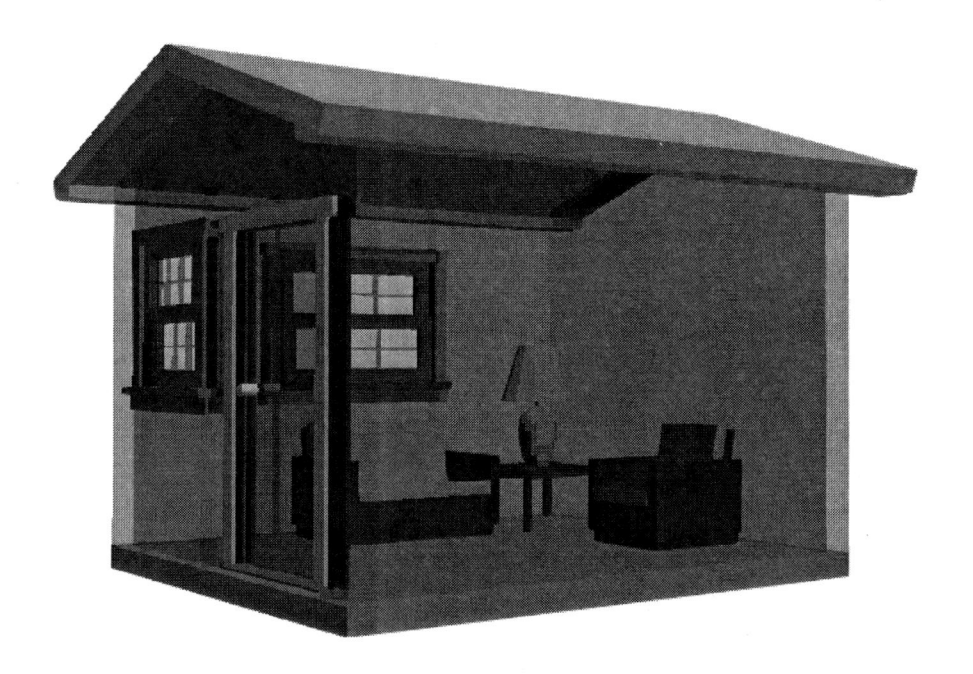

Project Goals

When you have completed this project, you will be able to:

1. Adjust the o2c **Rendered Color** for each DataCAD **Entity Color**.
2. Adjust the o2c **Opacity,** and **Reflection** for each color.
3. Use the **Object Viewer** to create rendered images of a model.
4. Use the **Object Viewer** to dynamically rotate and zoom on a shaded rendering of a model.

The DataCAD 11 Project Book

Prerequisite: DataCAD 11 interface experience, Chapter 10 (or Chapter 6, 9, 12, or 15)

Introduction

The o2c **Object Viewer** is a rendering program that can be run from within DataCAD to provide dynamic display of rendered views of DataCAD models. This exercise shows how to set the mapping of DataCAD colors to the **Object Viewer** colors and to create rendered images that include transparent objects.

INFO STOP

New in DataCAD 11 is the ability to render all materials of a particular color, or all materials by their layers, but not both. By selecting **Color** in the **Entity Color** area of the dialog, all **Red** entities, for instance, will be rendered alike, no matter which layers they are on. If you pick **Layer** in the **Entity Color** area, then all entities on **PLAN**, for instance, will be rendered alike, regardless of their colors. In this example we will stick to using the **Color** method, but just realize that you can do either.

••• **WARNING**••• Creating a rendering that has a lot of detail and uses a lot of different objects and colors can be time consuming. Remember that every object (or part of an object) that is to be rendered a different color must first be assigned a unique DataCAD color. Then each DataCAD color must be associated with the desired o2c color, and opacity and reflection settings have to be adjusted for each color.

•• **NOTE:** Most examples shown here use a modification of the **Play10** model from Chapter 10. (A floor slab and some furniture have been added to the Chapter 10 model.) You can substitute a model from Chapter 6, 9,12, 15, or any other 3D model file for use in this exercise.

1.__ Start the **DataCAD** software and open the **Play10** drawing (or your desired 3D model file).
2.__ From the **Menu Bar**, select **File, Save As** and save a new drawing file named **Appendix C**.
3.__ Add furniture, slabs, or whatever enhancements you would like to have in your rendered images.

In the example used here, the playhouse (Play10) model was opened and saved with a new file name. A floor slab was added using the horizontal slab **Base+Thk** option with the thickness set to - 6 inches. A table (from the Living Room/3D Tables symbol folder) was added. A lamp was added (from the Lights/3D Lighting Fixtures symbol folder) after using the **Identify** function to determine that the height of the table was 1'5", and adjusting the **Z-Offset** before placing the lamp.

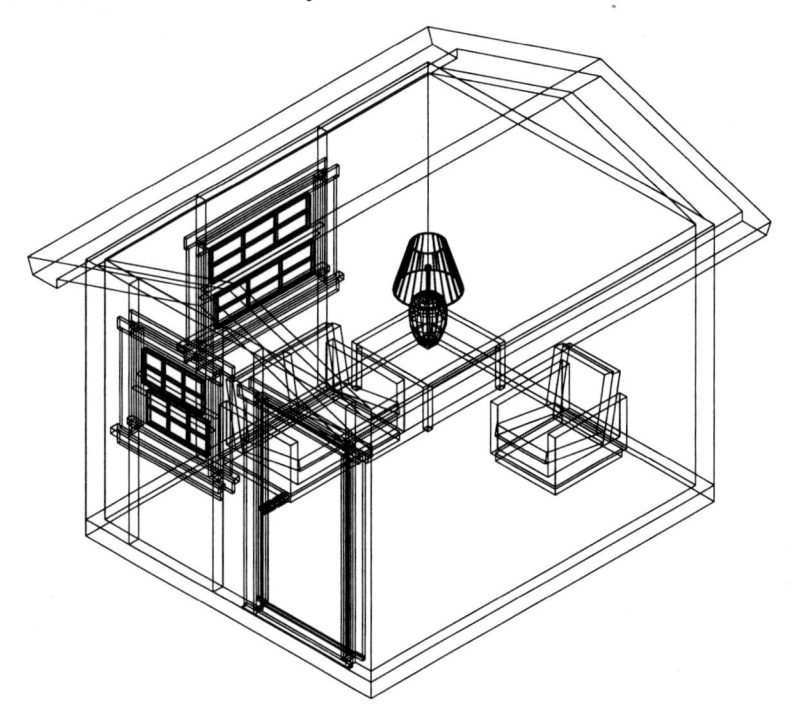

Figure C-1 shows the modified Play10 model.

Figure C-1, A model for rendering.

Hint: Be sure to isolate different types of objects on different drawing layers to make it easier to manage your model. If necessary, use the **Layer Manager** [**Ctrl+L**] to create new layers, Then use **Move** [press **m**], **To Layer** to move different objects to different layers.

4.__ Display an isometric view of your model.

Appendix C, Page 357

The DataCAD 11 Project Book

INFO STOP

It is possible to set any DataCAD color (or layer) to be rendered by the **Object Viewer** in any color found in the color palette. For example, all of the red *and* green entities in a DataCAD model could be rendered in brown by the **Object Viewer** (o2c). Also, any color (or layer) in the DataCAD model can be rendered in a different color by o2c.

Suppose that you have a love seat in the DataCAD model that has blue seat cushions. Also suppose that you have a window sill that is also blue. Both the sofa seat and the window sill will both be rendered the same color by o2c. If you want them to be rendered differently by o2c, you will have to first change the colors in the DataCAD model to be different from each other.

> •• **NOTE:** If your DataCAD model only contains entities that use the 15 DataCAD default colors, and if you want to have more than 16 different colors in the rendered image, you will have to change the color of entities to a color different from the original 15.

5.__ Adjust the colors of any DataCAD entities that you would like to change. Use the **Layer Manager** to help isolate different entities.

> •• **HINT:** Follow the instructions found in **Appendix B** and use Selection Sets or Masking to change the color of all entities that already have a particular color that you would like to change to a new color.

INFO STOP

If you have trouble setting the color for any part of a door or window modeled using the AEC_MODL macro, you might consider removing the door or window, and then recreating it after setting your own color for each component using the AEC_MODL macro color setting options. Some components (such as the jam or even the door slab itself) are hard to isolate and change after the model has been created.

Appendix C, Page 358

Setting the DataCAD Render Settings

Setting the DataCAD Render Settings

The next step (actually a lot of steps) is to map each DataCAD color in your model to an o2c color.

6.__ Adjust the display so that you can isolate an entity that has a DataCAD color that you want to map to an o2c color.

7.__ Hold the **Ctrl** key down and *right* click on the desired entity. Select **Rendering Settings** from the pop up menu.

This opens the **DcadWin.dmf** (DataCAD Materials File) dialog box.

8.__ Make sure **Color** is selected in the **Entity Color** area of the dialog.

9.__ Click on the **Rendered Color** color box, and then pick the desired color from the **Color Palette**.

10.__ Select the **OK** button in the **Color Palette** dialog box.

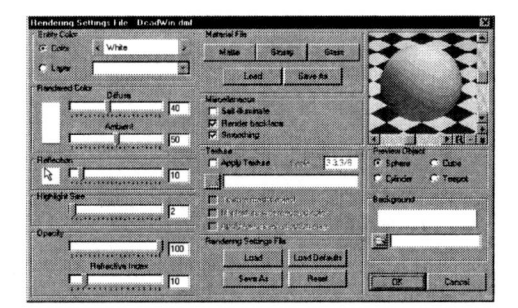

The **DcadWin.dmf** dialog box has an area named **Opacity** that contains a slider and a numeric value to the right of the slider. This is used to control the amount of transparency that objects with the current **Rendered Color** will have.

To set the **Opacity** for the current color, either click and drag the slider, or edit the numeric value in the text box to the right of the slider.

•• **HINT:** To create transparent walls, start with an **Opacity** setting of **40** for the wall color. After a test rendering, adjust this value until you get the effect you like. Observe that each time you look through another wall, the shading gets darker.

The **DcadWin.dmf** dialog box also has sliders and text boxes that can be used to control the **Reflection** characteristics for the current color. These include **Ambient, Diffuse,** and **Specular**.

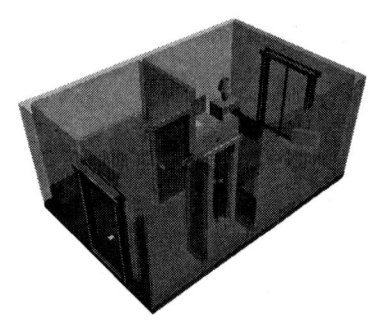

Appendix C, Page 359

The DataCAD 11 Project Book

•• **HINT:** If you want to experiment with the **Reflection** settings, you should do so on a simple 3D model. Some effects will be different on flat, cylindrical, and spherical objects, and will also depend on the angle of light falling on the object. For example, the highlights on the spherical dome and simulated trees in the adjacent detail of a rendering of Ferd City can be changed by adjusting the **Reflection** settings.

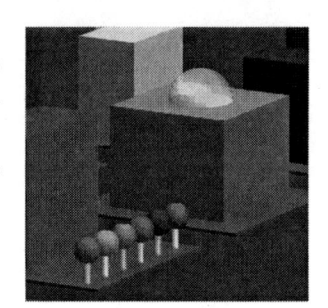

11.__ After making the desired o2c rendering settings, click **OK** in the **DcadWin.dmf** dialog box.

INFO STOP
If you want to open the **DcadWin.dmf** dialog box and set each color in the DataCAD palette directly, select (from the **Menu Bar**), **View, Rendering Settings**. When the **DcadWin.dmf** dialog box opens, click on the **Entity Color/Color** rectangle, pick a color from the **Color Palette**, select **OK**, and you can then click on the **Rendered Color** colored rectangle to map the selected DataCAD color to a new o2c render color.

12.__ Repeat the process of adjusting the DataCAD color to the o2c render color for each color in your model.

Using the Object Viewer

DataCAD 11 comes with a built-in 3D viewer called the **o2c Object Viewer** (o2c stands for Objects To See). This program allows for real-time zooming, rotating and panning of a shaded view.

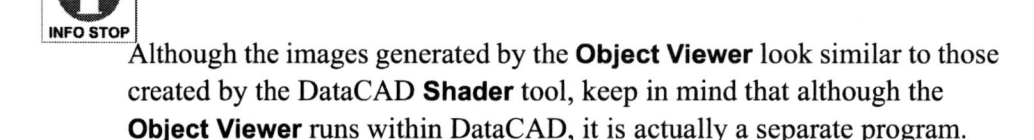

INFO STOP
Although the images generated by the **Object Viewer** look similar to those created by the DataCAD **Shader** tool, keep in mind that although the **Object Viewer** runs within DataCAD, it is actually a separate program.

1.__ With a file open in DataCAD, from the **MENU BAR** select **View, Object Viewer**.

Using the Object Viewer

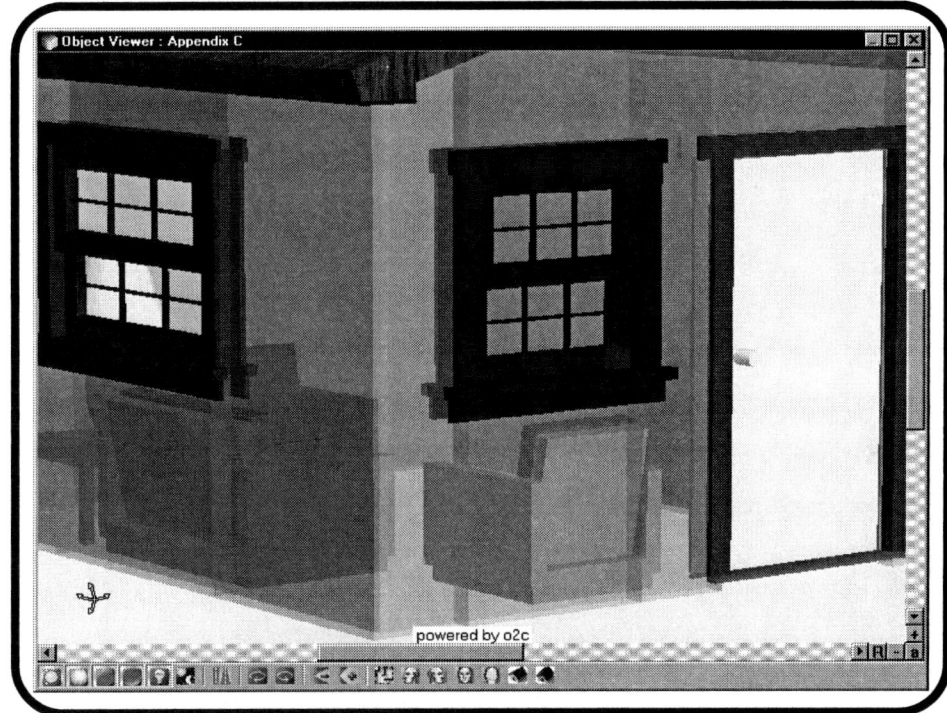

This opens a new window as shown in Figure C-12.

Figure C-2, The Object Viewer window.

The **Object Viewer** uses the mouse to dynamically control rotate and zoom.

2.__ Click and hold the *left* mouse button down and (while holding the left mouse button down) move the mouse left and right and up and down.

This rotates the display around horizontal and vertical axes.

3.__ Click and hold the *right* mouse button down and (while holding the *right* mouse button down) move the mouse up and down.

This zooms the display in and out. With just a little practice you will be able to travel anywhere around your model in a hurry. This is a great tool for previewing your design and can increase your design productivity.

The DataCAD 11 Project Book

4.___ Experiment with adjusting the display until you are comfortable with zooming and rotating.

Figure C-3, Dynamically adjusting the view.

There are a few more **Object Viewer** features to review.

Using the Object Viewer

5.__ Position the cursor in the **Object Viewer** and click the right mouse button once.

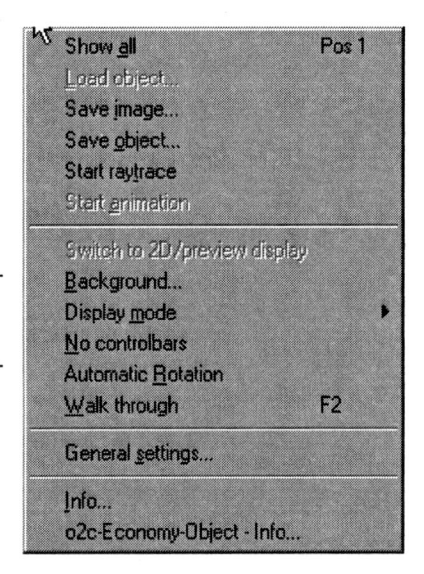

This opens the **Object Viewer** dialog box.

The **Show all** option returns the display to the initial view position.

The **Save image** option allows you to save the current image as a bmp format graphics file.

The **Save object** option allows you to save the current object as a o2c format graphics file. This file can be opened by whoever has the free o2c viewer program installed on their system (check the DataCAD web site, www.datacad.com). It is even possible to install an o2c viewer plug in to programs such as Microsoft Word, and use the viewer directly in the Word document.

The **Start raytrace** option will generate a high quality ray traced image of your view. This image will include shadows from a default light source.

The **Background** option will allow you to add a background image similar to the Shader function.

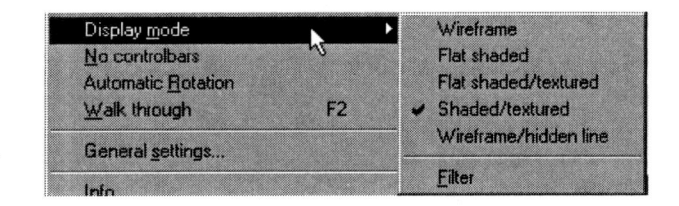

The **Display mode** option opens a pop out menu with additional viewing options.

The **No controlbars** option toggles scroll bars along the bottom and right edge of the **Object Viewer** window on and off.

The **Automatic Rotation** option can be toggled on and off. It allows you to start rotating the object with the mouse, and if you release the left mouse button while the mouse is moving, the object will continue to spin with the direction and speed the mouse was moving. Click the left mouse button to stop the rotation.

Appendix C, Page 363

The DataCAD 11 Project Book

The **Walk through** option is also an on off toggle. When it is on, you click and hold the left mouse button to pan the display. When it is off, you can use the arrow keys on the numeric keypad (the 4,8,6, and 2 keys) to pan the display.

The **General settings** option opens a dialog box that has options of interest to those who are communicating o2c objects over the internet, or embedding them in other applications.

6.__ Experiment with the various **Object Viewer** options.

APPENDIX D

Input Methods and Identify/Set All

Project Goals
When you have completed this project, you will be able to:
1. Use the 5 methods of drawing input for direction and distance.
2. Use **Identify/Set All**, one of DataCAD's most powerful features, to speed up your drawing, while maintaining coordination of layers and entity properties, including color, layer, line type, hatch pattern, weight, font, etc.

Prerequisite: DataCAD 11 interface experience

Introduction

In all the previous chapters we have stuck with the Relative Cartesian method of drawing. But there are 4 other methods available in DataCAD. The five methods are:

1. <u>Relative Cartesian</u>: Defines lines as increments of X and Y relative to the last point picked. This is the default entry method in DataCAD.
2. <u>Absolute Cartesian</u>: Defines lines as increments of X and Y relative to absolute 0 (zero).
3. <u>Relative Polar</u>: Defines lines as a distance from the last point picked at a specified angle.
4. <u>Absolute Polar</u>: Defines lines as a distance at a specified angle relative to absolute 0 (zero).
5. <u>Direction/Distance</u>: Defines a line in the direction that you move your mouse, and then by a distance that you type in.

You will probably never use Absolute Cartesian or Absloute Polar, but we'll describe them anyway. All the input methods create entities by defining DIRECTION and DISTANCE, just in different ways. Keep those two words in mind as you work through each one. Let's describe each method, and use them to draw a simple rectangle.

1. ___ Create a new drawing file by selecting **File, New**, then typing **Input** for the new file name. Click on **Create**.

Let's also make changes to some of our basic drawing settings.

The DataCAD 11 Project Book

2. __ Press the **g** [or the **G**] key to go to the **Grids** menu, and turn OFF **Grid Snap**, **Disp. Grid 1**, and **Disp. Grid 2**. Right-click once to exit the **Grids** menu.

Although drawing with grids and grid snapping on is good for some things, most things we draw don't fit within a grid of fixed dimensions. To get the most accurate drawings you are usually better off drawing using precise directions and distances (there are those words again) with one of the five input methods.

3. __ Press the **+** key [**Shift + =**] to give yourself a cursor that always extends from edge to edge of the Drawing Window.

4. __ Turn Walls OFF by either clicking on the **W** in SWOTHLUDFB to make it a lower case **w** (or you can press the **=** key).

Drawing With Relative Cartesian

Although you've already been drawing with this method in the previous chapters, here's the overview. In cartesian coordinates all X distances to the right of the Y-axis are described by positive numbers. All X distances to the left of the Y-axis are described by negative numbers. Likewise, all Y distances above the X-axis are described by positive numbers, and all Y distances below the X-axis are described by negative numbers. See Figure D-1.

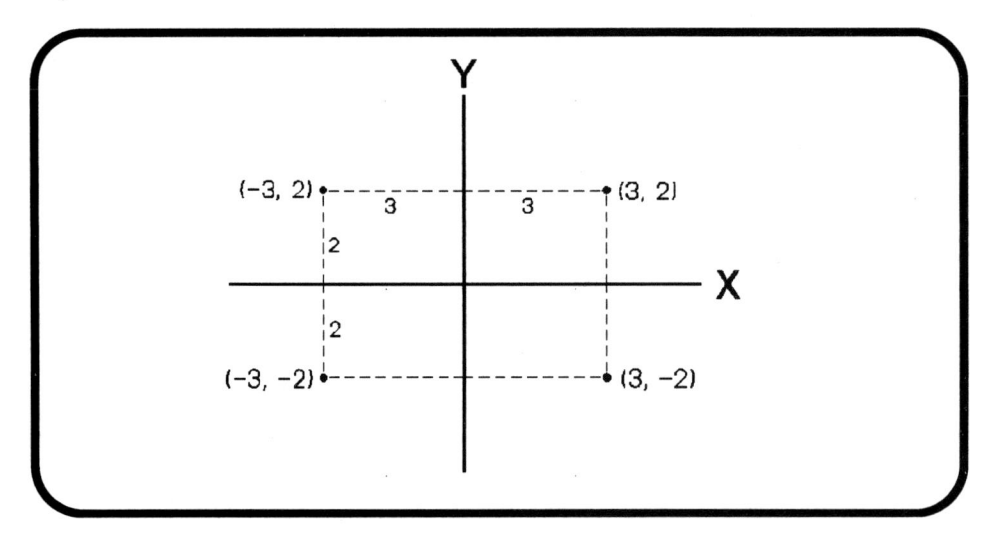

Figure D-1, X and Y coordinates.

Appendix D, Page 366

Drawing With Relative Cartesian

"Cartesian" means that the new line is drawn based on X (horizontal) and Y (vertical) coordinates. "Relative" means that a line is drawn relative to the last point picked. So if you click the mouse in the Drawing Window to define the first point of a line, that point becomes the last point picked. The second point of the line will then be drawn at the X and Y coordinates defined from that point. Here's how you would draw a 5'-0" line horizontally to the right.

1. ___ Set the current input mode to **Relative Cartesian** from the **STATUS AREA** dropdown (or select **Tools, Input Mode** from the **Menu Bar**; or press the **Insert** key until you see the **Relative Cartesian** input mode in the Message Line).

2. ___ Click in the Drawing Window to pick the first point of a line.

3. ___ Press the Space bar. This tells DataCAD you want to type in the second point of a new line.

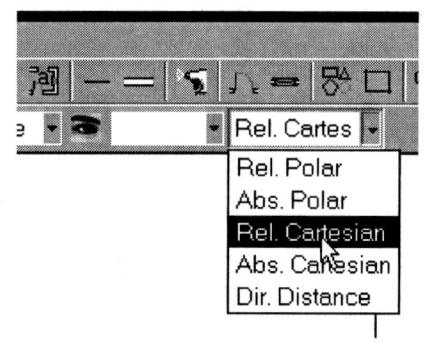

4. ___ You are prompted for an X distance. Type in a distance like 5'-0" (just type the number **5** or **5.0** and press **Enter**).

5. ___ Now you are prompted for a Y distance. Since the default distance is 0, just press **Enter** to accept it and the line is completed. See example at the top of Figure D-2.

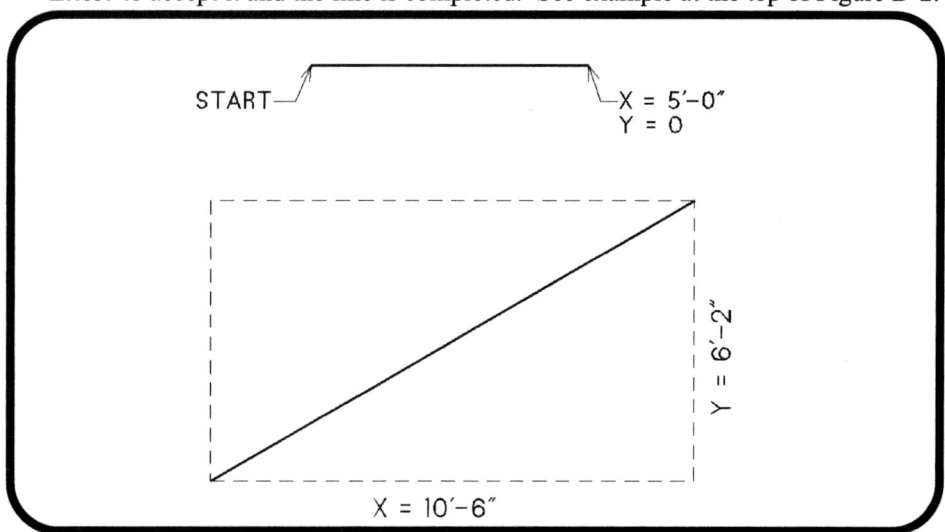

Figure D-2, A straight and angled line using Relative Cartesian.

Appendix D, Page 367

The DataCAD 11 Project Book

Let's say you want to draw an angled line. How you draw that line using the relative cartesian method depends on what kind of line you want to draw. If you want a line that is drawn to a specific point you need only know the X and Y coordinates of that point, without needing to know the actual length of the line. Let's say you want to draw a line from the lower left of your screen to the upper right corner. Since we know the X and Y distances this is a simple matter. Refer to Figure D-2.

6.__ Click the left mouse button once in the lower left corner of the Drawing Window.

7.__ Press the **Space** bar. This tells DataCAD you want to type in the second point of a new line.

8.__ You are prompted for an X distance. Type in the X distance of the line, which is 10'-6" (type **10.6** then press **Enter**).

9.__ Now you are prompted for a Y distance. Type in the Y distance of the line, which is 20'-2" (type **20.2** then press **Enter**), and the line is completed (Figure D-2).

But if you want a 45 degree line of a specific distance you would need to know, trigonometrically, what the X and Y distances would be for that line. You would probably be better off switching temporarily over to Relative Polar to do this. Otherwise, if you wanted to draw a 30 degree line 6'-6" long using the relative cartesian input method, first you would need to calculate the X distance, then the Y distance using trigonometry. The end result would be as seen in Figure D-3.

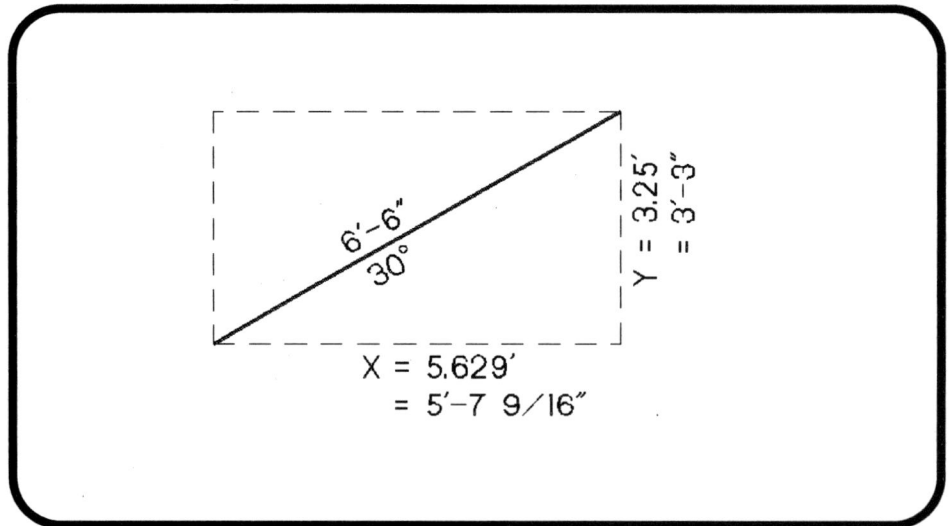

Figure D-3, A Relative Cartesian diagonal by trigonometry.

Drawing With Absolute Cartesian

After making those calculations you would then use the original four steps at the beginning of this section to input the X and Y distances for the line.

Drawing With Absolute Cartesian

Like Relative Cartesian, the Absolute Cartesian input method draws lines based on X (horizontal) and Y (vertical) coordinates. "Absolute" means that a line is drawn relative to the 0,0 (absolute zero) point of the drawing. Every CAD drawing has a 0,0 point which can not be moved or changed. With the absolute cartesian method if you click the mouse in the Drawing Window that point will be the first point of the line (Point 1). If you then press the **Space** bar you can input the second point (Point 2) of the line in reference to the absolute zero point, rather than in reference to Point 1.

1. ___ Set the current input mode to **Absolute Cartesian** from the **STATUS AREA** drop-down (or select **Tools, Input Mode** from the **Menu Bar**; or press the **Insert** key until you see the **Absolute Cartesian** input mode in the Message Line).

2. ___ Click the mouse anywhere in the Drawing Window (Point 1).

3. ___ Press the **Space** bar. This tells DataCAD you want to define the second point of the line off of the absolute 0 point.

4. ___ You are prompted for an X distance. Type in the X distance of the line, which is 7'-6" (type **7.6** then press **Enter**).

5. ___ Now you are prompted for a Y distance. Type in the Y distance of the line, which is 3'-2" (type **3.2** then press **Enter**), and the line is completed.

The line is drawn from your first mouse click (Point 1) to the absolute zero point (Point 2). See Figure D-4.

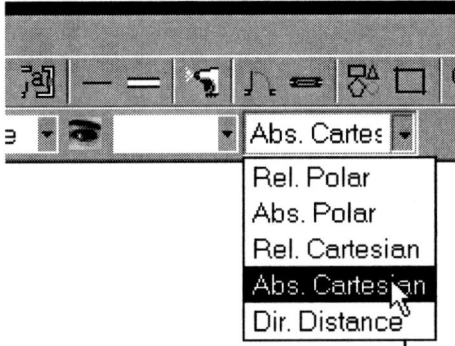

The DataCAD 11 Project Book

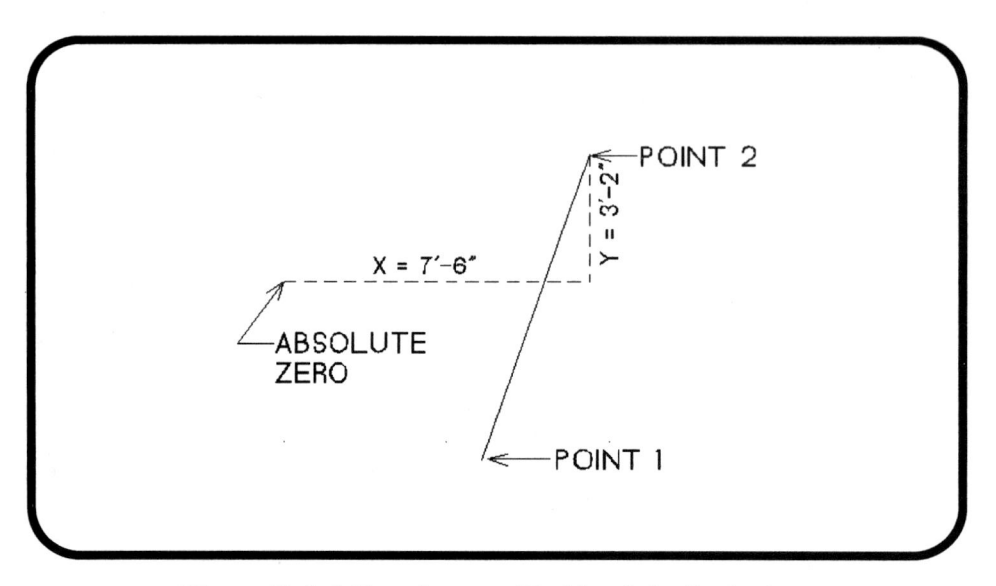

Figure D-4, A line drawn with Absolute Cartesian.

The Absolute Cartesian input method is very similar to the Absolute Polar method, except for using X/Y references, so for more examples on its use see the section regarding the Absolute Polar input method.

Drawing With Relative Polar

Although it is not the default DataCAD drawing method, this one is the most popular with most DataCAD users. "Polar" means that the new line is drawn based on the angles of a compass (the difference being that most CAD programs define 0/360 to the right, like the graphic here).

"Relative" means that a line is drawn relative to the last point picked. So if you click the mouse in the Drawing Window to define the first point of a line, that point becomes the last point picked. The second point of the line will then be drawn at an angle and distance from that point. Let's draw a line.

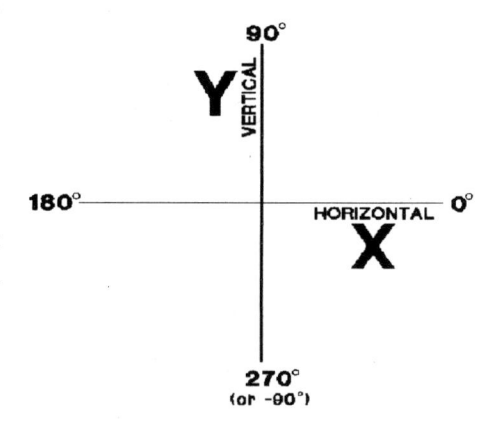

Drawing With Relative Polar

1. __ Set the current input mode to **Relative Polar** from the **STATUS AREA** drop-down (or select **Tools, Input Mode** from the **Menu Bar**; or press the **Insert** key until you see the **Relative Polar** input mode in the Message Line).

2. __ Click in the Drawing Window to pick the first point of the line.

3. __ Press the **Space** bar. This tells DataCAD you want to define a new line by typing something in.

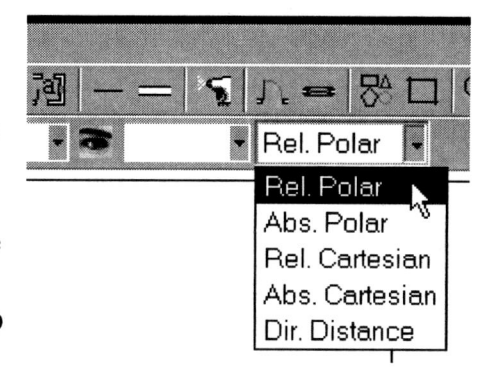

4. __ You are prompted for a distance. Type in a distance like 5'-0" (just type the number **5** or **5.0**, then press **Enter**).

5. __ Now you are prompted for an angle. Since the default angle is 0, just press **Enter** to accept it and the line is completed, as in Figure D-5.

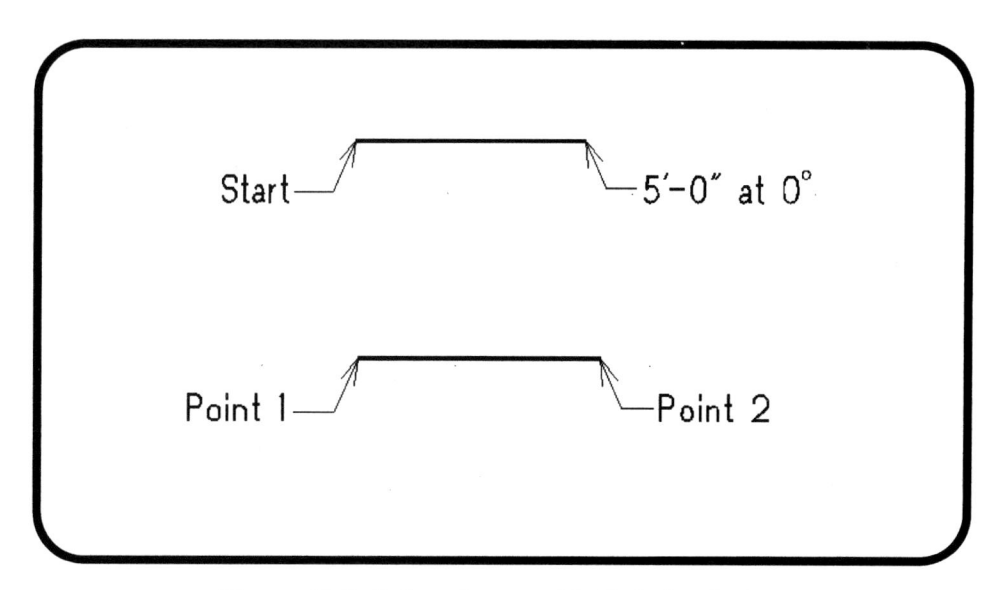

Figure D-5, A line drawn with Relative Polar.

If, instead, you wanted to draw the line at an angle of 30 degrees, as in Figure D-3, then you would enter an agle of 30 instead of 0.

6. Click in the Drawing Window to pick the first point of the line.

The DataCAD 11 Project Book

7. Press the **Space** bar. This tells DataCAD you want to define a new line by typing something in.
8. You are prompted for a distance. Type in a distance like 5'-0" (just type the number **5** or **5.0**, then press **Enter**).
9. Now you are prompted for an angle. Type **30** then press **Enter** and a 5 foot long line, at 30 degrees, is drawn. Simple!
10. To keep drawing additional lines from point 2, press the **Space** bar, type a distance, press **Enter**, type an angle, and press **Enter** again.

Drawing With Absolute Polar

"Polar" means that the new line is drawn based on the angles of a compass. "Absolute" means that a line is drawn relative to the 0,0 (absolute zero) point of the drawing. Every CAD drawing has a 0,0 point which can not be moved or changed. With the Absolute Polar method if you click the mouse in the Drawing Window that point will be the first point of the line (Point 1). If you then press the Space bar you can input the second point (Point 2) of the line in reference to the absolute zero point.

1. __ Set the current input mode to **Absolute Polar** from the **STATUS AREA** dropdown (or select **Tools, Input Mode** from the **Menu Bar**; or press the **Insert** key until you see the **Absolute Polar** input mode in the Message Line).
2. __ Click the mouse anywhere in the Drawing Window (Point 1).
3. __ Press the **Space** bar. This tells DataCAD you want to define the second point of the line from the absolute 0 point.
4. __ You are prompted to *Enter distance from the origin*. Press **Enter** to accept the default distance of 0.
5. __ Now you are prompted *Enter angle from the origin*. Press **Enter** to accept the default angle of 0.

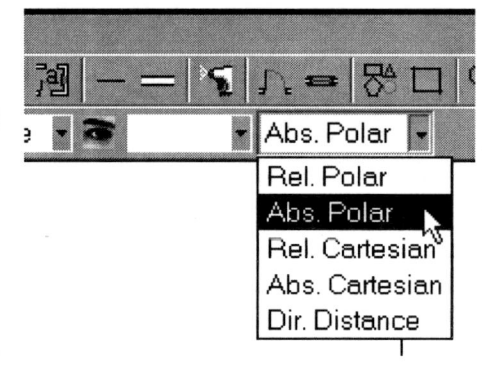

The line is drawn from your first mouse click (Point 1) to the absolute zero point (Point 2), as in the upper example in Figure D-6. Depending on how far away Point 1 was from the absolute zero point, the new line could be quite long.

Drawing With Absolute Polar

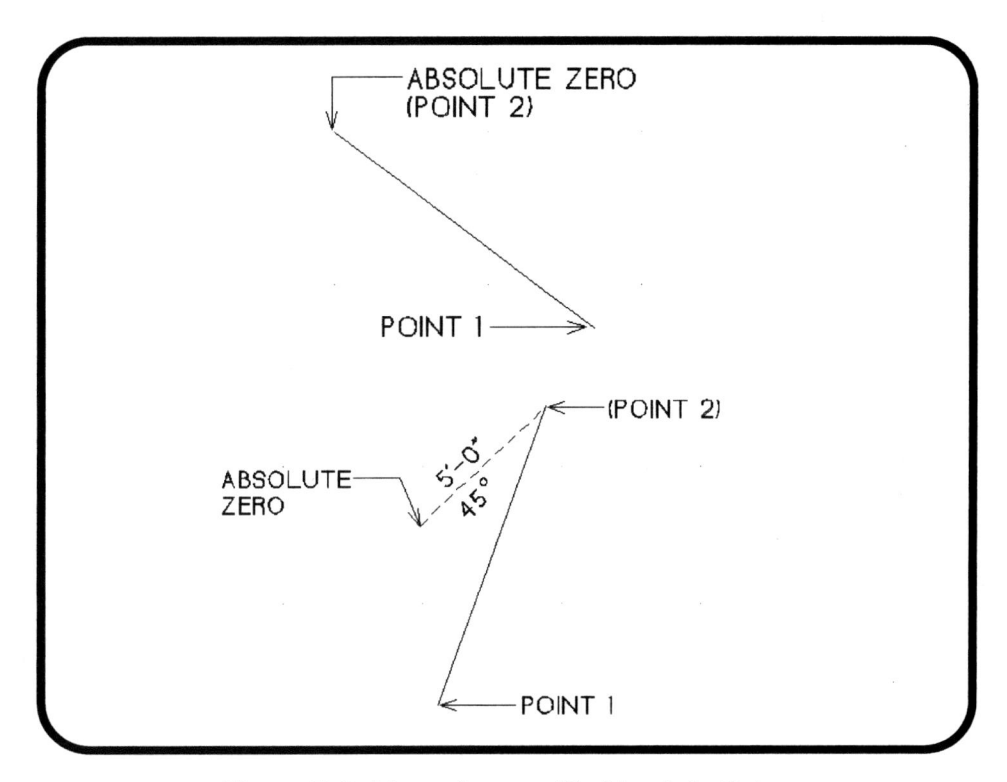

Figure D-6, Lines drawn with Absolute Polar.

You can also draw the second point of the line to any point referenced from, but not on, the absolute zero point, as in the lower example in Figure D-6.

6. ___ Click the mouse anywhere in the Drawing Window (Point 1).

7. ___ Press the **Space** bar. This tells DataCAD you want to define the second point of the line from the absolute 0 point.

8. ___ You are prompted to *Enter distance from the origin*. Press **5** (for 5'-0"), then press **Enter** to enter a distance of 5'-0" from absolute zero.

9. ___ Now you are prompted *Enter angle from the origin*. Press **45**, then **Enter**, to enter an angle of 45 degrees from absolute zero.

The line is drawn from your first mouse click (Point 1) to a point 5 feet and 45 degrees from absolute zero (Point 2).

The DataCAD 11 Project Book

If you do not right-click after drawing the first line then a second line will be displayed from Point 2 to the cursor. If you click the mouse the second line will be drawn from Point 2 to Point 3. But if you do not right-click, and instead press the **Space** bar, then you can define Point 3 by typing the distance and angle. So the location of Point 3 was "referenced" off of absolute zero, not off of Point 2.

If you again do not right-click, then continue drawing another line by pressing the space bar, the next line will be drawn from Point 3 to Point 4, with Point 4 being defined as a distance and angle from absolute zero. Let's try drawing 3 contiguous lines using this method (see Figure D-7).

10. __ Click the mouse anywhere in the Drawing Window (Point 1).

11. __ Press the **Space** bar. This tells DataCAD you want to define the second point of the line off of the absolute 0 point.

12. __ You are prompted to *Enter distance from the origin*. Press **5** then **Enter** to enter a distance of 5'-0" from absolute zero.

13. __ Now you are prompted *Enter angle from the origin*. Press **45** then Enter to enter an angle 45 degrees from absolute zero.

14. __ A line is drawn from Point 1 to Point 2.

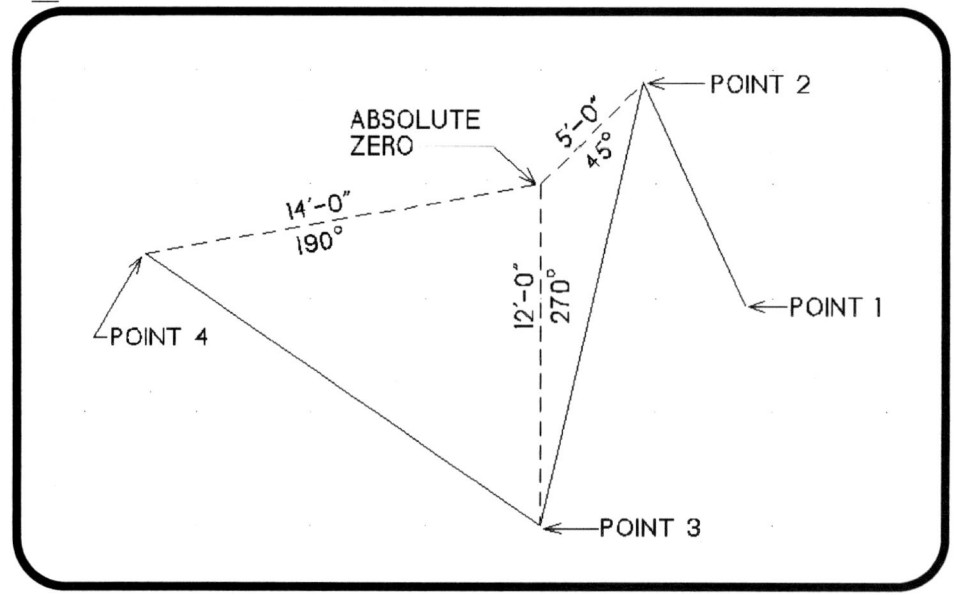

Figure D-7, 3 Consecutive lines with Absolute Polar.

Drawing With Direction/Distance

15.__ Press the **Space** bar.
16.__ You are prompted to *Enter distance from the origin*. Press **12** then **Enter** to enter a distance of 12'-0" from absolute zero.
17.__ Now you are prompted *Enter angle from the origin*. Press **270** then **Enter** to enter an angle 270 degrees from absolute zero.
18.__ A line is drawn from Point 2 to Point 3.
19.__ Press the **Space** bar.
20.__ You are prompted to *Enter distance from the origin*. Press **14** then **Enter** to enter a distance of 14'-0" from absolute zero.
21.__ Now you are prompted *Enter angle from the origin*. Press **190** then **Enter** to enter an angle 190 degrees from absolute zero.
22.__ A line is drawn from Point 3 to Point 4.

Whew! Now perhaps you can see why you probably won't use this input method too often.

Drawing With Direction/Distance

Direction/Distance is a relative newcomer to DataCAD, having been added in a free update in DataCAD 10. You define the direction by clicking in the Drawing Window, then moving your mouse in the direction of the line you want to create. The distance is then typed in to give you your line. This method works best with **Ortho** turned ON (the upper case **O** in SWOTHLUDFB).

1.__ Set the current input mode to **Dir. Distance** from the **STATUS AREA** drop-down (or select **Tools, Input Mode** from the **Menu Bar**; or press the **Insert** key until you see the **Direction, Distance** input mode in the Message Line).
2.__ Make sure **Ortho** is turned ON by clicking on the **O** in SWOTHLUDFB until it is an upper case letter, or by pressing the **o** or **O** key.
3.__ Click in the Drawing Window to pick the first point of the line.
4.__ Move the cursor up and to the right at a 45 degree angle. This will be the direction and angle of the new line.

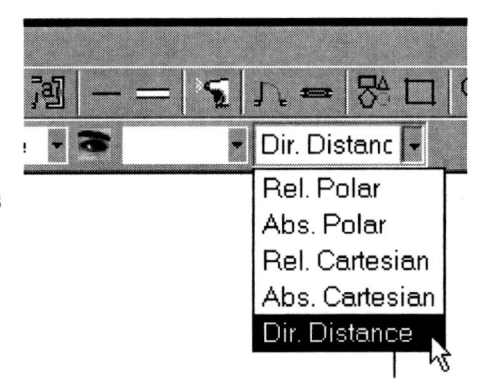

The DataCAD 11 Project Book

5.___ Press the **Space** bar. This tells DataCAD you want to define a new line by typing something in.

6.___ You are prompted for a distance. Type in a distance like 5'-0" (just type the number **5** or **5.0**, then press **Enter**). The line is drawn.

Since this input method was introduced some DataCAD users have fallen in love with it, so it's worth a try to see if it's a good fit for you.

Using Identify / Set All

This is unarguably one of DataCAD's best and most powerful features. Select the **Identify** option, pick any entity (line, arc, associative hatch, associative dimension, slab, etc.) in the Drawing Window, select the **Set All** option, and all the current settings will be changed to match those of the entity you picked, including linetype, layer, color, spacing, Z-heights, hatch settings, associative dimension settings, etc. Keep in mind that this command only matches entity properties, but not geometries. So picking a circle, for instance, will not cause an identical radius or diameter to be selected.

And since one of those matched settings is the layer that the entity was drawn on, DataCAD will make that the new active layer so that you are working on that same layer. Try these examples to see how this can save you a lot of time.

1.___ Create a new drawing file by selecting **File, New,** then typing **Identify** for the new file name. Click on **Create**.

2.___ Press **Ctrl+L** to open the **Layer Manager**.

3.___ Create one new layer. Click on **New**, enter **1**, then press **OK**.

4.___ Rename the first layer as **Plywood**. Click on the colored square to the left of the name, click on the **Red** color box, then click on **OK**.

5.___ Rename the second layer as **Insulation**. Click on the colored square to the left of the name, click on the **Magenta** color box, then click on **OK**.

6.___ Click on **OK** to close the **Layer Manager**.

7.___ If it is not already, make the **Plywood** layer the active layer by pressing the **Tab** key until that layer name appears in the **Status Area**.

8.___ From the **Edit** menu, select **Line Type** [or position the cursor over the **Line Type** button in the **Status Area** and click the left mouse button].

9.___ Select **Scroll Fwrd** several times until you see the **Plywood2** (or any other plywood line type). Click on **Plywood2** to select it.

Using Identify / Set All

10. ___ In the **Line Type** menu select **Spacing**. Type in 3/4" [**0.0.3/4**, or just **..3/4**], then press **Enter**.
11. ___ Right-click to back out of the **Line Type** menu.
12. ___ With your cursor in the Drawing Window draw several lines with the new **Plywood2** linetype. Your screen may look something like Figure D-8., with several red plywood lines, all on the **Plywood** layer.

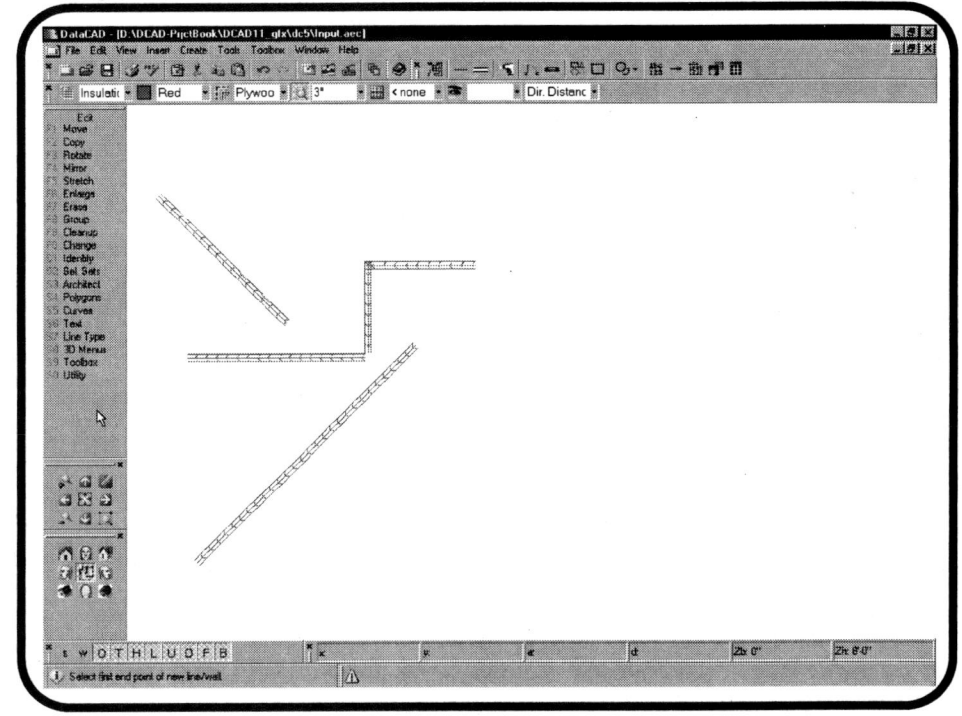

Figure D-8, Several plywood lines.

13. ___ Make the **Insulation** layer the active layer by pressing the **Tab** key until that layer name appears in the **Status Area**.
14. ___ From the **Edit** menu, select **Line Type** [or position the cursor over the **Line Type** button in the **STATUS AREA** and click the left mouse button].
15. ___ Select **Scroll Fwrd** several times until you see the **Insul1** (or any other insulation line type). Click on **Insul1** to select it.

Appendix D, Page 377

The DataCAD 11 Project Book

16. ___ In the **Line Type** menu select **Spacing**. Type in 2" [**0.2**, or just **.2**], then press **Enter**.

17. ___ Right-click to back out of the **Line Type** menu.

18. ___ With your cursor in the Drawing Window draw several lines with the new **Insul1** linetype. Your screen may look something like Figure D-9, with several **Magenta** insulation line types, all on the **Insulation** layer.

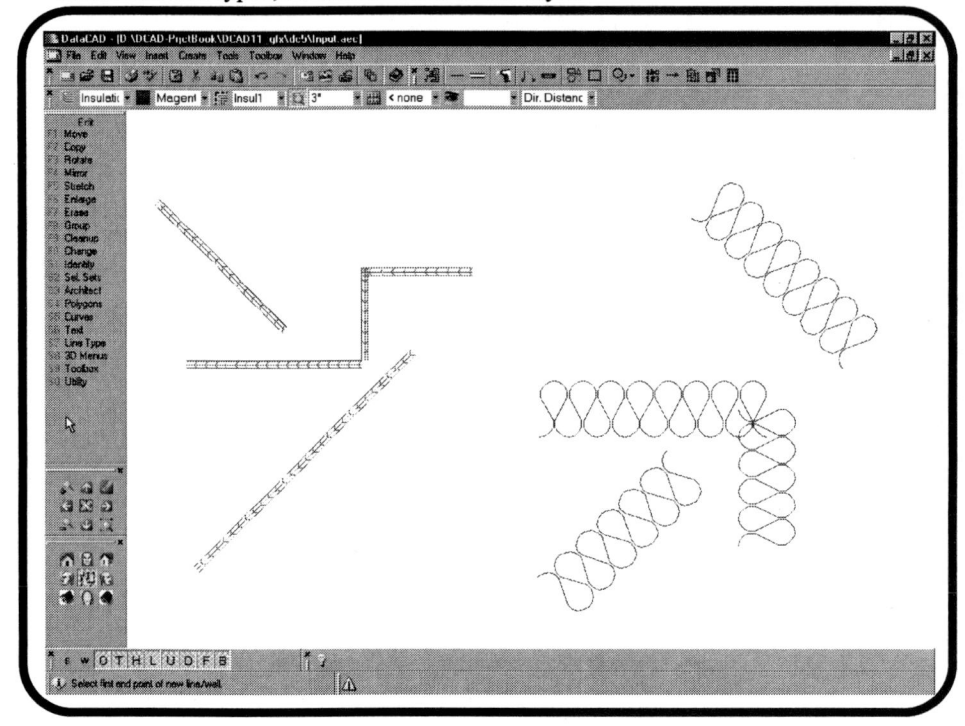

Figure D-9, Several added insulation lines.

You are currently on the Insulation layer, drawing magenta insulation line types. Now you want to continue drawing more red plywood lines on the **Plywood** layer. Here are two ways to do that:

Appendix D, Page 378

Using Identify / Set All

The Long Method:

1. ___ Press the **Tab** key to make the **Plywood** layer the active layer.
2. ___ From the **Edit** menu select **Line Type**.
3. ___ Click on **Scroll Fwrd** several times and select **Plywood2**.
4. ___ Select **Spacing**.
5. ___ Type in 3/4" [**0.0.3/4**, or just **..3/4**], then press **Enter**.
6. ___ Right-click out of the menu and you can start drawing red, 3/4" plywood lines again.

The Short Method:

1. ___ Select the **Identify** button in the **NAVIGATION PAD** (the one that looks like a finger), or press the **I** [not the **i**] key.
2. ___ Click on one of the plywood lines in the Drawing Window. It will appear gray and dashed.
3. ___ Select the **Set All** option. Done. You are drawinging red, 3/4" plywood lines again

Try drawing a few more lines to check it out.

Why is this so important? For three basic reasons: speed, accuracy and order. You can draw faster if you spend less time clicking through menus. Your drawings will be more accurate if you are actually drawing identical building elements that are really identical. And last, but perhaps least appreciated, is order. In CAD it is very important to maintain discipline with the entities on your layers. Door elements should only be on the Doors layer; walls should only be on the Walls layer; etc. You will thank yourself for this, as will other CAD drafters and engineers who you may share your drawings with.

Identify/Set All works for associative hatching and associative dimensions, too. If you open the **Detail13** file from Chapter 13 there are several associative hatches (two different Concrete hatches, and an Earth hatch), and several associative dimensions. Try using Identify/Set All on each associative hatch, and an associative dimension. Notice that when you do you will be able to immediately draw identical hatches of the correct scale and color, and on the correct layers.

And **Identify/Set All** works for 3D entities as well. If you draw several 3D slabs or 3D walls, all with different Z-bases and Z-heights, using **Identify/Set All** will allow you to immediately begin drawing identical 3D slabs and 3D walls with the correct Z-bases and Z-heights.

The DataCAD 11 Project Book